Concepts and Practices of STEM Education in Asia

May May Hung Cheng · Cathy Buntting ·
Alister Jones
Editors

Concepts and Practices of STEM Education in Asia

Springer

Editors
May May Hung Cheng
Department of Curriculum and Instruction
The Education University of Hong Kong
Hong Kong SAR, China

Cathy Buntting
Wilf Malcolm Institute of Educational Research
University of Waikato
Hamilton, New Zealand

Alister Jones
Office of the Vice-Chancellor
University of Waikato
Hamilton, New Zealand

ISBN 978-981-19-2598-6 ISBN 978-981-19-2596-2 (eBook)
https://doi.org/10.1007/978-981-19-2596-2

This Springer imprint is published by the registered company Springer Nature Singapore Pte Ltd.
The registered company address is: 152 Beach Road, #21-01/04 Gateway East, Singapore 189721, Singapore

Contents

Editors and Contributors

About the Editors

May May Hung Cheng obtained her Bachelor's and Master's degrees from the University of Hong Kong and her Ph.D. from the University of Waikato, New Zealand. She is currently Chair Professor of Teacher Education in the Department of Curriculum and Instruction at the Education University of Hong Kong. She was previously a Reader in Professional Education at the Department of Education, University of Oxford. Her main areas of research are teacher education and science education with a focus on teacher learning and teacher professional development. She was the President of the East Asian Association for Science Education (EASE) between 2016 to 2019.

Cathy Buntting is Director of the Wilf Malcolm Institute of Educational Research at the University of Waikato, New Zealand. She has a Master's degree in biochemistry and a Ph.D. in science education and has led multiple Government-funded projects in science education, technology education and STEM education. She is committed to research that contributes to the research-practice nexus and is Director of the New Zealand Science Learning Hub, a vast online portal making contemporary STEM research and development more visible and accessible to schools.

Alister Jones is Research Professor and Senior Deputy Vice-Chancellor at the University of Waikato, New Zealand. He has written extensively on curriculum development in both science and technology education in New Zealand and been consulted on educational development in New Zealand, Australia, the United Kingdom, USA, Hong Kong, China, Thailand and Chile, as well as Co-Chairing an APEC working group on science and mathematics education. He was Managing Director of the Australasian Science Education Research Association Limited for several years and is company director of a number of New Zealand-based education companies.

Contributors

May May Hung Cheng Department of Curriculum and Instruction, The Education University of Hong Kong, Hong Kong SAR, China

Meng-Fei Cheng Department of Physics, National Changhua University of Education, Changhua City, Taiwan

Hye-Eun Chu Macquarie University, Macquarie Park, Sydney, NSW, Australia

Chanyah Dahsah Science Education Center, Faculty of Science, Srinakharinwirot University, Bangkok, Thailand

Parbat Dhungana Department of Science and Environmental Studies, The Education University of Hong Kong, Hong Kong SAR, China

Yan Dong School of Education Technology, Faculty of Education, Beijing Normal University, Beijing, China

Hidayah Mohd Fadzil Faculty of Education, University of Malaya, Kuala Lumpur, Malaysia

Witat Fakcharoenphol Department of Teacher Education, Faculty of Education and Development Sciences, Kasetsart University, Nakhon Pathom, Thailand

Ei Seul Kim Seoul National University, Gwanak-gu, Korea

Sung-Won Kim Ewha Womans University, Seoul, Korea

Yoshisuke Kumano Faculty of Education, STEAM Education Institute, Shizuoka University, Shizuoka, Japan

Hyeong Moon Lee Seoul National University, Gwanak-gu, Korea

Yew-Jin Lee Natural Sciences & Science Education (NSSE), National Institute of Education, Nanyang Technological University, Singapore

Yu-Heng Lo Department of Physics, National Changhua University of Education, Changhua City, Taiwan

Sonya Martin Seoul National University, Gwanak-gu, Korea

Nanxi Meng Department of Learning Technologies, University of North Texas, Denton, TX, USA

Jiyeong Mun Ewha Womans University, Seoul, Korea

Yann Shiou Ong Natural Sciences & Science Education (NSSE), National Institute of Education, Nanyang Technological University, Singapore

Tepkanya Promkatkeaw Science Education Center, Faculty of Science, Srinakharinwirot University, Bangkok, Thailand

Pramudya D. A. Putra Science Education Programme, University of Jember, Kampus Tegalboto, Jember (East Jawa), Indonesia

Rohaida Mohd Saat Faculty of Education, University of Malaya, Kuala Lumpur, Malaysia

Navara Seetee Science Education Center, Faculty of Science, Srinakharinwirot University, Bangkok, Thailand

Nyet Moi Siew Faculty of Psychology and Education, Universiti Malaysia Sabah, Kota Kinabalu, Sabah, Malaysia

Nurul F. Sulaeman Physics Education Programme, Mulawarman University, Jalan Muara Pahu Gunung Kelua Campus, Samarinda, East Kalimantan, Indonesia

Siew Wei Tho Faculty Science and Mathematics, Sultan Idris Education University, Tanjung Malim, Malaysia

Tussatrin Wannagatesiri Department of Teacher Education, Faculty of Education and Development Sciences, Kasetsart University, Nakhon Pathom, Thailand

Yang Yang Center for Educational Science and Technology, Institute of Advanced Studies in Humanities and Social Sciences, Beijing Normal University, Zhuhai, Guangdong, China

Fang-Yin Yeh Department of Curriculum and Instruction, The Education University of Hong Kong, Hong Kong SAR, China

Yau Yuen Yeung Department of Science and Environmental Studies, The Education University of Hong Kong, Hong Kong SAR, China

Xinyu Zhou School of Information Technology in Education, South China Normal University, Guangzhou, Guangdong, China

Chapter 1
An Overview of STEM Education in Asia

May May Hung Cheng

Abstract This introduction chapter provides an overview of STEM education policies and curriculum initiatives in the Asian context. The three parts of this book are briefly introduced. There is then a summary and synthesis of relevant findings for each of the 14 chapters as well as a discussion linking the content of this book with the current development of STEM education research. Finally, implications, challenges and future directions for STEM education development and related research are proposed.

Keywords Conceptualising STEM education · Implementing STEM education · Teacher preparation · Teacher professional development

STEM Education in Asia: Contributing to an International Discourse

Concurrent with the global drive for STEM education, research in STEM education has shown exponential growth in the past decade (Brown, 2012; Lee et al., 2019; Mizell & Brown, 2016). Much attention has been given to integrating STEM education, shifting from previous STEM subject implementation that traditionally focused on science, technology, engineering, and mathematics as isolated disciplines (Brown, 2012; Kelley & Knowles, 2016; Lee et al., 2019; Wang et al., 2011) or which used examples of technology to contextualize science education (science-technology-society, STS) (see Jones & Buntting, 2015). However, implementing integrated STEM is not straightforward. Some of the challenges arising from achieving integration in STEM education, documented in recent research, include areas of concern such as curriculum enrichment of existing subjects like science or restructuring STEM-based discipline curricula (Nadelson &

M. M. H. Cheng (✉)
Department of Curriculum and Instruction, The Education University of Hong Kong, 10 Lo Ping Road, Hong Kong SAR, China
e-mail: maycheng@eduhk.hk

M. M. H. Cheng et al. (eds.), *Concepts and Practices of STEM Education in Asia*,
https://doi.org/10.1007/978-981-19-2596-2_1

Seifert, 2017) and teacher preparation and professional development (Al Salami et al., 2017; Asghar et al., 2012; Lee et al., 2019). Researchers have also called for a better understanding of STEM curricula as well as classroom practices and operations due to a lack of empirical evidence discerning elements crucial for a successful implementation of STEM education (Honey et al., 2014; Kang, 2019; Thibaut et al., 2018).

Despite global attention to STEM policy and an increase in STEM education research, Asian-based STEM education studies are still in an emerging state in the international literature. According to Lee et al.'s (2019) editorial article, Asian-based STEM education studies comprised only 8.5% of the total examined international STEM education publications from 2013 to 2017 ($N = 665$), compared to around 65% from the United States. There are also increasing calls for schools to design integrated STEM units that improve, or at least preserve, students' examination results while concurrently fostering the knowledge, skills, and attitudes needed for the 21st century. Little is known about the best practices for implementing or integrating STEM in ways that will do this (English, 2016; Herschbach, 2011; Kelley & Knowles, 2016), especially in the Asian context.

After more than a decade of STEM education promotion in many Asian countries, an analysis of the current status and practices in the region is not only timely, but it provides a reference to inform future STEM education and research initiatives in the Asian region. Thus, the purpose of this edited book is to enrich the literature related to STEM education at primary and secondary levels in Asia, with particular attention given to the analysis of the educational context in a number of Asian countries, including STEM-related policies, pedagogical practices, and the design and evaluation of STEM programmes, including impacts on student learning outcomes and the ways in which STEM education is catering for students' interests and needs. Together, the contributions from different Asian regions invite researchers and educators to learn from effective STEM practices, and point out areas for further development.

The Asian Region is well-known for its wide range of geographical, political, economic, and religious diversity between and within the countries in the region. Given these variations, and the insights arising from the differing contexts, international readers will be able to gain insights from the ways in which STEM learning and teaching initiatives have been tested, modified, and developed in a range of highly complex and diversified contexts. Further, the specific resources and strategies outlined in this book can be a useful reference for all curriculum developers and STEM educators when they design school-based curricula and STEM teacher education or development programmes. Graduate students and researchers will find in this book contextual information related to STEM education in Asia, directions for further investigation, and insights for conducting studies in this region.

STEM Education Policies and Curriculum Initiatives in the Asian Context

A brief introduction of STEM policies and curriculum practices in the Asian context provides a useful backdrop for understanding the content of this book. In a comprehensive review of international STEM education policy, Marginson et al. (2013) categorized different STEM policy narratives into four groups, including Post-Confucian East and Southeast Asia, English-speaking countries, Europe, and the emerging economies and education systems (e.g. South Africa, Brazil, and Argentina). Their review identified a common economic discourse on the contribution of STEM education to individual nations' economic innovation in response to the challenges of the 21st century, and highlighted diverse narratives embedded in different discursive terrains. In the Asian region, more emphasis has been placed on improving teacher quality to meet higher standards, having high standard provision of STEM education for all (specifically in Science and Mathematics), and driving comprehensive reforms towards pedagogical approaches that emphasize student-centred learning, inquiry-based learning, problem solving, and creativity. Comparatively, a widespread 'at risk' narrative underpinned by student under-performance and long tails of low-achievers in international achievement tests, declining student participation in STEM subjects, and concerns related to the general quality of the teachers are more pervasive among the majority of English-speaking countries. These findings suggest a difference in focus and concerns related to STEM education in different parts of the world.

To further support readers' understanding of the development of STEM education in Asia, a brief overview is provided for Mainland China, Japan, Singapore, Thailand, Taiwan, Hong Kong, Malaysia, Korea, and Indonesia.

Within Asia, the Korean Ministry of Education, in 2011, reconstructed STEM education to incorporate 'Arts,' making STEAM an interdisciplinary construction for 'creative convergence education' (Marginson et al., 2013; Park et al., 2016). The Korean STEAM initiatives were spurred by continuous government projects for human development (e.g. Educating Talented and Gifted Youth in Science; see Jon & Chung, 2013), and recent international comparison studies indicating students' high performance yet low-level of interest and enjoyment in science and mathematics (Organisation for Economic Co-operation & Development, 2010). At the school level, individual components of STEAM are required to be taught to primary and secondary school students (Kang, 2019). Furthermore, key objectives of 'creative convergence education' are materialized in the course of 'creative experiences' (taking up to 9–13% of total class hours) in the National Guidelines of the 2015 Revised National Curriculum (Cho & Huh, 2017; Kang, 2019). Nationally, the Korea Institute/Foundation for the Advancement of Science and Creativity (KOFAC), as the most representative government agency for STEAM, has played a major role in organizing STEAM programmes and initiatives. KOFAC defines STEAM as education based on scientific technology, and has driven STEAM initiatives through supporting the science culture and maker cultures; supporting professional development of teachers and resource development and distribution in creative convergence education; and supporting gifted youth (Hong, 2017; KOFAC, n.d.).

In Japan, policies and programmes that affect various aspects of STEM education are addressed under broader discussions of science, technology, and human resources development, the synthesis of broad-based education and the 'STEM for STEM people' elite track, the 'PISA shock', and the re-evaluation of a 'low pressure and relaxing' curriculum characterizing the 2000s (Ishikawa et al., 2013; Marginson et al., 2013; Takayama, 2010). Within the context of the national curriculum, the 2008 Revision of Courses of Study, or the national curriculum guidelines implemented by the Japanese Ministry of Education, Sports, Science, and Technology stipulated increased contact hours as well as content in Science and Mathematics subjects for primary and lower secondary schools. In addition to the nationwide curriculum provision, other STEM promotion strategies included the launch of specialist or 'elite' science programmes (for secondary and tertiary level), addressing gender disparity in STEM participation and occupation, and STEM-discipline student internationalization initiatives (Ishikawa et al., 2013; Marginson et al., 2013).

In Singapore, STEM education has long been part of the framing of national policy on education and industry. Its perceived relevance to the industry is reflected in the current movement of the 'innovation-driven' economy since 2000 (Idris et al., 2013). Within the national curriculum context, STEM education in formal schooling in Singapore is embedded in Mathematics Education and Science Education in both primary and secondary schools. In addition to discipline-based integration, a state-wide STEM education programme, mostly in the form of extracurricular activities, is provided by the Agency for Science under the supervision of the Ministry of Education to promote STEM education in Singapore (Idris et al., 2013).

In Mainland China, school-, city-, and provincial-based STEM programmes for primary and secondary schools represent a bottom-up approach to STEM education in Mainland China involving different stakeholders (e.g. universities, research centres, industry, etc.). Interest in STEM education and research, initiated by the introduction and analysis of STEM education and policy in the country, started to take root in 2007. The 'China STEAM Development Report', co-authored by the Ministry of Education People's Republic of China (MOE P. R. of China) and other educational and research institutes, pointed out several characteristics of STEAM education in Mainland China. First, around 72% of STEAM-related courses are embedded in extracurricular activities or elective courses. Second, within the formal curriculum context, the current implementation of STEM education in Mainland China is primarily school-based and embedded in non-core subjects (e.g. information technology), with engineering as a core aspect, and strong application features (e.g. coding, robotics, and media production) (MOE P. R. of China et al., 2017).

In Hong Kong, a Special Administrative Region of China, the government officially addressed the importance of STEM education and started promoting

STEM education in primary and secondary schools in 2015 (Education Bureau of Hong Kong [EDBHK], 2016). Within the Hong Kong curriculum context, the elements of STEM education are embedded in the three key learning areas (KLAs) of Science, Technology, and Mathematics Education, with key strategies focusing on the curriculum renewal of the KLAs enriching learning activities, and enhancing teacher education for strengthening STEM education (EDBHK, 2016).

In Taiwan, the K-12 curriculum guidelines issued by the Ministry of Education in 2018 (Ministry of Education of Taiwan, 2018) integrate the STEM subjects into the science and technology curriculum. There are efforts to further promote the core literacies of systematic thinking and problem solving for all citizens in order to promote competition and enhance expertise in the technology industry.

In Malaysia, an 'at-risk' mindset was prevalent in relation to low intakes of science and technology courses and programmes despite the implementation of the national enrolment policy since the 1970s (Shahali et al., 2017). Additionally, the crisis of an insufficient STEM workforce for the future and a downward trend of test scores in TIMSS and PISA have dominated the official STEM narratives in the country (Marginson et al., 2013; Shahali et al., 2017). This perception of national crisis drove the Malaysian Ministry of Education to stress the quality of STEM education for enhancing students' STEM knowledge and interest in science subjects under the broad *Malaysia Education Blueprint* promulgated in 2013 (Ministry of Education Malaysia, 2013). In the context of the Malaysian primary and (lower) secondary education curricula, STEM education initiatives are embedded primarily in the 'Core Science Subjects' through the planning of an integrated STEM approach in teaching and learning (Shahali et al., 2017). In the upper secondary curricular context, the nomenclature 'STEM stream' has replaced the term 'Science & Technology Stream' (previously renamed from 'Science Stream') since 2020 in the Malaysian school system.

In Thailand, STEM education was on the agenda of the Ministry of Education in 2015, with the inclusion of STEM education in the Thai National Education plan in 2017 (Office of the Education Council Thailand, 2017). Efforts to promote STEM education start from the Early Childhood Education sector, and include science, mathematics, and technology in the curriculum. The primary and secondary curricula for mathematics, science, and geography were revised, while design and technology and information and communication technology were moved into the science curriculum (Ministry of Education Thailand, 2017). As a result, the technology strand in the science learning area includes design and technology as well as computational thinking. STEM education in Thailand is driven and promoted by a number of organizations, including large corporations and non-government organizations, in addition to schools and universities.

In Indonesia, according to the curriculum goals of science education for K-12, there is an emphasis on integrating science concepts with technology, environment, and society (Indonesia Ministry of Education and Culture, 2020). Research efforts to promote teacher education for STEM education have found that Indonesian pre-service teachers have low understanding of integrated STEM approaches (Putra & Kumano, 2018).

Having provided an overview of macro-level STEM education policies and curriculum initiatives in the Asian contexts presented in this book, the structure of the book is now considered, including an introduction to the individual chapters.

The Three Sections of This Book

This book has a further 14 chapters, written by contributors from nine countries or regions in Asia. It is divided into three sections focusing on the conceptualization of, implementation of, and teacher preparation for STEM education. The contributors are experts in STEM education or are leading major research and development projects in STEM in their own regions. The first section is focused at the macro-level on the conceptualization and formulation of STEM education policies in different regions, contributing to our understanding of the current status of STEM education in Asia. The second section examines some features of STEM learning and teaching at the classroom level and includes studies on student learning in STEM programmes. Pedagogical innovations implemented in different parts of Asia are also reported and discussed. The third section moves to professional teacher development. It discusses practices of teacher professional development in the region and reports on current provisions as well as challenges.

This introduction chapter provides below a synthesis of the relevant findings reported in the remaining chapters, as well as linking the content of this book with the current development of STEM education research. Finally, some directions for future research on STEM learning and teaching in Asia are proposed.

Conceptualizing STEM Education in Asia

The three chapters in this section cover discussions of STEM policies, directions, and programmes across different regions in Asia. The section begins with Chap. 2, a review of STEM strategies and programmes in secondary education in Asia. The review draws on 32 papers written in English published between 2011 and 2020 and written in the Asian context. Guided by descriptive frameworks drawn from analysis of good practices of integrated STEM, STEM courses and programmes were examined in terms of the nature and scope of STEM programmes, instructional practices adopted in the STEM programmes, and the evidence of the impact of the STEM programmes on student learning in the knowledge, skills, and attitude domains. This review characterizes different STEM enactments in Asian secondary education, contributing to identifying and discussing effective STEM programmes and strategies.

In Chap. 3, STEAM education is reported to be adopted widely in theoretical research and educational practices in Mainland China by colleagues, Meng et al. STEAM education in Mainland China is informed by the Core Competencies and

Values (CCV) for Chinese Students (MOE P. R. of China, 2016), in which there are four specific aspects: scientific concepts and applications, scientific thinking and innovation, scientific exploration and communication, and scientific attitudes and responsibilities. The chapter explains the alignment of policies at the provincial and regional level as well as the formulation of STEM learning and teaching strategies, characteristics of STEM learners in Mainland China, research collaborations in STEAM learning, teaching approaches, and the practice of teacher professional development related to STEAM education.

In Chap. 4, Yoshisuke Kumano positions the discussion of STEM education in Japan in the context of the third largest economy in the world. He draws on the implications of the developments of STEM/STEAM learning in the United States and policy directions by the Japanese government, which chart the role of STEM education for the future society, that is, 'Society 5.0' as defined by the Cabinet of Japan (Cabinet Office of Japan, 2021). Kumano shares how policy is translated into action with the Shizuoka STEM Academy and concludes the chapter by calling for joint effort from schools, institutions, companies, and policymakers in addressing the needs for the future society.

Implementing STEM Education in Asia

The eight chapters in this section cover contributions from six different regions in Asia: Singapore, Thailand, Taiwan, Hong Kong, Malaysia, and Korea. Together, the chapters also consider STEM education across the early childhood, primary, and secondary education levels and urban and rural settings. A range of initiatives is reported, including considering formal and informal learning opportunities, adopting an engineering design process, implementing online experiments, engaging in scientist-teacher-student partnerships, promoting mentor–mentee learning within university school partnerships, and using the scientific research process to promote STEM learning.

Yann Shiou Ong and Yew Jin Lee analyse the implementation of STEM education in the Applied Learning Programmes (ALP) in Singapore by reviewing the websites of 15 secondary schools in Chap. 5. They use a theoretical model of relevance of STEM education, defining the relevance of STEM education for students in relation to fulfilling intrinsic or extrinsic needs in the present or in the future, and individual, societal, and vocational needs. The chapter concludes with three implications, including the need for consideration of student choices, awareness and discussion of the undesirable impacts of STEM solutions, and an emphasis on the epistemic aspects of STEM.

In Chap. 6, Promkatkeaw et al. discuss education policies in Thailand and the rationale for emphasizing STEM at the Early Childhood Education (ECE) level. There are clear directions for integrating science, mathematics, and technology in ECE in the Thai curriculum with an aim of promoting inquiry, communication, and thinking skills at an early age as described in the curriculum standards. The implementation is supported by professional development for ECE teachers, provision of STEM activities or lesson plans adopting an engineering design process and learning resources, as well as research efforts focusing on the impact of STEM learning on the cognitive development of young children. The chapter concludes with a call for the development of learning materials and lesson activities, and embedding the United Nations' sustainable development goals into STEM learning.

With an aim of promoting STEM learning to all students including girls in Taiwan, Meng-Fei Cheng and Yu-Heng Lo (Chap. 7) designed a 3-hour course to enhance students' creativity and engage them in an engineering design process. Cheng and Lo analysed the need for girl-friendly STEM education, and adopted various strategies in the course including providing a low-stress environment, creating a cooperative rather than competitive learning environment, and open-ended assessment. Findings suggest that the course reduced the gender gap in terms of enhancing the engineering design ability and attitudes towards technology among all students.

The online experiments (OEs) or remotely-controlled experiments introduced in Chap. 8 by Yau Yuen Yeung and his colleagues won a special prize in an international innovation and invention competition in 2018. The OEs which involve the use of technology and knowledge in science experiments were introduced to four secondary classes of students in Hong Kong and Mainland China. Findings suggest that students of both genders generally gained good science or STEM learning experiences. The chapter concludes with recommendations for designing OEs and the application of OEs, for example, during times of crisis like the COVID-19 pandemic, and where schools are not able to offer laboratory practice or for experiments that require extra precautions or extended time periods.

In Chap. 9, Rohaida Mohd Saat and Hidayah Mohd Fadzil explain how STEM is reflected in Malaysian education policies and school curriculum choices. There are defined STEM subject areas such as Science and Computer Science, STEM Core Subject Packages, STEM elective subjects, and Applied Science and Technology subjects for the STEM stream. Teaching and learning approaches such as inquiry-based and project-based learning are identified as being relevant for STEM. The three waves of initiatives to promote STEM are elaborated, and a detailed analysis of the Scientist-Teacher-Student Partnership (STSP) approach is analysed, with discussion of the roles of the different parties and the factors contributing to the success of the approach.

The STEM initiative in Chap. 10 is contextualized in the rural areas of East Malaysia, Sabah. Nyet Moi Siew reports on the implementation of a mentor–mentee programme involving secondary students, as well as in-service and pre-service teachers. The STEM activities were implemented using an engineering design process, aiming to develop the students' 21st century skills. Quantitative and qualitative data were collected over a five-year period to evaluate the outcomes of the programme and to identify challenges. Findings suggest benefits for developing students' thinking and creativity, problem solving skills, and skills involved in sketching, designing, and constructing models. Two challenges for the mentees or students were time constraints for completing their plans and limited understanding of scientific concepts.

In Chap. 11, Jiyeong Mun and Sung-won Kim elaborate on the development of STEM education in Korea and support from the Ministry of Education as well as the KOFAC. Details are provided about a STEM R&E (research and education) programme funded by the KOFAC, in which students were involved in a scientific research process including drafting proposals, applying for grants, consulting experts, and making presentations. Involving students as future scientists facilitates students' development of research skills, problem-solving abilities, and collaboration and peer cooperation skills. Students reflected that they learned both from their achievement and failures in the research process.

In Chap. 12, Hye-eun Chu et al. report how STEAM education is realized in Korean classrooms, using three school-based projects as illustrations. Teachers in both Korea and Australia were implementing the projects and testing out various teaching approaches to realize STEAM education. The first project illustrates a four-phased teaching design for adopting the STEAM approach. The second project involves an arts activity for designing a quasi-3D hologram using a mobile phone or iPad and a pyramid made of clear plastic. Interview findings about students' perceptions of learning science are summarized in relation to this project. The third project looks into the impact of STEAM on scientific creativity by involving students in designing a sustainable school building. The experience of these projects provides reference for teachers who are teaching in traditional classroom settings while attempting to implement STEAM education.

Preparing Teachers and Teacher Professional Development for STEM Education in Asia

The three chapters in this section together address teacher preparation for STEM education at both the pre-service and in-service levels. The contributions from Indonesia (Chap. 13) and Thailand (Chap. 14) are based on analysis of existing provision and information from websites or university programme materials. The chapter from Hong Kong (Chap. 15) reports on data collected from in-service teachers, recommending directions for professional development to facilitate STEM implementation at the school level.

Thus, in Chap. 13, Sulaeman et al. provide background to K-12 science education in Indonesia, pathways to become a science teacher, and the curriculum of science teacher preparation programmes. Drawing on information from the websites of three universities, Sulaeman et al. identify several strategies adopted by the universities in re-designing their programmes to address STEM teacher preparation needs, including designing a new compulsory course in STEM education, integrating STEM into other pedagogical courses, and providing elective courses and extra-curricular activities. Further opportunities to develop STEM pedagogical content knowledge through pre-service science teacher education programmes are proposed.

Chapter 14 provides a background on the Thailand 5-year STEM Master Plan from 2015 to 2019 for increasing the number of trained personnel to serve the STEM education policies. In this context, STEM education is promoted in the core curriculum, requiring urgent teacher development. Specifically, Fakcharoenphol et al. argue for strengthening support for both pre-service and in-service teachers in implementing STEM education, including covering different subject areas, not limited to science and mathematics. The chapter reports on the findings of a meta-analysis of research related to STEM teacher education in Thailand. Implications indicate directions for further support and development, including the setting up of communities of practice, developing teaching materials or resources, reforming student assessment, and reminding teachers to focus more on capturing students' interest and engaging them in evaluating their design or invention.

In Chap. 15, the findings are reported for a STEM education project that was conducted in collaboration with Hong Kong primary schools to enhance primary school teachers' professional ability to develop STEM activities. The project adopted a practitioner research approach in collaboration with the university research team, and teachers participating in the project were interviewed. The findings revealed participating primary school teachers' perceived challenges in implementing STEM education, the support they received, and their future needs in STEM teacher professional development. Challenges reported during the preparation and teaching phases were related to STEM instruction and lesson planning, limited resources, and other concerns embedded in broader contextual considerations. Participating teachers also indicated a preference for future STEM teacher professional development activities that address lesson content, pedagogical approaches and lesson planning; provide opportunities for authentic learning and hands-on experiences; and consider the linkages between STEM lessons and even the linkage or development of a framework describing STEM learning between the primary and secondary levels.

Implications, Challenges, and Future Directions

These chapters provide evidence for Asian-based STEM education studies within the worldwide landscape calling for and supporting widening participation in STEM education. Additionally, school-level STEM education has been much less studied compared to that in the higher education sector (Lee et al., 2019). The chapters in this book therefore add to the literature related to current practices in STEM education in schools around Asia.

Importantly, the chapters reflect some of the tremendous efforts underpinning and driving STEM education in Asia. For example, STEM education is seen as a meaningful endeavour informed by a vision to prepare for a future society in Japan (Chap. 4), and it is supported by organizations and foundations investing in the next generation, such as KOFAC in Korea (Chaps. 11 and 12), and non-government organizations in Thailand (Chap. 6). There are efforts in curriculum revision and development in Indonesia, Taiwan, and Thailand. Further, the benefits of university-school partnership projects are reported in the contexts of Hong Kong (Chaps. 8 and 15), Thailand (Chaps. 6 and 14), and Singapore (Chap. 5), with benefits of involving the research community in mentor–mentee programmes described in the context of Malaysia (Chap. 10).

Together, the chapters in this book reveal that different regions in Asia have been implementing STEM education in new ways at different levels of education. For example, initiatives are described for early childhood education in Thailand, the primary level in Hong Kong, and secondary schools in Japan, Korea, Thailand, Indonesia, and Singapore. In defining STEM education practices, the chapters in this book have consistently employed the term STEM in their analysis and as the focus of their study. Among the chapters included in this book, efforts to integrate STEM with other subject areas and STEAM initiatives are described in the context of Korea and Mainland China. Further efforts are needed to examine the implications of STEM education for early childhood and primary schools. In addition, the impacts of STEAM education as compared with STEM education on student learning warrant further attention.

Researchers contributing to this book were keen to identify changes in students' learning and attitudes towards STEM. Two of the Chaps. (7 and 8) specifically identify changes in students' attitudes towards STEM learning at school, or how STEM learning experiences may impact on their choices for future study or careers. The focus of Cheng's study in Chap. 7 on the impact of STEM initiatives on female students to enhance the future pool of human capital for the technology or engineering sector reflects international interest in gendered responses to STEM education initiatives. For example, Kang (2019) investigated the usefulness and usability of online media in technology-supported STEM learning, finding that having STEM experiences during elementary school had a long-term impact on university students' affective learning. In addition, STEM initiatives for female students may also be interpreted as a move towards nurturing all students to achieve basic STEM literacy and ensure equal learning opportunities. Continued research into the impacts and ways of enhancing the impacts of STEM education on students' learning is important.

Some of the approaches to delivering STEM education at the school level include partnerships with experts and organizations beyond schools. For example, there are efforts to involve scientists and engineers in industry to mentor students through industry-school partnerships in Japan and Korea (Chaps. 4 and 11). Specifically, students in these examples were offered opportunities to carry out authentic research projects guided by teachers and working with scientists or engineers. One of the many benefits of this approach is that students may identify role models for their future careers, thus influencing their perceptions of or attitudes towards choosing technology or engineering related jobs in their future careers.

While many of the chapters identify positive impacts of integrating STEM into an existing subject or curriculum, this is nonetheless challenging, as discussed in

Chaps. 12, , 14, and 15. The fact that engineering and technology components are often not explicit or visible in the curriculum core components suggests that the STEM acronym more frequently references science in the actual implementation of STEM curricula and classroom practices in the Asian context. According to the review by Marginson et al. (2013), many Asian regions refer to the role of STEM education as one that fosters 'broad-based scientific literacy' (p. 70). In order to implement STEM education as an integrated learning experience for all students, teachers need to understand its interdisciplinary nature, realize the differences in its implementation as compared to implementing traditional (siloed) subject learning, and have the experience of STEM learning themselves (as suggested in Chap. 15).

It remains challenging for the many aims of STEM education to be fully achieved. As demonstrated across this book, the challenges include curriculum design, teacher preparation and development, implementation at the school and classroom level, resource support, and assessing the impacts of the initiatives. Further studies are needed to identify solutions to address the challenges. For example, it is worth investigating the impacts of providing teachers with STEM learning experiences (as suggested in Chap. 15), constructing a teacher education curriculum to prepare teachers who have not experienced STEM education in their previous education (Chaps. 13 and 14), supporting teachers to develop a different mindset as compared with views in which traditional teaching rewards correct and uniform solutions (Chaps. 12 and 15), facilitating collaborations between teachers and industry or the research community (as discussed in many of the chapters), and enhancing teachers' capacity to use technology in their teaching (Chap. 8).

Researchers may facilitate teachers' professional development by engaging teachers in collaborations with universities, for example, in practitioner research that involves developing STEM lessons that address students' needs, such as the needs of female students; building coherence between STEM topics; and developing teachers' ability and confidence in relation to teaching unfamiliar topics, for example, science teachers teaching coding or engineering.

At the macro-level, the effective embedding of integrated STEM education will require administrators and education officials to support the implementation of STEM education at both the school and the system level, including providing time and opportunities for professional development, funding the development and sharing of teaching resources, creating curriculum time for STEM lessons or projects, and valuing new approaches to assessment. While the pursuit of STEM may mark a new era for education and offer new opportunities for students, the education community in its broadest sense needs to be prepared to embrace the changes as well as the challenges. It is our collective hope that this book provides some inspiring insights into ways in which this might be done.

References

Al Salami, M. K., Makela, C. J., & de Miranda, M. A. (2017). Assessing changes in teachers' attitudes toward interdisciplinary STEM teaching. *International Journal of Technology and Design Education, 27*(1), 63–88. https://doi.org/10.1007/s10798-015-9341-0.

Asghar, A., Ellington, R., Rice, E., Johnson, F., & Prime, G. M. (2012). Supporting STEM education in secondary science contexts. *Interdisciplinary Journal of Problem-Based Learning, 6*(2), Article 4. https://doi.org/10.7771/1541-5015.1349.

Brown, J. (2012). The current status of STEM education research. *Journal of STEM Education: Innovations and Research, 13*(5), 7–11. https://www.jstem.org/jstem/index.php/JSTEM/article/view/1652.

Cabinet Office of Japan. (2021). *Kagakugijyutsu Innovation Kihon Keikaku* [Science, technology, and innovation basic plan]. Government of Japan. https://www8.cao.go.jp/cstp/kihonkeikaku/6gaiyo.pdf.

Cho, J., & Huh, J. (2017, December 20). *New education policies and practices in South Korea.* UNESCO. https://bangkok.unesco.org/content/new-education-policies-and-practices-south-korea.

Education Bureau of Hong Kong. (2016). *Report on promotion of STEM education: Unleashing potential in innovation.* Education Bureau. https://www.edb.gov.hk/attachment/en/curriculum-development/renewal/STEM%20Education%20Report_Eng.pdf.

English, L. D. (2016). STEM education K-12: Perspectives on integration. *International Journal of STEM Education, 3*(3), 1–8. https://doi.org/10.1186/s40594-016-0036-1.

Herschbach, D. R. (2011). The STEM initiative: Constraints and challenges. *Journal of STEM Teacher Education, 48*(1), 96–122. https://doi.org/10.30707/JSTE48.1Herschbach.

Honey, M., Pearson, G., & Schweingruber, H. (Eds.). (2014). *STEM integration in K-12 education: Status, prospects, and anagenda for research*: National Academies Press. https://doi.org/10.17226/18612.

Hong, O. (2017). STEAM education in Korea: Current policies and future directions. *Asian Research Policy, 8*(2), 92–102.

Idris, N., Daud, M. F., Meng, C. C., Eu, L. K., & Ariffin, A. D. (2013). *Country comparisons: Country report Singapore STEM.* Australian Council of Learned Academies (ACOLA). https://acola.org/wp-content/uploads/2018/12/Consultant-Report-Singapore.pdf.

Indonesia Ministry of Education and Culture. (2020). *Fokus Pembelajaran SD/MI, SMP/MTs, SMA/MA.* [Learning focus for elementary school and middle school]. Badan Standar Nasional Pendidikan. https://bsnp-indonesia.org/fokus-pembelajaran/.

Ishikawa, M., Fujii, S., & Moehle, A. (2013). *Country comparisons: Japan.* Australian Council of Learned Academies (ACOLA). https://acola.org/wp-content/uploads/2018/12/Consultant-Report-Japan.pdf.

Jon, J.-E., & Chung, H.-I. (2013). *Country comparisons: Country report—Republic of Korea.* Australian Council of Learned Academies (ACOLA). https://acola.org/wp-content/uploads/2018/12/Consultant-Report-Korea.pdf.

Jones, A., & Buntting, C. (2015). Technology education and science education. In R. Gunstone (Ed.), *Springer Encyclopedia in Science Education* (pp. 1049–1053). SpringerReference.

Kang, N. (2019). A review of the effect of integrated STEM or STEAM (science, technology, engineering, arts, and mathematics) education in South Korea. *Asia-Pacific Science Education, 5*(6). https://doi.org/10.1186/s41029-019-0034-y.

Kelley, T. R., & Knowles, J. G. (2016). A conceptual framework for integrated STEM education. *International Journal of STEM Education, 3*(1), 1–11.

Korea Institute/Foundation for the Advancement of Science & Creativity. (n.d.). *Spreading science culture and maker culture.* https://www.kofac.re.kr/wp-content/uploads/2018/KOFAC_BROCHURE.pdf.

Lee, M. H., Chai, C. S., & Hong, H. Y. (2019). STEM education in Asia Pacific: Challenges and development. *Asia-Pacific Education Researcher, 28*(1), 1–4. https://doi.org/10.1007/s40299-018-0424-z.

Marginson, S., Tytler, R., Freeman, B., & Roberts, K. (2013). *STEM: Country comparisons. International comparisons of science, technology, engineering and mathematics (STEM) education: Final report.* Australian Council of Learned Academies (ACOLA). https://acola.org/wp-content/uploads/2018/12/saf02-stem-country-comparisons.pdf.

Ministry of Education Malaysia. (2013). *Malaysia education blueprint 2013–2025*. (*Preschool to post-secondary education*). Kementerian Pendidikan Malaysia. https://www.moe.gov.my/menumedia/media-cetak/penerbitan/dasar/1207-malaysia-education-blueprint-2013-2025/file.

Ministry of Education People's Republic of China. (2016). *Jiàoyù bù guānyú yìnfā "jiàoyù xìnxī huà "shísānwǔ" guīhuà" de tōngzhī* [The 13th five-year plan for educational informatization]. http://www.moe.gov.cn/srcsite/A16/s3342/201606/t20160622_269367.html.

Ministry of Education People's Republic of China, Beijing Normal University, & Beijing World Education and InformationTechnology Research Institute. (2017). *Zhōngguó STEAM jiàoyù fāzhǎn bàogào* [China STEAM development report]. Beijing Normal University.

Ministry of Education of Taiwan. (2018). *Shí'èr nián guómín jīběn jiàoyù kèchéng gāngyào—zìrán kē lǐngyù* [Curriculum Guidelines of 12-Year Basic Education—Science Education]. The Ministry of Education of Taiwan.

Ministry of Education Thailand. (2017). *Tạw chī̂ wạd lœa s̄āra kār reīyn rū̂ kæn klāng klùm s̄āra kār reīyn rū̂ withyāṣ̄āstr̒ (c̄hbạb prạbprung ph.Ṣ. 2560) Tām h̄lạks̄ūtr kæn klāng kār ṣ̄ụks̄'ā k̄hận phụ̄nṭhān phuthṭhṣ̄ạkrāch 2551* [Indicators and core learning areas science (Revised B.E. 2560) based on the basic education core curriculum B.E. 2551]. Office of the Basic Education Commission, MOE. https://drive.google.com/file/d/1_ALwE9xuCL3Fjet3XI4gYjBj8p_1zLaA/view.

Mizell, S., & Brown, S. (2016). The current status of STEM education research 2013–2015. *Journal of STEM Education, 17*(4), 52–56. https://www.jstem.org/jstem/index.php/JSTEM/article/view/2169.

Nadelson, L. S., & Seifert, A. L. (2017). Integrated STEM defined: Contexts, challenges, and the future. *The Journal of Educational Research, 110*(3), 221–223. https://doi.org/10.1080/00220671.2017.1289775.

Office of the Education Council Thailand. (2017). *Phænkār ṣ̄ụks̄'ā h̄æ̀ng chāti ph.Ṣ. 2560–2579* [Thai national education plan A.D. 2017–2036]. https://planipolis.iiep.unesco.org/sites/default/files/ressources/thailand_educationplan.pdf.

Organisation for Economic Co-operation and Development. (2010). *PISA 2009 results: What students know and can do–student performance in reading, mathematics, and science* (Vol. I). OECD Publishing. https://doi.org/10.1787/9789264091450-en.

Park, H., Byun, S.-Y., Sim, J., Han, H.-S., & Baek, Y. S. (2016). Teachers' perceptions and practices of STEAM education in South Korea. *Eurasia Journal of Mathematics, Science and Technology Education, 12*(7), 1739–1753. https://doi.org/10.12973/eurasia.2016.1531a.

Putra, P. D. A., & Kumano, Y. (2018). Energy learning progression and STEM conceptualization among pre-service science teachers in Japan and Indonesia. *New Educational Review, 53*(3), 153–162. https://doi.org/10.15804/tner.2018.53.3.13.

Shahali, E. H. M., Ismail, I., & Halim, L. (2017). STEM education in Malaysia: Policy, trajectories and initiatives. *Asian Research Policy, 8*(2), 122–133.

Takayama, K. (2010). Politics of externalization in reflexive times: Reinventing Japanese education reform discourses through "Finnish PISA success." *Comparative Education Review, 54*(1), 51–75. https://doi.org/10.1086/644838.

Thibaut, L., Ceuppens, S., De Loof, H., De Meester, J., Goovaerts, L., Struyf, A., Boeve-de Pauw, J., Dehaene, W., Deprez, J., De Cock, M., Hellinckx, L., Knippraath, H., Langie, G., Struyven, K., Van de Velde, D., Van Petegem, P., & Depaepe, F. (2018). Integrated STEM education: A systematic review of instructional practices in secondary education. *European Journal of STEM Education, 3*(1), 2. https://doi.org/10.20897/ejsteme/85525.

Wang, H. H., Moore, T. J., Roehrig, G. H., & Park, M. S. (2011). STEM integration: Teacher perceptions and practice. *Journal of Pre-College Engineering Education Research (J-PEER)*, *1*(2), Article 2. https://doi.org/10.5703/1288284314636.

May May Hung Cheng obtained her Bachelor's and Master's degrees from the University of Hong Kong and her Ph.D. from the University of Waikato, New Zealand. She is currently Chair Professor of Teacher Education in the Department of Curriculum and Instruction at the Education University of Hong Kong. She was previously a Reader in Professional Education at the Department of Education, University of Oxford. Her main areas of research are teacher education and science education with a focus on teacher learning and teacher professional development. She was the president of the East Asian Association for Science Education (EASE) between 2016 to 2019.

Part I
Conceptualising STEM Education in Asia

Chapter 2
Identifying Effective STEM Programmes and Strategies in Asia

May May Hung Cheng and Fang-Yin Yeh

Abstract This chapter reports a systematic review of existing literature related to the study of STEM in Asian secondary education. Thirty-two studies published in English from 2011 to 2020 were selected and reviewed. Guided by descriptive frameworks drawn from analysis of good practices of integrated STEM, STEM courses and programmes in Asia were examined in terms of firstly, the nature and scope of STEM programmes, secondly, instructional practices adopted in the STEM programmes, and thirdly, the evidence of the STEM programme impact on student learning in the knowledge, skills, and attitude domains. The findings showed that in addition to the frequent emphasis on science learning, an emerging trend of integrating engineering design at the secondary school level was observed among the existing Asian-based STEM programmes and courses. Additionally, project-based learning and engineering design-based learning were two main instructional designs used among the selected studies that demonstrated the flexibility in meeting different teaching and learning purposes. The review also suggests that crosscutting concepts are neglected among Asian-based STEM programmes and courses. This review characterized different STEM enactments in Asian secondary education, contributing towards identifying and discussing effective STEM programmes and strategies. Finally, this chapter points out areas for further development and invites researchers to further examine good practices in STEM education in the Asian context.

Keywords STEM studies · Asian classrooms · Primary school and secondary school · Pedagogical approaches in STEM studies · Student performance · STEM programme effectiveness

---

M. M. H. Cheng (✉) · F.-Y. Yeh
Department of Curriculum and Instruction, The Education University of Hong Kong, 10 Lo Ping Road, Hong Kong SAR, China
e-mail: maycheng@eduhk.hk

F.-Y. Yeh
e-mail: fyyeh@eduhk.hk

M. M. H. Cheng et al. (eds.), *Concepts and Practices of STEM Education in Asia*,
https://doi.org/10.1007/978-981-19-2596-2_2

Introduction

In recent years, especially in the second decade of the 21st Century, STEM education has gained more prominence in government policy and education circles in Asia. Concurrent with the global drive for STEM education, research in STEM education has shown exponential growth in the past decade. Much attention has been given to integrated STEM in recent STEM education research (Honey et al., 2014; Kelley & Knowles, 2016; Lee et al., 2019). Amongst the increased focus on integrated STEM, different challenges arising from its implementation have been documented such as the lesson enhancement or curriculum restructuring in the established discipline-based structure (Nadelson & Seifert, 2017), teacher preparation and professional development in STEM integration (Asghar et al., 2012; Lee et al., 2019), while one of the profound constraints for STEM teaching was teachers' lack of knowledge on embedding the four STEM components into the teaching of the subject (Kelley & Knowles, 2016). Research has also called attention to the need to better understand instructional practices and classroom operations due to a lack of empirical evidence of discerning elements crucial to the successful implementation of STEM education (Honey et al., 2014; Kang, 2019; Thibaut et al., 2018; Wahono et al., 2020).

Compared to the global STEM education scholarship, Asian-based STEM education studies are still in an emerging state, as Lee et al. (2019) highlighted that Asian-based STEM education studies comprised only 8.5% of the total examined international STEM education publications from 2013 to 2017 ($n = 665$), compared to around 65% from the United States. Thus, focusing on the Asian contexts, and with specific attention to the enactment of STEM education in Asian secondary schools, this chapter systematically reviews the enactment of STEM education with particular attention given to the nature and scope of STEM programmes or courses, instructional practices, and evidence reflecting the effectiveness of the instructional approaches in terms of students' learning.

Building a Framework for Reviewing STEM Programmes

There is in general a lack of consensus on how STEM education should be approached, largely due to the different experiences that involve different types of integration across STEM disciplines and school subjects (Sanders, 2009), or experiences that involve different forms, for example, single courses, school programmes, or extracurricular activities (Honey et al., 2014). A few recent studies have formulated different frameworks to describe the approaches and conditions of integrated STEM that most likely lead to positive outcomes of STEM education (Honey et al., 2014; Statauskiene & Mazgelyte, 2016; Thibaut et al., 2018). These frameworks provide useful references for reviewing a myriad of STEM programmes and courses implemented across Asian curriculum and classroom contexts.

Honey et al. (2014) developed a descriptive framework of integrated STEM education for a narrative review of 28 selected good practices in K-12 education in the United States. The descriptive framework identified four higher level features (and various illustrative subcomponents) of (1) goals (different statements of goals); (2) nature and scope (different types of STEM integration, disciplinary emphasis, duration, size); (3) outcomes (different outcomes in learning, schooling, and employment); and (4) implementation (different instructional designs, supports, partnership).

Second, in a European Commission-supported project, Statauskiene and Mazgelyte (2016) devised a top-down, theory-based four-dimensional framework to analyse 22 selected good practices in integrated STEM education with the aim of understanding the recurrent features of good practices in secondary schools across seven European countries. The framework integrated (1) learning theory (behaviourism, cognitivism, constructivism, humanism); (2) type of intellect (Gardner's multiple intelligences); (3) type of integration; and (4) competences (according to the European Reference Framework of Key Competences for lifelong learning).

Third, Thibaut et al.'s (2018) study reviewed 23 international STEM studies that provided a clear description of instructional practices for integrated STEM in secondary education. The study formulated, from the bottom-up, a five-principle framework, including (1) integration of content, (2) problem-centred learning, (3) inquiry-based learning, (4) design-based learning, and (5) cooperative learning.

With a specific focus on identifying features of STEM classroom practices and effective STEM programmes and strategies, components relating to the integration, implementation (of instructional practices), and the outcomes (of student learning) from the three integrated STEM education frameworks mentioned above are referenced in developing the framework for this review. Table 2.1 presents the framework formulated for this review. Features from Honey et al.'s (2014) descriptive framework provided the main structure of this review, and subcomponents from Honey et al. (2014) and Thibaut et al. (2018) were referenced to assist the illustrative procedure, which allows other subcomponents to be included in the review process.

Table 2.1 Framework for STEM programme review

Main features informed by Honey et al. (2014)	Subcomponents
Nature and scope of STEM programme integration	E.g. types of STEM connection and disciplinary emphasis; duration; size; formal or after-/out-of-school settings (see Honey et al., 2014)
Implementation of STEM programme: instructional practices	E.g. Problem-centred learning, project-based learning, problem-based learning, inquiry-based learning, design-based learning, cooperative learning (see Thibaut et al., 2018)
Outcomes of STEM programme for student learning	E.g. STEM literacy; 21st Century competences; interest, and engagement (see Honey et al., 2014)

This review aimed for a systematic result pertinent to characterizing existing STEM classroom practices and effective STEM programmes and strategies in Asian secondary education. The key questions guiding the study are as follows:

(1) Nature and scope of STEM programme integration:
RQ1: What is the nature and scope of the Asian STEM programme in terms of types of integration, disciplinary emphasis, duration, and scale?
(2) Implementation of STEM programme: Instructional practices
RQ2: What are the main instructional practices implemented in the Asian STEM programmes or courses?
(3) Outcomes of STEM programmes for student learning
RQ3: What are the domains of student learning outcomes examined in the selected Asian STEM programmes?
RQ4: What is the effectiveness of promoting different aspects of student learning in relation to the STEM instructional practices reported?

Methodology

A systematic review was conducted to identify, evaluate, and summarize the findings of all relevant literature on the enactment of STEM education in secondary schools in Asia. Literature sources include international STEM-related studies written in English with the publication timeframe ranging from 2011 to 2020. An article search was conducted using the academic databases of the Web of Science, Science Direct, and Google Scholar. Databases were browsed using a combination of keywords: 'STEM' plus related words such as education, discipline, programme, course, or integration, Asian country/region names, and school level. The selected studies had to be journal articles (thus conference papers were excluded) that described the instructional practices for the STEM courses and programmes and presented results for student learning. Articles were checked by one rater regarding whether inclusion criteria were met. When doubtful, papers were discussed by the two researchers until consensus was reached.

A total of 32 articles representing qualitative, quantitative, and mixed methods studies were selected. The review covered eight countries/regions in East and Southeast Asia, including Hong Kong (1 study), South Korea (1 study), Japan (2 studies), China (3 studies), Taiwan (11 studies), Thailand (1 study), Indonesia (2 studies), and Malaysia (11 studies) (see Appendix). The student participant sample size ranged from 24 to 411 in the individual studies. The 32 studies were reviewed to answer research questions 1, 2, and 3. For research question 4, 27 out of 32 studies that employed the pretest–posttest or the nonequivalent groups designs were reviewed to make claims for the effectiveness of the STEM courses for student learning.

Findings

The Nature and Scope of STEM Programmes

The results of the review identified different STEM connections planned in the selected STEM courses in terms of (1) integration of concepts from different STEM disciplines such as the applications of science and mathematics concepts to scientific inquiry (e.g. Wang et al., 2015); (2) connecting concepts in subjects to the practice of another such as the applications of science and mathematics concepts to engineering design practices (e.g. Anwari et al., 2015; Lou et al., 2011); and (3) combining practices such as integrating scientific inquiry and engineering design/technological practices (e.g. Siew, 2017; Tsai et al., 2018). Commonly found features among these different types of STEM connection were aspects of learning-by-doing (e.g. project tasks, artefact design, experiments) and multiple representations (e.g. the integration of medical knowledge and skills in the form of spatial visualisation in Hsu et al., 2017).

Different subject contents were emphasised among the selected STEM programmes and courses. As shown in Fig. 2.1, focus on science (13 studies) and engineering (15 studies) learning was common in secondary education, especially at the senior secondary school level in Asia. STEM education with an emphasis on science learning primarily focuses on topics such as physics, chemistry, and ecology/biology (see Appendix). STEM education that highlights engineering learning incorporates the engineering design process and encompasses topics relating to mechanical engineering (e.g. solar automatic trolley, CaC_2 steamship, self-propelling sailboat robot). Additional attention on environmental engineering (e.g. clean water treatment system) and civil engineering (e.g. irrigation system, water-level warning

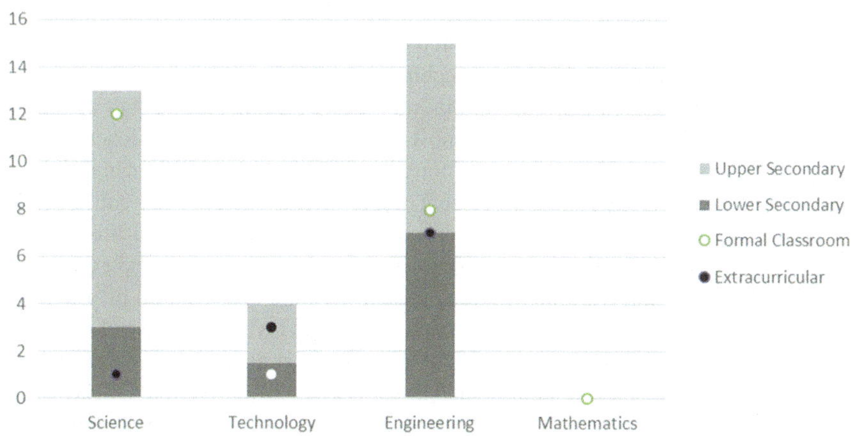

Fig. 2.1 STEM programme by emphasis of subject areas, school levels, and settings (number of studies)

system) were also observed. STEM education that emphasised technology learning was less common (4 studies). However, engineering design is an integral component in technology learning, primarily through emerging technologies in education such as the microcontroller unit, 3D printing technologies, and robotics. STEM education emphasising mathematics was not recorded among the studies on secondary STEM education in Asia.

In terms of the scope, STEM education with an emphasised on science learning was primarily implemented in formal classroom settings. Courses emphasising engineering and technology learning were commonly implemented across various extracurricular or co-curricular settings (e.g. outreach programmes, school clubs, workshops) (see Fig. 2.1). Courses vary in their implementation timeframes in terms of duration, with the shortest comprising a 10-min virtual reality immersion activity (Huang, 2019) to the longest, an action research project that lasted a whole school year (Chen & Lin, 2019). The majority of the STEM courses involved either a small group (fewer than 60 participants, 13 studies) or a large group (above 100 participants, 13 studies) of participants from a single class, multiple classes within one school, or across different schools.

The Implementation of STEM Programmes

This section reports the main and the supporting instructional designs identified in the selected Asian STEM courses and programmes. The main instructional designs specified in the 32 studies are project-based learning (17 studies), technology-supported learning (6 studies), engineering design-based learning (5 studies), inquiry-based learning (3 studies), and problem-based learning (1 study). Lesser mentioned instructional designs such as the inventive thinking approach, critical thinking process, or the scientific imagination embedded in the design process were employed by one or two studies (see Appendix). In the following section, the four instructional designs: project-based learning, technology-supported learning environment, engineering design-based learning, and inquiry-based learning are further detailed.

Project-based learning (PBL) is the most common instructional approach (17 out of 32 STEM studies, or 53%) adopted among the STEM courses. PBL is highlighted as a student-centred instructional model that organizes learning around creating open-ended projects (e.g. authentic prototypes, products, or artefacts) with real-life connections. PBL promotes the diverse specific skills, group work, social, and communication skills within the learning process, and through the construction of products, showcases students' mastery of concepts in different STEM subjects (see Lou et al., 2017; Pang et al., 2019; Saleh et al., 2020; Wan Husin et al., 2016). Apart from the conceptual definitions, a few studies such as Lou et al. (2017) viewed 'STEM PBL' as a teaching model defined to 'specify vague work tasks in the interdisciplinary structure' (Stearns et al., 2012 as cited in Lou et al., 2017, p. 2391).

Around half of the STEM courses that employed PBL (8 out of 17) were supported by instructional models with clear teaching phases. An emphasised on the E (engineering) component is observed in the six-phase educational model of 'iSTEM (imagination STEM) Learning' (Tsai et al., 2018), the Maker approach (see Chen & Lin, 2019; Lou et al., 2011) and the 6E inquiry-based learning (see Hashim et al., 2020). The engineering design process is also used in implementing STEM PBL, such as in Wan Husin et al. (2016) that merged a three-step engineering design process (i.e. think, make, and improve) into their 'Project-oriented Problem-based Learning' instructional model. Technology-supported learning and inquiry-based learning are two other supporting instructional designs commonly used to support PBL among the selected studies.

Technology-supported learning (TSL) is an umbrella term for teaching practices that utilize educational technology to enhance learning. A total of 14 out of 32 STEM courses (44%) indicated TSL either as their main instructional feature (6 STEM courses, all at the senior secondary school level) or as the supporting instructional design (8 STEM courses). Technology was incorporated into the learning environment through the use of internet technology (5 studies), computer application (3 studies), or the more advanced computer technology (6 studies) to support technology-enabled collaboration, virtual/online laboratories, remedial learning, educational gaming, real-time formative assessment, or skill-based curricula.

Internet technology such as the use of learning management systems (Moodle, Edmodo, or the self-developed 'STEM interactive internet platform' in Tsai et al., 2018), information communication technology (Facebook) and the real-time internet database (for information and knowledge integration in Jeong & Kim, 2015) all supported an environment for collaborative, interactive, and inquiry-based activities. The use of computer applications for STEM courses includes computer adaptive remedial learning systems (Chang et al., 2015), software for physics laboratory simulation (Wang et al., 2015), and the mobile application-based educational game (Wang & Chiang, 2020). Recent STEM courses also utilized more advanced computer technology such as virtual reality for immersive experiences (Huang, 2019), augmented reality (Hsu et al., 2017; Majid & Majid, 2018) and 3D printing (Chien, 2017; Lin et al., 2018) for skills training (e.g. medical surgery, digital modelling).

Engineering design-based learning (EDBL) is an umbrella term for STEM courses that incorporate engineering design tasks. A total of 15 STEM courses (47%) employed EDBL either as the main pedagogical anchor (5 STEM courses, all at the senior secondary school level) or as the supporting instructional design (10 STEM courses). Different learning purposes were discerned among the programmes that utilised EDBL. One aspect of learning emphasised applying STEM disciplinary knowledge and skills in the engineering design cycle to solve problems. Examples include Chien's (2017) STEM-oriented pre-engineering curriculum on digital modelling and manufacturing (concerns the design, output and revision decisions) grounded in mathematics and science principles. Another type of EDBL used engineering or technological design tasks to facilitate student learning and to develop integrated STEM knowledge (e.g. Changtong et al., 2020; Hafiz & Ayop, 2020;

Khozali & Karpudewan, 2020), skills (e.g. Anwari et al., 2015; Hafiz & Ayop, 2020; Wan Husin et al., 2016), interest and engagement in STEM (e.g. Chen & Chang, 2018; Hafiz & Ayop, 2020; Shahali et al., 2017; Wang & Chiang, 2020).

The implementation of EDBL also varies among the programmes and courses. Most of the programmes adopted engineering design process (EDP) developed by U.S.-based institutions (e.g. National Research Council, 2013; Massachusetts Department of Education, 2006; and Museum of Science-Boston, 2009 as cited in Fan & Yu, 2017; Hafiz & Ayop, 2020; Shahali et al., 2017; Siew, 2017) and scholars (e.g. Atman et al., 2007; Martinez & Stager, 2013; and Moore et al., 2016 as cited in Chien, 2017; Wan Husin, et al., 2016; Huri & Karpudewan, 2019). A few EDP models were developed by local educational institutions to stress course alignment with their respective national curricula (e.g. Changtong et al., 2020). EDBL was also implemented through an emphasis on engineering design activities. Examples include the 'task-centred teaching model' (e.g. building robotic sailboat in Chen & Chang, 2018), or the 'jigsaw learning approach' in Changtong et al. (2020) that engages students in the engineering design activity through collaborative learning and inquiry-based investigation.

Inquiry-based learning (IBL) was employed in nine STEM courses (28%) either as the main pedagogical anchor (3 STEM courses, all at the lower secondary school level) or as the supporting instructional design (6 STEM courses). The 5E Instructional Model (e.g. Bybee et al., 2006) and the 6E Instructional Model (e.g. Burke, 2014) were commonly adopted and integrated (in 2 studies each) among the selected STEM courses. The 5E Instructional Model comprised engagement, exploration, explanation, elaboration, and evaluation. It was integrated into Wang et al.'s (2015) 'model-based inquiry' approach that emphasized the construction and reconstruction of models based on scientific investigation, and in Lay and Osman's (2017) study that combined a creative design spiral, inquiry/discovery activities, and teamwork for educational media design. The 6E Instructional Model, adopted in Lin et al. (2020) and Hashim et al. (2020), added an 'engineering' element to the 5E model in the inquiry cycle to emphasize the 'purposeful design and inquiry' in STEM education (Lin et al., 2018, p. 2).

Other STEM courses that implemented IBL without a guiding inquiry cycle viewed it as a broad pedagogical approach with the overall goal for students to make meaning, more often through hands-on activities that incorporated technologies or educational kits. Examples include Hsu et al.'s (2017) 'authentic inquiry' that engages students in computer-simulated experiments to develop inquiry skills, and in Majid and Majid's (2018) 'guided discovery learning' through mobile AR that facilitated more engagement in the learning process and promoted analytic learning through imposing questions.

The Outcomes of STEM Programmes and Courses for Student Learning

This section reports the outcomes of STEM programmes and courses for student learning. It first presents an overview of student learning by three domains of knowledge/practices, skills, attitudes, and evaluation methods, followed by reporting the effectiveness of the STEM programmes for student learning.

Domains of Student Learning and Measurement

In reviewing STEM knowledge/practices, three dimensions were examined. First, the 'disciplinary core ideas' include motions, energy, ecosystem, evolution, engineering design. Second, the 'crosscutting concepts' include concepts that bridge disciplinary core boundaries, including patterns, cause and effect, proportion and quantity, energy and matter, structure and function, stability and change. Third, the 'scientific and engineering practices' include asking questions (for science) and defining problems (for engineering), developing and using models, planning and carrying out investigations, using mathematical and computational thinking, constructing explanations (for science) and designing solutions (for engineering), obtaining, evaluating and communicating information) (National Research Council, 2013).

For the skills domain, 'peer learning' and 'cognitive thinking skills' were inductively formulated from the analysis to represent the learning emphasis reported in the selected STEM programmes and courses. First, peer learning includes social skills such as collaboration skills, communication skills, respecting different points of view, and so forth. Second, cognitive thinking skills group higher thinking skills such as creativity, critical thinking, or inventive thinking skills.

The attitude domain examines students' interest in, perceptions of, and the degree of acceptance of the instructional content. Two categories were inductively formulated from the analysis. First, 'interest in STEM' examines students' interests, motivation, or confidence in the STEM disciplines and careers. Second, 'engagement in STEM activities' looks at student engagement in STEM activities.

Figure 2.2 presents the patterns of student learning reported in the 32 studies by learning domains (knowledge, skills and attitude) and instructional practices (PBL, TSL, EDBL, and IBL). The dotted line indicates the total number of studies that reported student learning outcomes. From the figure, disciplinary core ideas and scientific and engineering practices in the knowledge/practice domain, and interests in STEM in the attitude domain remained the focus for student learning across different STEM courses.

STEM courses that adopted PBL, TSL, and IBL tended to report student learning outcomes across different learning domains. Courses that incorporated EDBL were more focused on the effect of the teaching methods on students' knowledge/practice learning. Courses that adopted PBL were interested primarily in students' attitudes in STEM disciplines and/or career orientations (e.g. Chen & Chang, 2018). Courses

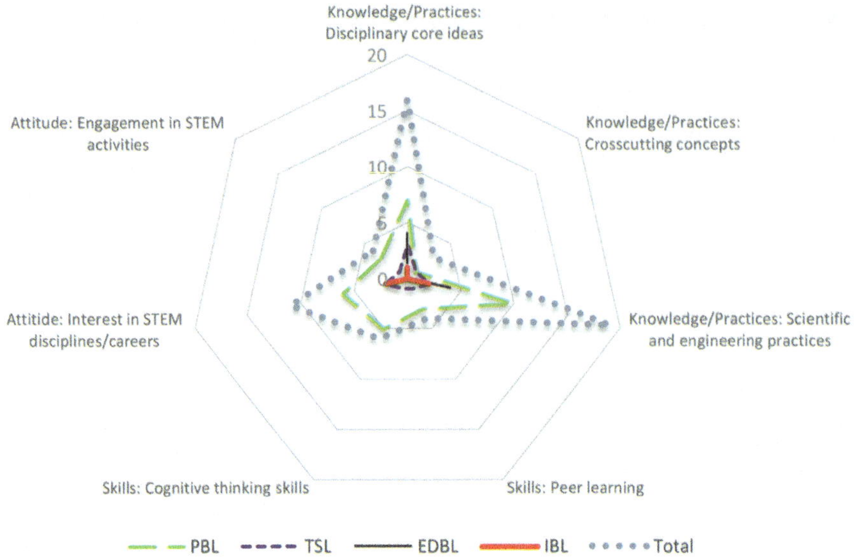

Fig. 2.2 Patterns of student learning by learning domains and instructional practices

incorporating TSL reported a closer examination of students' confidence in technological skills (e.g. Tam et al., 2020) or the perceived value of technology in enhancing their learning (e.g. Chang et al., 2015).

The learning of crosscutting concepts was embedded in different STEM courses that connect different domains of science (e.g. physical science and earth science in Jeong & Kim, 2015; physical science, life science, and engineering design in Mutakinati et al., 2018). However, crosscutting concepts received the least attention for student learning outcomes among the STEM intervention courses.

In terms of the evaluation methods used to measure student learning outcomes, most studies used multiple approaches, categorized into three groups in this review:

(1) *Student self-reported information* collects data on student learning outcomes through the primary use of a questionnaire, survey, or tendency scales (e.g. the 21st Century Skills Questionnaire in Wan Husin et al., 2016);
(2) *Teacher subjective assessment* uses a pre-determined rubric, criteria, or scale for teachers to evaluate student production, skills, or behaviours (e.g. the Critical Thinking Rubric in Mutakinati et al., 2018)
(3) *Objective assessment* refers to tests mainly in the forms of true–false or multiple-choice questions for students to complete (e.g. Industrial Electronics C Technician Skills Certification Test in Chang et al., 2015).

Figure 2.3 presents an overview of the methods used to evaluate students' STEM learning outcomes across the three learning domains. From the figure, *student self-reported information* and *teacher subjective assessment* were commonly used in

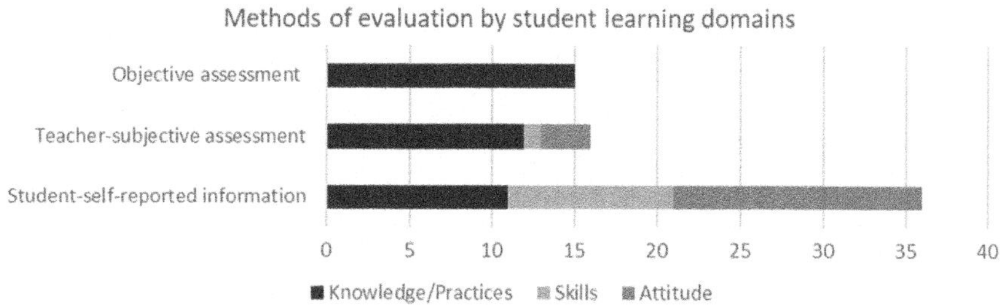

Fig. 2.3 Methods of evaluation by student learning domains

assessing student learning across the three learning domains. *Objective assessment* primarily targeted evaluation of students' conceptual and procedural knowledge.

The Effectiveness of the Instructional Practices in Terms of Student Learning

To further study the effectiveness of the instructional practices, a group of 27 studies that employed quasi-experimental designs were examined. Based on the intervention results as reported in individual studies, Table 2.2 presents student learning performance results by ' + ' (study reported instructional practices having a significant impact on student learning), '–' (study reported that the control group performed better than the experimental group) and 'n/s' (study reported instructional practices having no significant impact on student learning).

Knowledge and Practices

Regarding student learning outcomes in the knowledge and practices domain, STEM studies that integrated IBL and EDBL more often showed positive impacts on students' learning in disciplinary core ideas and scientific and engineering practices. Common traits among these STEM courses are the explicit and purposeful planning of the teaching and learning of disciplinary core ideas (e.g. chemistry-related in Lay & Osman, 2017; Majid & Majid, 2018, or physics-related in Balakrishnan et al., 2019; Saleh et al., 2020) and the purposeful engagements of science and mathematics principles in engineering design tasks (e.g. Fan & Yu, 2017; Huri & Karpudewan, 2019; Tsai et al., 2018).

A variety of scientific and engineering practices were examined in the STEM studies, including scientific process skills (Wang et al., 2015), the scientific imagination process (Siew, 2017), productivity skills and product creativity (Chien, 2017; Lay & Osman, 2017), and problem prediction and analysis (Chien, 2017; Fan & Yu, 2017; Tam et al., 2020). Common characteristics among the STEM courses that reported positive impacts of the teaching interventions on different scientific and engineering practices is the incorporation of a clear inquiry cycle or a clear EDP model in the courses (e.g. the 5-phase inquiry model in Tam et al., 2020; the 5E Instructional

Table 2.2 Student performance results: knowledge/practices, skills, attitudes

Main instructional design	Supporting instructional design	Study conducted by	Knowledge and practices domain			Skills		Attitude	
			Disciplinary core concept	Crosscutting concept	Scientific and engineering practices	Peer learning	Cognitive thinking skills	Interest in STEM	Engagement in STEM activities
PBL	EDBL	Anwari et al. (2015) *			(n/s)				
PBL	EDBL	Chen and Chang (2018) **	(+)		(+)			(+)	
PBL	EDBL	Shahali et al. (2017) *						(+)	
PBL	EDBL	Wan Husin et al. (2016) *			(+)	(n/s)	(n/s)		
PBL	TSL; EDBL	Lin et al. (2018) **-; **	(-)		(n/s)				
PBL	TSL; EDBL	Tsai et al. (2018) *	(+)						
PBL	TSL; EDBL	Wang and Chiang (2020) *				(+)		(+)	(n/s)
PBL	TSL	Jeong and Kim (2015) **						(+)	
PBL	TSL	Purwaningsih et al. (2020) **			(+)				
PBL	IBL	Hashim et al. (2020) *					(+)		
PBL	IBL	Lay and Osman (2017) **	(+)		(+)	(n/s)	(n/s)		
PBL	n/a	Lou et al. (2017) *					(+)		
PBL	n/a	Pang et al. (2019) **-						(+)	
PBL	n/a	Saleh et al. (2020) **	(+)				(n/s)		
TSL	EDBL	Khozali and Karpudewan (2020) *		(+)					
TSL	IBL	Majid and Majid (2018) *	(+)						
TSL	IBL	Wang et al. (2015) **	(n/s)		(+)	(+)			(+)
TSL	n/a	Chang et al. (2015) *	(+)						

(continued)

Table 2.2 (continued)

Main instructional design	Supporting instructional design	Study conducted by	Knowledge and practices domain			Skills		Attitude	
			Disciplinary core concept	Crosscutting concept	Scientific and engineering practices	Peer learning	Cognitive thinking skills	Interest in STEM	Engagement in STEM activities
TSL	(immersive)	Huang (2019) **-						(n/s)	
EDBL	TSL	Chien (2017) **-	(n/s)		(+)				
EDBL	n/a	Siew (2017) *			(+)				
EDBL	n/a	Changtong et al. (2020) **	(+)		(n/s)				
EDBL	n/a	Fan and Yu (2017) **	(+)		(+)				
EDBL	n/a	Huri and Karpudewan (2019) *	(+)						
IBL	(hands-on)	Balakrishnan et al. (2019) **-	(+)						
IBL	EDBL	Lin et al. (2020) **			(n/s)			(n/s)	
IBL	TSL	Tam et al. (2020) *			(+)			(+)	

* Single group pretest–posttest design; ** Nonequivalent group design; ** - Posttest-only control group design

Model integrated into Lay & Osman, 2017; Wang et al., 2015; integrated EDP in Chien, 2017; Fan & Yu, 2017; Siew, 2017; Wan Husin et al., 2016).

Skills

Programmes that integrated a clear cognitive thinking skills module (e.g. creativity in Lou et al., 2017) often reported having a positive impact on the respective skills measured. On the other hand, in studies without purposeful incorporation of the cognitive thinking modules, when cognitive thinking skills were measured (e.g. inventive thinking in Lay & Osman, 2017; Saleh et al., 2020; Wan Husin et al., 2016), the impact of the instructional designs on students' skills were often insignificant. The possibility of the short duration of the STEM programmes in limiting students' performance results was also identified in Wan Husin et al. (2016). In which a more extended period of the programme was recommended.

Peer learning was not a focal evaluation aspect for student learning among the selected secondary school level STEM courses. However, pedagogical interventions with collaborative components (e.g. the collaborative online environment in Wang et al., 2015) reported students' significant improvement in collaboration skills, expressing ideas, and making arguments. Some PBL studies reported no significant statistical differences in communication skills (e.g. Lay & Osman, 2017; Wan Husin et al., 2016) were observed to have high baseline scores in students' peer learning performances.

Attitude

The positive impact of the instructional designs on students' interest in STEM and engagement in STEM activities was reported in different studies. Specifically, studies that employed PBL reported favourable outcomes in students' interest in STEM disciplines (Chen & Chang, 2018; Jeong & Kim, 2015; Pang et al., 2019; Shahali et al., 2017; Wang & Chiang, 2020) and students' interest in STEM careers (Chen & Chang, 2018; Shahali et al., 2017). Those courses that implemented PBL shared a strong feature of engineering and/or technological design components for engaging students in practical activities.

One study (Tam et al., 2020) that employed technology-supported learning reported a positive impact of the instructional design on students' ICT self-efficacy. Two studies, including Huang (2019) that examined the impact of technology-supported learning on students' science self-efficacy, and Lin et al. (2020) that examined the effectiveness of inquiry-based learning on students' attitude towards technology, both reported insignificant impact of the teaching interventions on the students attitudinal learning. Similar to the findings in previous sections, identification of meaningful student learning outcomes, purposeful planning of instructional designs, and assessment planning could be central to preparing students to achieve the course outcomes.

Two STEM courses (Wang & Chiang, 2020; Wang et al., 2015) reported a positive impact of the instructional designs to at least retained students' engagement in STEM. The two STEM courses integrated computer applications (i.e. simulation software, mobile application) into their student-centred STEM courses and shared a strong peer collaboration component.

Discussion and Conclusion

The review identified existing STEM classroom practices in secondary education in the Asian contexts in terms of the nature of integration, instructional practices, and the evidence of the STEM programmes' impact on students' learning.

The review reflects that the efforts for STEM education provision in the Asian context are more frequently implemented with an emphasis on science and engineering learning. Previous literature such as Lee et al. (2019) has expressed concerns over a focus on science and the inequitable attention given to other STEM components (i.e. T. E. M) in school learning in the Asian contexts (see also English, 2016 for a similar concern in Western-based studies). The current review pointed to an emerging trend of integrating the E (engineering design) among the existing Asian-based STEM programmes and courses at the secondary education level.

While STEM courses emphasising technology learning were less common among the selected studies, they shared engineering design as an integral element of technology learning. Similar to Chien's (2017) approach to technology education, the review observed a direction in viewing engineering-oriented technology education as a pathway to pre-engineering education in secondary education in the Asian contexts. Concurring with Lee et al. (2019), the review continued to show mathematics and technology as subject areas, and mathematics as components needing more explicit representation and elevation in school-level STEM education and research.

In terms of instructional design, PBL was the main instructional design adopted by most studies to facilitate core curriculum content learning and the development of interdisciplinary skills. Compared to PBL in science learning that emphasized scientific practices, Kang (2019) highlighted that PBL in integrated STEM exhibits additional features originating from engineering and technological education. From the review, the diverse 'STEM PBL' models –which are student and design-tasked centred, with clear instructional stages informed by IBL, engineering design principles, or other teaching designs—illustrated the flexibility in designing STEM PBL for meeting different teaching and learning purposes.

The implementation of EDBL among the selected STEM programmes and courses also exemplified its flexibility for different teaching and learning objectives, e.g. learning the engineering design cycles or advancing integrated STEM knowledge. As stated by Householder and Hailey (2012), the distinctiveness of the engineering design challenge depended upon the implicit reliance upon analysis and close adherence to the application of science in the product design. A few STEM courses that adopted EDBL (Changtong et al., 2020; Chien, 2017; Fan & Yu, 2017) have emphasized the applications of science and mathematical knowledge and explicitly highlighted science and mathematics knowledge to support the design process.

STEM courses that adopted TSL illustrated an emerging trend in technology-supported STEM learning in Asian classrooms. The review partly concurs with Lee et al.'s (2019) observation that some technology-supported STEM lesson designs may still be inclined towards didactic teaching (e.g. lectures). However, the review illustrated other utilization of technologies (e.g. internet technology, computer application, more advanced computer technology) in Asian classrooms to support technology-enabled collaboration (e.g. knowledge sharing, ideation, task production), learning (e.g. engaging with multiple representations, virtual/online laboratories, remedial education, skill-based learning, gaming), and assessment (e.g. formative assessment).

When interpreting the effectiveness of the instructional practices for student learning, caution is needed due to the limited number of studies reviewed, the varied programme durations, and the focus of and approach to the evaluation of different STEM learning domains. However, some common characteristics of the STEM courses that reported positive student learning outcomes in the knowledge and practices domain may offer some insights in future planning. The common characteristics include explicit planning of disciplinary core ideas in the lesson content, purposeful engagement of science and mathematical principles in engineering design activities, and incorporating the clear inquiry cycle and/or EDP models.

Furthermore, the review suggests that the development of crosscutting concepts is severely neglected among the Asian-based STEM programmes and courses. As a way to better integrate crosscutting concepts learning, future STEM programmes and courses could make explicit the performance expectations of crosscutting concepts for student learning in STEM courses. Future researchers may consider putting more effort into developing objective tests for measuring the different types of knowledge and skills involved in STEM.

Lastly, little is known about the best practices for implementing or integrating STEM (English, 2016; Kelley & Knowles, 2016). This systematic review of existing STEM studies in the Asian contexts is guided by frameworks derived from the previous meta-synthesis of good practices in STEM education showcased in the United States and Europe (Honey et al., 2014; Thibaut et al., 2018). Findings from this review may contribute to an initial discussion of the effective or good practices and the variant conditions that most likely lead to positive outcomes in relation to teaching, learning, and assessment in the Asian context. With the continual efforts of the government and educational bodies to promote STEM in Asia, effective or good practices in STEM education will likely grow. The findings of similar meta-synthesis studies and further meta-analysis studies in the Asian contexts will be enriched when a more extensive collection of effective or good practices in STEM education in Asia are showcased in the future.

Appendix: Basic Information of STEM Programmes and Courses (in Alphabetical Order)

Study conducted by	Location	Grade level	Subject area emphasis	Duration	Size	Formal/Extra-curricular	Task/topics	Instructional practices
Anwari et al. (2015)	Japan	Grade 9	Engineering	3 lessons	147	Formal	Design enhanced DC motor	1. PBL 2. EDBL 3. Scaffolding instruction
Balakrishnan, et al. (2019)	Malaysia	Form 2, 3	Science (physics)	/	255	Formal	Build card structure; magnobolt kit	1. IBL 2. Hands-on
Chang et al. (2015)	Taiwan	Grade 11	Science (electronics)	10 weeks	32	Formal	Operate multimeter to measure voltage and current	1.TSL
Changtong et al. (2020)	Thailand	Grade 9	Engineering	5 days	115	Formal	Racing car challenges	1. EDBL 2. Collaborative learning (jigsaw learning)
Chen and Chang (2018)	Taiwan	Grade 10	Technology	1 semester	82	Extracurricular	Design a self-propelling Sailboat robot	1. PBL 2. TSL: Arduino 3. EDBL
Chen and Lin (2019)	Taiwan	Grade 8 and 9	Science (physics, engineering)	2 years	24	Formal	NBA Ocean Park; cross-sea bridge etc.	1. PBL (Maker-centred)
Chien (2017)	Taiwan	Grade 10	Technology (3D printing)	8 weeks	182	Formal	CO2 dragster design	1. EDBL
Fan and Yu (2017)	Taiwan	Grade 10	Engineering	10 weeks	332	Formal	Mechanism toy design	1. EDBL
Hafiz and Ayop (2020)	Malaysia	Form 4	Engineering	4 sessions	32	Extracurricular	Water rocket	1. PBL 2. EDP 3. Inquiry-based
Hashim et al. (2020)	Malaysia	Form 1	Engineering	/	44	Extracurricular	Create water level warning prototypes	1. PBL 2. 6E
Hsu et al. (2017)	Taiwan	Grade 10	Science (medical science)	2 lessons	32	Formal	Laparoscopic surgery and cardiac catheterization simulation	1. TSL: AR simulator 2. IBL
Huang (2019)	China	Grade 11	Science (biology)	1 lesson	66	Formal	VR experiences	1. TSL: VR head-mounted display

(continued)

(continued)

Study conducted by	Location	Grade level	Subject area emphasis	Duration	Size	Formal/Extra-curricular	Task/topics	Instructional practices
Huri and Karpudewan (2019)	Malaysia	Form 4	Science (chemistry)	5 weeks	50	Formal	Laboratory activities	1. EDBL 2. Problem-based learning
Jeong and Kim (2015)	South Korea	Grade 7	Science (environmental studies)	3 months	145	Formal	Global climate change monitoring activities	1. PBL 2. TSL (real-time database)
Khozali and Karpudewan (2020)	Malaysia	Form 4	Science (ecology)	7 weeks	64	Formal	Design terrarium model	1. TSL: Facebook 2. EDBL
Lay and Osman (2017)	Malaysia	Form 4	Science (chemistry)	N/A	138	Formal	Create an educational media	1. PBL (MyKimDG) 2. IBL: 5E 3. Creative design process
Lin et al. (2018)	Taiwan	Grade 10	Engineering	10 weeks	43	Formal	Bridges	1. PBL 2. EDBL 3. TSL:3D printing)
Lin et al. (2020)	Taiwan	Grade 7	Engineering	8 weeks	139	Formal	Egg protection devices	EDBL
Lou et al. (2011)	Taiwan	Grade 10	Engineering	2 months	40	Extracurricular	Design solar automatic trolley	1. TSL: ICT platform 2. Problem-based learning
Lou et al. (2017)	Taiwan	Grade 9	Engineering	6 weeks	60	Formal	CaC2 Steamship design	1. PBL 2. Creative Designing Process
Majid and Majid (2018)	Malaysia	Form 4	Science (chemistry)	/	25	Formal	AR chemistry kit	1. TSL: AR 2. Scaffolding
Mutakinati et al. (2018)	Japan	Grade 7	Engineering	6 lessons	160	Formal	Simple water treatment system design	1. PBL: (Critical Thinking)
Pang et al. (2019)	Malaysia	Grade 7–12	Technology (Robotic)	/	400	Extracurricular	Robot racer, bowler, archer, sumo, robot tug of war	1. PBL

(continued)

(continued)

Study conducted by	Location	Grade level	Subject area emphasis	Duration	Size	Formal/Extra-curricular	Task/topics	Instructional practices
Purwaningsih et al (2020)	Indonesia	Grade 9–12	Science (physics)	/	53	Formal	Constructing a water rocket	1. PBL 2. TSL: Edmodo platform
Saleh et al. (2020)	Malaysia	Form 4	Science (physics)	/	70	Extracurricular	Egg-drop challenge	1. PBL
Shahali et al. (2017)	Malaysia	Lower secondary	Engineering	5 days	242	Extracurricular	Solar car, home automation, earthquake town	1. PBL 2. EDBL
Siew (2017)	Malaysia	Grade 10	Engineering	10 h	50	Extracurricular	Irrigation system design	1. EDBL 2. Imagination Training
Tam et al. (2020)	Hong Kong	Grade 7 & 8	Technology	3 days	411	Extracurricular	Produce wearable items with LED light ribbons	1. TSL (micro:bit) 2. IBL
Tsai et al. (2018)	Taiwan	High school	Engineering	3 months	39	Extracurricular	Solar pneumatic ship design	1. TSL: Moodle 2. PBL
Wan Husin et al. (2016)	Malaysia	Grade 7	Engineering	6 days	125	Extracurricular	Circuit, flash LED; water turbine	1. PBL 2. Problem-based learning 3. EDBL
Wang and Chiang (2020)	China	Grade 7	Engineering	4 weeks	72	Formal	Solve engineering problems for Robinson Crusoe	1. PBL 2. EDBL: novel engineering
Wang et al. (2015)	China	Grade 11	Science (physics)	1 semester	145	Formal	Hands-on and virtual physics experiment	1. IBL: 5E 2. TSL: simulation software

References

Asghar, A., Ellington, R., Rice, E., Johnson, F., & Prime, G. M. (2012). Supporting STEM education in secondary science contexts. *Interdisciplinary Journal of Problem-Based Learning, 6*(2), Article 4.

Burke, B. N. (2014). The ITEEA 6E learning by design model: Maximizing informed design and inquiry in the integrative STEM classroom. *Technology and Engineering Teacher, 73*(6), 14–19.

Bybee, R. W., Taylor, J. A., Gardner, A., Van Scotter, P., Powell, J. C., Westbrook, A., & Landes, N. (2006). *The BSCS 5E instructional model: Origins and effectiveness.* Biological Sciences Curriculum Studies (BSCS). https://media.bscs.org/bscsmw/5es/bscs_5e_full_report.pdf.

English, L. D. (2016). STEM education K-12: Perspectives on integration. *International Journal of STEM Education, 3*(1), Article 3.

Honey, M., Pearson, G., & Schweingruber, H. (Eds.). (2014). *STEM integration in K-12 education: Status, prospects, and an agenda for research.* National Academies Press.

Householder, D. L., & Hailey, C. E. (Eds.). (2012). *Incorporating engineering design challenges into STEM courses.* National Center for Engineering and Technology Education. https://digitalcommons.usu.edu/ncete_publications/166/.

Kang, N. (2019). A review of the effect of integrated STEM or STEAM (science, technology, engineering, arts, and mathematics) education in South Korea. *Asia-Pacific Science Education, 5*(6). https://doi.org/10.1186/s41029-019-0034-y.

Kelley, T. R., & Knowles, J. G. (2016). A conceptual framework for integrated STEM education. *International Journal of STEM Education, 3*(1), Article 11. https://doi.org/10.1186/s40594-016-0046-z.

Lee, M. H., Chai, C. S., & Hong, H. Y. (2019). STEM education in Asia Pacific: Challenges and development. *Asia-Pacific Education Researcher, 28*(1), 1–4. https://doi.org/10.1007/s40299-018-0424-z.

Nadelson, L. S., & Seifert, A. L. (2017). Integrated STEM defined: Contexts, challenges, and the future. *The Journal of Educational Research, 110*(3), 221–223. https://doi.org/10.1080/00220671.2017.1289775.

National Research Council. (2013). *Next generation science standards: For states, by states.* The National Academies Press. https://doi.org/10.17226/18290.

Sanders, M. E. (2009). STEM, STEM education, STEMmania. *The Technology Teacher, 68*(4), 20–26.

Statauskiene, L., & Mazgelyte, R. (2016). *Defining a good practice in STEM education within a framework of March Project.* Education Development Centre.

Thibaut, L., Ceuppens, S., De Loof, H., De Meester, J., Goovaerts, L., Struyf, A., Boeve-de Pauw, J., Dehaene, W., Deprez, J., De Cock, M., Hellinckx, L., Knipprath, H., Langie, G., Struyven, K., Van de Velde, D., Van Petegem, P., & Depaepe, F. (2018). Integrated STEM education: A systematic review of instructional practices in secondary education. *European Journal of STEM Education, 3*(1), Article 2. https://doi.org/10.20897/ejsteme/85525.

Wahono, B., Lin, P. L., & Chang, C. Y. (2020). Evidence of STEM enactment effectiveness in Asian student learning outcomes. *International Journal of STEM Education, 7*(1), Article 36. https://doi.org/10.1186/s40594-020-00236-1.

Studies Reviewed

Anwari, I., Yamada, S., Unno, M., Saito, T., Suwarma, I. R., Mutakinati, L., & Kumano, Y. (2015). Implementation of authentic learning and assessment through STEM Education approach to improve students' metacognitive skills. *K-12 STEM Education, 1*(3), 123–136.

Balakrishnan, B., Azman, M. N. A., Sharif, A. M., Yaacob, M. I. H., Zain, H. H. M., & Parmin. (2019). The effectiveness of hands-on modules in learning science: A case study in Malaysia. *The International Journal of Science, Mathematics and Technology Learning, 26*(1), 21–29.

Chang, S.-H., Ku, A.-C., Yu, L.-C., Wu, T.-C., & Kuo, B.-C. (2015). A science, technology, engineering and mathematics course with computer-assisted remedial learning system support for vocational high school students. *Journal of Baltic Science Education, 14*(5), 641–654.

Changtong, N., Maneejak, N., & Yasri, P. (2020). Approaches for implementing STEM (science, technology, engineering & mathematics) activities among middle school students in Thailand. *International Journal of Educational Methodology, 6*(1), 185–198. https://pdf.ijem.com/IJEM_6_1_185.pdf.

Chen, C.-S., & Lin, J.-W. (2019). A practical action research study of the impact of maker-centered STEM-PjBL on a rural middle school in Taiwan. *International Journal of Science and Mathematics Education, 17*(Suppl. 1), 85–108. https://doi.org/10.1007/s10763-019-09961-8.

Chen, Y., & Chang, C. C. (2018). The impact of an integrated robotics STEM course with a sailboat topic on high school students' perceptions of integrative STEM, interest, and career orientation. *EURASIA Journal of Mathematics, Science and Technology Education, 14*(12), em1614. https://doi.org/10.29333/ejmste/94314.

Chien, Y. H. (2017). Developing a pre-engineering curriculum for 3D printing skills for high school technology education. *Eurasia Journal of Mathematics, Science and Technology Education, 13*(7), 2941–2958. https://doi.org/10.12973/eurasia.2017.00729a.

Fan, S.-C., & Yu, K.-C. (2017). How an integrative STEM curriculum can benefit students in engineering design practices. *International Journal of Technology and Design Education, 27*(1), 107–129. https://doi.org/10.1007/s10798-015-9328-x.

Hafiz, N. R. M., & Ayop, S. K. (2020). The effect of an integrated-STEM water rocket module: SEMARAK towards STEM elements among form four students. *International STEM Journal, 1*(1), 1–16.

Hashim, H., Ali, M. N, Samsudin, M. A., & Shodiqi, H. A. (2020). Nourishing students' creative thinking through exemplar STEM lesson. *International STEM Journal*, 1(1), 27–37. https://ejournal.um.edu.my/index.php/STEM/article/view/27200.

Hsu, Y. S., Lin, Y. H., & Yang, B. (2017). Impact of augmented reality lessons on students' STEM interest. *Research and Practice in Technology Enhanced Learning, 12*, Article 2. https://doi.org/10.1186/s41039-016-0039-z.

Huang, W. (2019). Examining the impact of head-mounted display virtual reality on the science self-efficacy of high schoolers. *Interactive Learning Environments*. https://doi.org/10.1080/10494820.2019.1641525.

Huri, N. H. D., & Karpudewan, M. (2019). Evaluating the effectiveness of Integrated STEM-lab activities in improving secondary school students' understanding of electrolysis. *Chemistry Education Research and Practice, 20*(3), 495–508. https://doi.org/10.1039/c9rp00021f.

Jeong, S. K., & Kim, H. (2015). The effect of a climate change monitoring program on students' knowledge and perceptions of STEAM education in the Republic of Korea. *Eurasia Journal of Mathematics, Science and Technology Education, 11*(6), 1321–1338. https://doi.org/10.12973/eurasia.2015.1390a.

Khozali, N. B., & Karpudewan, M. (2020). An interdisciplinary Facebook incorporated STEM education strategy in teaching and learning of dynamic ecosystems. *Eurasia Journal of Mathematics, Science and Technology Education, 16*(11), em1902. https://doi.org/10.29333/ejmste/8704

Lay, A.-N., & Osman, K. (2017). Developing 21st century skills through a constructivist-constructionist learning environment. *K-12 STEM Education, 3*(2), 205–216. https://www.learntechlib.org/p/209542/.

Lin, K.-Y., Hsiao, H.-S., Chang, Y.-S., Chien, Y.-H., & Wu, Y.-T. (2018). The effectiveness of using 3D printing technology in STEM project-based learning activities. *Eurasia Journal of Mathematics, Science and Technology Education, 14*(12), em1633. https://doi.org/10.29333/ejmste/97189.

Lin, K.-Y., Hsiao, H.-S., Williams, P. J., & Chen, Y.-H. (2020). Effects of 6E-oriented STEM practical activities in cultivating middle school students' attitudes toward technology and technological inquiry ability. *Research in Science & Technological Education, 38*(1), 1–18. https://doi.org/10.1080/02635143.2018.1561432.

Lou, S.-J., Chou, Y.-C., Shih, R.-C., & Chung, C.-C. (2017). A Study of Creativity in CaC2 steamship-derived STEM project-based learning. *EURASIA Journal of Mathematics Science and Technology Education, 13*(6), 2387–2404. https://doi.org/10.12973/eurasia.2017.01231a.

Lou, S.-J., Shih, R.-C., Diez, C. R., & Tseng, K.-H. (2011). The impact of problem-based learning strategies on STEM knowledge integration and attitudes: An exploratory study among female Taiwanese senior high school students. *International Journal of Technology and Design Education, 21*, 195–215. https://doi.org/10.1007/s10798-010-9114-8.

Majid, N. A. A., & Majid, N. A. (2018). Augmented reality to promote guided discovery learning for STEM learning. *International Journal on Advanced Science, Engineering and Information Technology, 8*(4–2), 1494–1500.

Mutakinati, L., Anwari, I., & Kumano, Y. (2018). Analysis of students' critical thinking skill of middle school through STEM education project-based learning. *Jurnal Pendidikan IPA Indonesia, 7*(1), 54–65. https://doi.org/10.15294/jpii.v7i1.10495.

Pang, Y. J., Hussin, H., Tay, C. C., & Ahmad, S. S. S. (2019). Robotics competition-based learning for 21st century stem education. *Journal of Human Capital Development (JHCD), 12*(1), 83–100. https://journal.utem.edu.my/index.php/jhcd/article/view/5389/3854.

Purwaningsih, E., Sari, S. P., Sari, A. M., & Suryadi, A. (2020). The effect of STEM-PjBL and discovery learning on improving students' problem-solving skills of impulse and momentum topic. *Jurnal Pendidikan IPA Indonesia, 9*(4), 465–476. https://doi.org/10.15294/jpii.v9i4.26432.

Saleh, S., Muhammad, A., & Abdullah, S. M. S. (2020). STEM project-based approach in enhancing conceptual understanding and inventive thinking skills among secondary school students. *Journal of Nusantara Studies, 5*(1), 234–254. https://doi.org/10.24200/jonus.vol5iss1pp234-254.

Shahali, E. H. M., Halim, L., Rasul, M. S., Osman, K., & Zulkifeli, M. A. (2017). STEM learning through engineering design: Impact on middle secondary students' interest towards STEM. *EURASIA Journal of Mathematics, Science and Technology Education, 13*(5), 1189–1211. https://doi.org/10.12973/eurasia.2017.00667a.

Siew, N. M. (2017). Fostering students' scientific imagination in STEM through an engineering design process. *Problems of Education in the 21st Century, 75*(4), 375–393. https://doi.org/10.33225/pec/17.75.375.

Tam, H.-L., Chan, A. Y.-F., & Lai, O. L.-H. (2020). Gender stereotyping and STEM education: Girls' empowerment through effective ICT training in Hong Kong. *Children and Youth Services Review, 119*, Article 105624. https://doi.org/10.1016/j.childyouth.2020.105624.

Tsai, H.-Y., Chung, C.-C., & Lou, S.-J. (2018). Construction and development of iSTEM learning model. *EURASIA Journal of Mathematics, Science and Technology Education, 14*(1), 15–32. https://doi.org/10.12973/ejmste/78019.

Wan Husin, W. N. F., Mohamad Arsad, N., Othman, O., Halim, L., Rasul, M. S., Osman, K., & Iksan, Z. (2016). Fostering students' 21st century skills through Project Oriented Problem Based Learning (POPBL) in integrated STEM education program. *Asia-Pacific Forum on Science Learning and Teaching, 17*(1), Article 3.

Wang, J., Guo, D., & Jou, M. (2015). A study on the effects of model-based inquiry pedagogy on students' inquiry skills in a virtual physics lab. *Computers in Human Behavior, 49*, 658–669. https://doi.org/10.1016/j.chb.2015.01.043.

Wang, L., & Chiang, F. K. (2020). Integrating novel engineering strategies into STEM education: APP design and an assessment of engineering-related attitudes. *British Journal of Educational Technology, 51*(6), 1938–1959. https://doi.org/10.1111/bjet.13031.

May May Hung Cheng obtained her Bachelor's and Master's degrees from the University of Hong Kong and her Ph.D. from the University of Waikato, New Zealand. She is currently Chair Professor of Teacher Education in the Department of Curriculum and Instruction at the Education University of Hong Kong. She was previously a Reader in Professional Education at the Department of Education, University of Oxford. Her main areas of research are teacher education and science education with a focus on teacher learning and teacher professional development. She was the president of the East Asian Association for Science Education (EASE) between 2016 to 2019.

Fang-Yin Yeh obtained her Bachelor's degree from the National Taiwan University of Science and Technology, Master's degree from the University of Kansas, USA, and Ph.D. from the University of Hong Kong. She is currently a Post-Doctoral Fellow in the Department of Curriculum and Instruction at the Education University of Hong Kong. Her research focuses on school curriculum in East Asia. She is also interested in teacher professional development.

Chapter 3
STEM Education in Mainland China

Nanxi Meng, Yang Yang, Xinyu Zhou, and Yan Dong

Abstract Despite global attention on STEM policy and STEM education research, Asian-based STEM education studies are still in an emerging state. In the current chapter, we introduce the development of STEM education in Mainland China from its historical context, the understanding of the term STEM education and its variants, and the research and practice of STEM education. Specifically, we reviewed literatures on the STEM learner and learning, STEM teaching approaches and STEM teacher development. It is found that the theoretical research has been carried out thoroughly with the foundation of studying the mature experience globally, most of the research results about STEM education and its implementation published in Chinese are still theoretical models based on the existing problems. In STEM education implementation, a few actions have been taken to promote STEM in greater educational communities, including exploring the feasible model through international education exchange among schools and teachers, adopting the comprehensive practical activity as the course format in K-12 institutes, taking problem-based and project-based learning as the basic teaching approaches, integrating formal and informal education, and building the STEM education ecology. We briefly introduce these practices in the chapter, in order to share the experience and lessons learned.

N. Meng
Department of Learning Technologies, University of North Texas, 1155 Union Circle #311127, Denton, TX 76203, USA
e-mail: nanxi.meng@unt.edu

Y. Yang
Center for Educational Science and Technology, Institute of Advanced Studies in Humanities and Social Sciences, Beijing Normal University, No.18, Jinfeng, Road, Zhuhai 519087, Guangdong, China
e-mail: yyang@bnu.edu.cn

X. Zhou
School of Information Technology in Education, South China Normal University, No. 55, Zhongshan road, Guangzhou 510631, Guangdong, China

Y. Dong (✉)
School of Education Technology, Faculty of Education, Beijing Normal University, 19, Xinwai ST, Beijing 100875, China
e-mail: yan.dong@bnu.edu.cn

M. M. H. Cheng et al. (eds.), *Concepts and Practices of STEM Education in Asia*,
https://doi.org/10.1007/978-981-19-2596-2_3

With the ongoing education reform, it is expected that more empirical evidence will emerge along with the wide and in-depth implementation of STEM education in mainland China.

Keywords STEM education · Teaching strategy · Peer collaboration · Teacher professional development

Historical Context of STEM Education in China

On September 5, 1988, Deng Xiaoping put forward the famous theory: "Science and technology are primary productive forces". Since the reform and opening up in 1978, the government of mainland China has been prioritizing the development of science and technology, in order to support the modernization and rapid growth of the economy. The traditional education served its purpose of supplying a large number of skilled workers to develop internationally competitive manufacturing industry and build a well-to-do society in the last decades, yet the undergoing of the Fourth Industrial Revolution featured by the blurring of boundaries between the physical, digital, and biological worlds has changed the way of human existence and development, and the demand for future talents. Foreseeing the challenges, the State Council has implemented a series of policies to improve science and technology (S&T) innovation (The State Council, P. R. of China, 2016).

The transformational demand on talents leads to the change of the educational model. STEM and STEAM education was introduced and adopted by China, with the focus on developing literacy and competence related to science. Delving into the questions of "What kind of talents to cultivate?" and "How to cultivate them?", researchers and scholars in mainland China designed the *Core Competencies and Values (CCV) for Chinese Student* (Ministry of Education [MOE], P. R. of China, 2016), to serve as the framework of cultivating the talents with morality, adjusting to the development of world educational reform and improve the global competitiveness.

The research and practice of STEM education is still in its infancy in China. Despite the obstacles brought by the traditional rote-based, exam-oriented concepts, teaching models and approaches, STEM education has been showing strong vitality, the inherent exploratory characteristics of integrated curriculum and STEM as an innovative model and pedagogy is stimulating the interest and enthusiasm of students, teachers, researchers and other stakeholders which will be elaborated in the following sections.

The Understanding of STEM Education in Mainland China

STEM education was originally defined by the National Research Council (NRC, 1996) and Sanders (2009) as a teaching and learning method that integrates the content and skills of science, technology, engineering and math. With decades of practical exploration, this definition was expanded beyond the four disciplines by the National Science Teaching Association (NSTA) of the U.S., and defined STEM education as an interdisciplinary approach to learning. Specifically, NSTA (2012) described STEM education in K-12 context as:

> 'STEM education is an experiential learning pedagogy in which the application of knowledge and skills are integrated through in-context projects or problems focused on learning outcomes tied to the development of important college and career readiness proficiencies' (p. 1).

Compared to STEM education, STEAM education, among many variants, was adopted widely in theoretical research and educational practices in China. STEAM education is regarded as one of the essential approaches to cultivate talents with core competencies and values and develop comprehensively in the future social environment. Scholars argued that STEAM education shares the same value with the CCV (Zhu & Lei, 2018; Hu et al., 2016), as they both emphasize fostering the character and ability to use the integrated knowledge from multiple disciplines to solve complex problems. The CCV provides theoretical guidance for STEAM educational practices, with special attention to the development of humanistic literacy, physical health, and many other aspects that contribute to the holistic development of students. Specifically, the CCV embodied STEAM capabilities in four aspects: scientific concepts and applications, scientific thinking and innovation, scientific exploration and communication", and scientific attitudes and responsibilities (Hu et al., 2016).

Therefore, reforming the STEAM education with the guidance of CCV to continuously enrich the design curriculum and learning objectives, teaching projects, content and teaching practices is the foundation of STEAM education in mainland China.

STEM Education Research and Practices

In this chapter, we first provide an overview of the theoretical research related to STEM education in China, followed by zooming into the practices at the school level, in order to provide a holistic and in-depth understanding of STEM education development in China.

The theoretical research of STEM education in mainland China were conducted in five concentrations:

(1) The analysis of STEM education policies internationally and domestically.
(2) the analysis of STEM education and learning models abroad.

(3) case studies of the best practices abroad.
(4) the analysis of teaching and learning materials abroad, and
(5) proposing new STEM education models for the given contexts in China.

For the purpose of reflecting the features, experiences and lessons learned of developing STEM in China and the structure of the chapter, we focus only on the analysis of publications writing about STEM or STEAM policies and practices in China. Literature sources include STEM-related studies written in English and Chinese with the publication timeframe ranging from 1998 to 2021. An article search was conducted using the academic databases of the Web of Science, Science Direct, and Google Scholar. Databases were browsed using a combination of keywords: 'STEM' plus related words such as education, discipline, programme, course, or integration, and school level. The selected studies included journal articles, books and doctoral dissertation that provide information in the five aspects listed above. A total of 31 articles representing qualitative, quantitative, and mixed methods studies were selected.

STEM Education Related Policies

Through literature review, we found that the research on practice and experience of formulating STEM education related policies in China shows three key features. First, international comparative research serves as the foundation of policy development. Through the thorough understanding of mature experiences gathered by educational institutions in the U.S. and other countries, the education community in China formulated the national policies to guide the practices. Second, the studies demonstrated active exploration on the connotation and extension of STEM education. Third, there were a lot of efforts spent on the integration of STEM education and the construction of national development strategies.

In 2014, Shanghai Municipal Education Commission first proposed the concept of "STEM +" education. The "+" indicated the integration of science, technology, engineering, mathematics and humanities, artistic accomplishment and social values, with special emphasis on the cultivation of science, humanities and social values. The "+" is not merely the expansion of subject knowledge, but an upgraded notion of providing holistic education. In 2015, the Chinese government issued "Guiding Opinions on Comprehensively and Deeply Promoting Educational Informatization During the 13th Five-Year Plan Period" and other related policies, clearly supporting the development STEM education, marking that STEM education became part of the national education development plan.

2016 could be regarded as the nationwide initial year of STEM education in China. The State of Council released the *Implementation Plan for the Outline of the National Scientific Literacy Action Plan* (The State Council, P. R. of China, 2016), in which advocated the interdisciplinary scientific inquiry for the first time. Since the releasing of *the Implementation Plan*, the concepts like "STEM", "STEAM" or "STEM + education" and their paraphrases had emerged in the policy documents of the national and provincial education departments, and start serving as the essential reference. In the *13th Five-Year Plan for Education Informatization proposed by* the Center for Educational Information Management (MOE, P. R. of China, 2016), STEAM education was mentioned. The FYP stated:

> 'In areas where conditions permit, innovative education models such as "makerspace", "interdisciplinary learning (STEAM education)" and "maker education" should actively explore the application of ICT'.

2017 was viewed as a big movement of STEM education in China (Yao & Guo, 2017), STEM education was officially integrated in the K-12 curriculum standards, marked by three key events. First, the STEM Education Research Center was established under the National Institute of Education Sciences (NIES, P. R. of China, 2017), to provide decision-making services for the Ministry of Education, Ministry of Science and Technology and other ministries, enrich and improve the theoretical basis of STEM education, lead and promote the in-depth development of STEM education practice, and explore and build a collaborative innovation mechanism for STEM education. Second, Artificial intelligence (AI) has officially included in the curriculum of K-12 education. Given the rapid development of AI technology and its abundance of application, the State Council issued the New Generation of the Artificial Intelligence Development Plan (The State Council, P. R. of China, 2017), which specifically listed the strategies to implement the national AI education project, including setting up AI-related courses in elementary and secondary schools, promoting programming education, and engaging social forces in the development and promotion of educational programming software and games. AI textbooks for K-12 education have been developed and put into use. Third, engineering as the traditionally neglected subject in K-12 education was introduced into the curriculum system, and the interconnectivity among subjects was emphasized. The Ministry of Education promulgated the revised *Science Curriculum Standards for Primary Schools, which officially added* "technology and engineering" as major components into the science courses in primary schools to enrich the science curriculum (MOE, P. R. of China, 2017a), which advocated interdisciplinary learning, and encouraged teacher to adopt STEM education pedagogy in their teaching. In 2018, the *Ordinary High School Curriculum Standards* for each subject was promulgated, STEM/STEAM and STEM + education emerged in multiple related curriculum standards. Specifically, in the High School Information Technology Curriculum Standards, the concept of STEAM education was adopted as the guidance, to "bring students the interest and joy of research and creation, and foster the consciousness and ability of problem solving and innovative designing with information technology". In the optional compulsory module *High School General Technical Curriculum Standards*, the "Science, Technology and Humanities Integration and Innovation Topics" under the module of Technology and Innovation", it was mentioned that "Comprehensively apply the knowledge, methods and skills of science, technology, engineering, art, mathematics, society ("STEAMS") and other disciplines to carry out problem solving and technological innovation in the form of theme-based learning or project-based learning". In the *Ordinary High School Biology Curriculum Standard*, students were required to "pay attention to the connection among different subjects", due to the "interconnectivity of Biology and mathematics, technology, engineering, and information science." (MOE, P. R. of China, 2017b).

The NIES announced the *White Paper on STEM Education in China* (NIES, P. R. of China, 2017). This document analyzed the historical context and the current status of STEM education in China. Meanwhile the *China STEM Education 2029 Innovation Action Plan* was released. In addition to formal STEM education at school, the plan emphasized the participation of the informal education institutes, and encouraged social resources (i.e. related government departments, research institutes, high-tech enterprises, educational foundations etc.) to engage in STEM education innovation. It is projected that in the next decade, STEM education should benefit all students, especially those from special groups, to foster their innovative thinking and scientific inquiry skills via focusing on learning process, apply innovative evaluation strategies and provide innovative training models to teachers. There are mainly six objectives for this plan: promoting the design of national STEM education policy; implementing a STEM personnel training plan; building a platform for resource integration and teacher training; constructing STEM education standards and an evaluation system; building an integrated STEM Innovation Ecosystem; and exploring the strategies of education and talent cultivation for economic development. The plan also announced building a number of STEM education model schools, providing STEM classrooms, project-based learning classrooms and maker spaces to supply the innovative talents that are imperatively needed for the development of China (NIES, 2017).

Driven by the national policy guidance and the upsurge of educational practice, provinces and regions have explored and issued guiding documents of their regional STEM practices. For example, Shandong province released the first provincial STEM guiding document, followed by Shenzhen and Guangdong province issuing guidelines for curriculum construction, project-based learning guidelines and teacher professional development. Jiangsu province took the lead in developing the syllabi for STEM curriculum. Many schools have established STEM professional classrooms or makerspaces to promote project-based learning in classroom teaching.

To sum up, despite that there are no specialized or comprehensive STEM curriculum standards at the national level in mainland China, the policies supporting STEM education have shaped the development of STEM education by referring to the mature experiences of other countries and trial areas in China (i.e. Shanghai). With the guiding plans and documents promulgated from the Ministry of Education, K-12 education systems started to include STEM/STEAM education as an innovative and interdisciplinary model in the formulation of curriculum policy at the national level. Policies at the provincial and regional level were aligned with national policy to guide STEM practice with regional advantages. Social educational institutes were involved to explore STEM practice in informal education. At the national level, STEM is introduced and integrated into the curriculum via a top-down educational reform.

STEM Learning and Teaching

STEM education in China first emerged in selected areas and provinces as school or extra-curriculum practices as an experiment, then gradually became a paradigm of educational integration. Despite ample theoretical exploration, the discrepancy between policy and practice still exists, which makes describing the learners' behavior and innovative teaching approaches challenging due to lack of empirical evidence. Therefore, in the current chapter, we try to describe the features of STEM learners in mainland China, followed by the proposed models and approaches of STEM education.

In order to conclude the features of STEM learners and learning in mainland China, the authors of this chapter conducted a brief literature review of studies written in both Chinese and English on this topic. We conducted the search on cnki.net under the database of Chinese Social Sciences Citation Index.[1] In total 14 articles on STEM learner and learning were identified when searching the term "China STEM learning" and "China STEM learner", 14 articles on STEM teaching approaches, and 17 articles on STEM teacher professional development, for the period of 2011–2021. Our analysis is based on the articles selected above, synthesized with the literatures published in English, and the findings were presented below.

Three Characteristics of STEM Learners in Mainland China

First, despite the outstanding grasp of scientific content knowledge, the effectiveness of science education in elementary and secondary education in mainland China has been questioned by scholars and educational researchers (Yan et al., 2018). According to the PISA test, Chinese students generally lack creativity, their scientific interest and cognitive science knowledge level were also lower than the international average (Yan et al., 2018). Second, mixed results were shown about students' STEM learning attitude, performance and early career attitude between two genders. Unlike the predominant results shown in western cultures that girls presented less interest and self-efficacy in learning STEM related subjects, studies have shown contradictory results of the STEM learning interest and capacity for students in mainland China. For example, Zhang and colleagues (2021) conducted a survey among 566 primary and secondary students, and found that boys' STEM interests and STEM career interest are significantly higher than those of girls, yet Zhan et al.'s study (2021) showed that there was no significant difference between the genders on STEM learning attitude. The social and cultural differences might lead to the discrepancy, which implies that teachers and researchers should take the regional culture and stereotypes into consideration when designing STEM courses and learning activities for students. Third, the

[1] The Chinese Social Sciences Citation Index (CSSCI) covers about 500 Chinese academic journals of humanities and social sciences. Now many leading Chinese universities and institutes use CSSCI as a basis for the evaluation of academic achievements and promotion.

cohort education model and student's friendship contributed to collaborative learning in STEM classrooms. Wei et al. (2021) conducted an empirical research in primary programming class with 84 students. In the study, the students were asked to pair up either with their friends or a random partner. The study found that, in general, peer collaboration improved boys' programming self-efficacy, but not girls; both boys in boy-boy pairs and girls in the boy-girl pairs significantly improved their computational thinking skills. This study was one of the first large-scale programming/STEM empirical studies conducted with students sampled in mainland China.

Research on Collaboration in STEAM Learning

As an interdisciplinary learning model, STEAM education emphasizes peer collaboration, which is a hot topic for STEAM researchers in mainland China. Addition to the traditional empirical studies that survey and interviewing served as the main channels to collect data, researchers have started to explore the feasibility of applying biological data to study the STEAM learning process. In the interdisciplinary STEM interaction and collaboration, learners generally form a study group of two or more to complete common learning goals in different interactive ways (English & Mousoulides, 2015). Dong et al. (2019) found that multi-person peer interaction could provide ample opportunities for STEM interdisciplinary problem solving and foster high-order thinking skills, which were the core elements in promoting active learning to achieve greater results. Effective multi-person peer interaction produces synergistic effects through group cooperation, promotes interaction and negotiation between learners, promotes knowledge construction and generation, and enables learners to achieve higher learning achievements (Zhai & Shu, 2018). Given the initial stage of exploring and implementing STEAM education, it is urgent to deepen the understanding of the learning mechanism in STEM from the perspective of cognitive science, to provide theoretical guidance to the design and implementation of STEM learning (Dong et al., 2020b).

Exploring the students' cognitive state in the process of peer interaction in STEM learning groups, especially the coordination mechanism happened during the collaboration in their mind, is an important foundation for scientifically developing STEM education. The current research of interpersonal brain synchronization based on ultra-scanning technology mostly focuses on the process of peer interaction with two people, while the actual cooperative learning in STEM classroom teaching is mostly based on multi-person interaction. The increase in the number of members in a group will cause more complicated peer relationships among members. Although studies have found that the transition from two-person to multi-person cooperation will fundamentally change the way individuals think and the process of peer interaction (Xie et al., 2019), it is still difficult to explain the common multi-person peer interaction process in high school STEM education. In addition, in order to improve the quality and educational effects of high school STEM activities, it is urgent to reveal the internal structural factors that affect learners' interactive behavior and cognitive rules. With more understanding related to multi-person interaction, researchers may use scientific and effective strategies to design interventions in STEM lessons. Existing studies have explored how factors or strategies such as group gender ratio, knowledge and experience, and communication patterns affect the frequency of peer interaction and the final cooperative learning effect (Lu, 2020). However, it has failed to clearly reveal the brain coordination mechanism in the process of multi-person peer interaction in the STEM context. It remains challenging to provide scientific evidence and practical basis for the scientific design and effective implementation of STEM education.

Conducting STEAM learning research with the support of biological data through the lens of natural science can provide profound impact on the current paradigm. From a theoretical perspective, through the use of ultra-scanning technology, the data collected were from the large-scale experimental data in real educational situations, which were more authentic compared to the traditionally collected data, which contributed to deepen the understanding of the cognitive mechanism of the interaction process of multiple peers in the STEM education context, and to enrich interdisciplinary learning related theories of cognitive neuroscience. From the practical level, the research result could inform the design of intervention of STEM education, STEAM learning experiences, and provide guidance to the effective intervention and evaluation of STEM activities for primary and secondary school teachers. Informed by the understanding of cognitive mechanism, researchers may collect relevant scientific evidence during the implementation. With the success of STEAM research and practice in mainland China, the experiences and lessons learned could be shared with other countries in the world.

Teaching Approaches

Interdisciplinary, collaborative, and project-oriented are the important characteristics of STEM education activities (Yu & Hu, 2015; Li et al., 2020). Project-based science learning contributes to the learning of middle school science and other subjects and benefits all learners (Saavedra et al., 2021). In interdisciplinary collaborative learning, the establishment of learners' perspectives and the development of competence literacy are gradual, dynamic and continuous (Yu & Wu, 2019). To improve the learning performance with the consideration of the characteristics of students in mainland China, we introduce the practices and experiences from five aspects: STEAM education model exploration through international exchange, creating the course of comprehensive practical activity, adopting problem-based and project-based learning as the teaching approaches, integrating formal and informal education, and establishing STEAM education ecosystem.

Model Exploration Through International Education Exchange

In order to establish initial understanding of STEAM practices, researchers were informed by experience from STEAM leading countries like the U.S. In September 2017, the Department of Education of Zhejiang Province (DEZP) launched the Zhejiang–Indiana Parallel Classroom Program on STEAM Courses (DEZP, 2017), among the 15 selected seeding and 15 breeding primary and secondary schools from both Zhejiang Province and K-12 schools from Indiana. Classroom teachers from Indiana were invited to teach STEAM courses to Chinese students, and share their teaching insights and experiences with the Chinese peer teachers. Witnessing the passion and success of teachers from both countries, the DEZP organized 15 teacher training programs abroad. Teachers in these programs covered the entire K-12 spectrum, and provided well-developed education programs to learn their unique educational concepts and teaching practices, as well as the leadership skills towards building a world-leading STEAM teaching system.

Comprehensive Practical Activity as the Course Format

Different from STEAM education, which was considered as an integrated approach or experiential learning pedagogy (NSTA, 2012), STEAM education was implemented in the format of a course in China. In September 2017, the Ministry of Education of China issued the "Comprehensive Practice Activity Curriculum Guidelines for Primary and Secondary Schools" to implement the STEAM education. This guideline required all the regional educational departments to fully understand the importance of the comprehensive practical activity curriculum, and guarantee that comprehensive practical activity courses were fully available and well-organized. Specifically, the comprehensive practical activity course (CPAC) is an interdisciplinary course that inspires students to identify and discover problems and questions in their real-life context and transform them into activities related to researchable themes. Through exploration, service, production, experience and other methods, the CPAC is designed to cultivate comprehensive qualities of students, i.e. comprehensively use knowledge of various disciplines to recognize, analyze and solve practical problems, improve the students' overall quality and focus on the development of core literacy, especially social responsibility, innovative spirit and practical ability, to meet the needs of rapidly changing social life, professional world and individual development. The CPAC is one of the compulsory components of the national compulsory education and general high school curriculum. It is set in parallel with subject courses and it is an important part of the basic education curriculum system. This course is managed and guided by the regional educational department. The course content is mainly developed by teachers of local schools, and it is fully implemented throughout the K-12 education system.

Teachers and scholars have been actively implementing and monitoring the implementation of STEAM education in CPAC. Du (2020) explored the CPAC in 54 elementary schools in Zhejiang Province prior to integrating STEAM education, and found the key challenges reported by teachers in teaching the course include: (1) the lack of corresponding curriculum training, (2) the lack of guidance in curriculum designing, (3) the lack of ready-made and age appropriate teaching materials for students, and (4) the lack of corresponding standards when compiling textbooks. In addition to practicing the STEAM education pedagogy, adopting relevant teaching materials, teachers with different subject backgrounds could develop school-based curriculum and seminars collaboratively, in order to guide student learning aiming to achieve better performance. In addition, the establishment of a STEAM facilitation center on campus facilitated the collaboration among multiple departments including curriculum center, teaching and research group, teaching office, and academic affairs office, and helps to guarantee the quality of the implementation of CPAC.

Problem-Based and Project-Based Learning as Teaching Approaches

Problem- (PrBL) and project-based learning (PjBL) are the common approaches of learning STEAM in mainland China, as both involve a set of instructional strategies that empower learners to engage in the classroom learning and increase student achievement. Research has revealed that PrBL and PjBL worked particularly well when teaching STEAM subjects as they increase engagement and enjoyment for all student groups, especially for girls and underrepresented groups (Han et al., 2015). The constructivist nature in both models was intended to replace the traditional lecture-based teaching approaches to actively engage students in interdisciplinary learning through real-world problem solving with autonomy.

Based on existing research, researchers in mainland China explored the above approaches, and proposed new approaches according to the characteristics of STEAM learners and the national development strategies. For example, China was known for its emphasis on math education, in order to practice the requirement of infiltrating the mathematics culture in mathematics education, an alternative understanding of STEAM education as "reading into arts and engineering with science, based on mathematical elements" (Yakman & Lee, 2012) was adopted. Therefore, as one of the integrated educational resources, mathematics culture plays the adhesive role for integrating the related subjects in PrBL and PjBL classrooms. Specifically, in mathematics cultural projects, teachers design open and operational project tasks with mathematics cultural materials, guiding students to discover, ask questions, analyze problems, integrate multidisciplinary knowledge to solve problems, and promote students' interdisciplinary learning and innovative practice.

Additionally, Dong and Sun (2019) integrated the PrBL and PjBL approaches and the elements of design thinking, and proposed the production-based learning (DoPBL) model to promote interdisciplinary learning. DoPBL model emphasized that students physically and mentally equally participate in the learning process, to solve problems with the thorough consideration of the end users, created the tangible/intangible learning products with appropriate level of usability and feasibility. This model was proposed due to the recent focus on production-based education, which encourages students to participate in the meaningful real-world production related activities.

Within the PrBL and PjBL, the 5E instructional model and its variants were the most commonly adopted teaching model. First proposed by Bybee & Landes in the early 90 s, the 5E provides a carefully planned sequence of instruction that places students at the center of learning, and it comprises engage, explore, explain, elaborate, and evaluate stages. This model encourages all students to explore, construct understanding of scientific concepts, and relate those understandings to phenomena or engineering problems (Bybee et al., 2006). Based on this model, researchers in mainland China developed multiple variants according to the needs of students and the innovative implementation of STEAM education. For example, Li and Li (2019) proposed the 5EX model, which expanded the additional 5E model and put emphasis on enhancing the mathematics foundation of students. Specifically, the 5EX model includes Enter and Questions (EQ), Exploration and Mathematics (EM), Engineering and Technology (ET), Expansion and Creativity (EC), and Evaluation and Reflection (ER).

The Integration of Formal and Informal Education

Addition to school education, the development of STEAM education also relies on integrating the informal educational institutes and learning environments. Currently, two types of social forces have been integrated in the STEAM education. The first type is the non-profit educational institutes like science museums and exhibition centers. These institutes expanded the learning environment of STEAM education, offered students more possibilities to participate beyond the classroom or schools. Wu and colleagues (2019) proposed the STEAM education model of "learning, research, career development", in which schools in Shanghai collaborated with Shanghai Natural History Museum to design learning materials and courses. Compared to the traditional in-school learning, this model advanced the depth of learning in research activities to enhance the problem-solving skills, and promoted career-oriented learning, which encouraged students to establish career goals and actively engage in learning activities. More importantly, this model extended STEAM education into informal learning time and occasions like summer and winter camps, which was one step closer to holistic education.

A number of for-profit and non-profit educational organizations are also active in China's STEM education arena, promoting the development of China's STEM education in various ways. For instance, the Lego Group launched the *Lego Basic Course of STEAM* jointly with East China Normal University Press in 2018, which is the first set of Lego STEM solutions for the Chinese curriculum system. Based on STEM-related disciplines, this course covers the four major themes of physical science, earth and space science, engineering and technology, and life science (Wu, 2018). This course is a good example of curriculum localization, it is not the simple introduction of the existing materials abroad, but a new creation based on the learning needs and characteristics of the Chinese curriculum. Additionally, a number of educational corporates contribute to programming education, robotic education, and maker education in K-12, to develop curriculum systems and teaching tools based on the characteristics of the Chinese learners and teachers, making reference to the teaching models adopted by developed countries (Liu, 2020).

Building STEAM Education Ecology

STEAM education has become an important approach in the global scientific and technological talent training and education reform, building a sustainable STEM ecosystem has never become more important than before to ensure its healthy development. Gao et al. (2020) proposed the STEM Ecosystem Construction Framework (STEM-ECF) and its evolution strategies, in order to improve the structure of the ecosystem, and systematically support the sustainable development of STEM education. The six factors include environmental facilities, policy funds, teachers, curriculum resources, method and practice, and achievement transformation.

The STEM-ECF highlights the characteristics of students in each education stage (preschool, elementary school, junior high school, and high school), aiming to achieve the goal of fostering scientific literacy required in the CVV, forming the platform to improve the implementation of STEAM education. Particularly, pre-school education intends to cultivate the learning interest and stimulate the development of various parts of the brain. Primary education serves students with certain learning skills, the STEM-ECF should focus on the comprehensiveness of students. Middle school and high school are the essential linkage between elementary and higher education, the STEM-ECF should focus on fostering the scientific inquiry skills and pertinence. Besides, school education, family education and social education should also be engaged. The frame is shown as below in Fig. 3.1.

The STEM-ECF is centered on students and with students as the recipients of the material, energy, and information flow. Among these elements, the learning facilities are the foundation and facilitator, STEAM related policies and funds provide the external security of its operation, as the policies guide the development of the STEAM education domestically and the funds predict the timing of a given project. The qualified teachers are the core component and the essential driving force of the STEM-ECF to promote the operation. Curriculum resources provide scientific support of the STEM ecosystem, they also differentiate the STEM-ECF system from other education ecosystems. When constructing the curriculum system, it is necessary to interpret the differences across multiple disciplines in depth in order to better carry out cross-disciplinary integrated learning. Teaching practice is the prerequisite to promote the healthy development of the STEM ecosystem, promoting and theorizing the best teaching practices are the vital actions of forming the STEM-ECF. Last but not least, a healthy STEM-ECF in operation would bring numerous excellent products, actively facilitated research on the transformation and application of the products, and it can further strengthen the connection between the STEM education and social and economic systems.

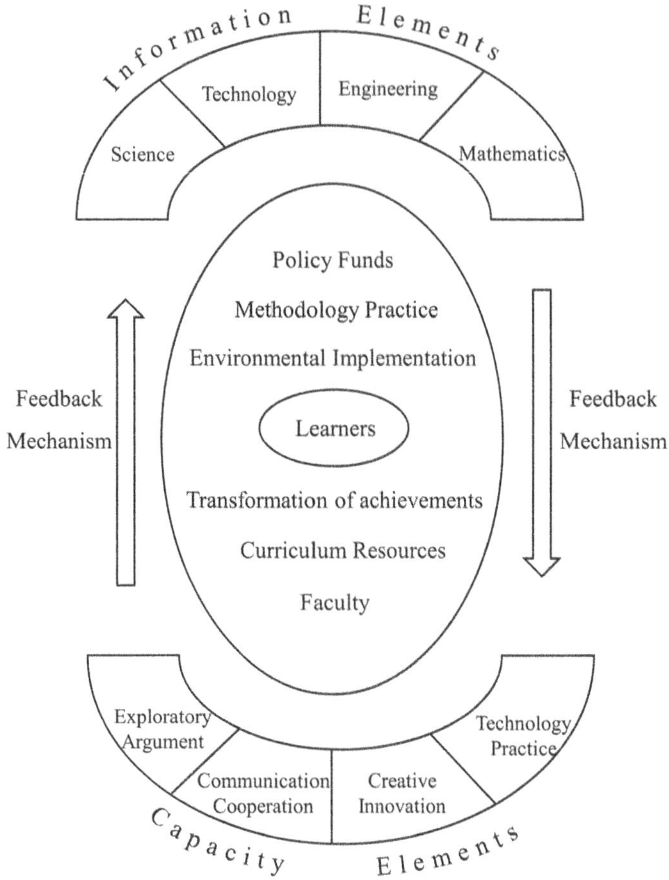

Fig. 3.1 STEM ecosystem building framework, adapted from Gao et al. (2020, p. 83)

The Practice of Teacher Professional Development

As the core driving force in the implementation of STEAM education, the quality of teachers could affect the success of STEM education. The "passion and enthusiasm in teachers" is recognized as the key to the national STEAM education strategy, yet a lack of ICT literacy formed an obstacle for teachers to practice STEAM teaching. Information literacy is a must-have for STEAM teachers in China besides the good knowledge and skills on STEAM pedagogy and related teaching approaches. Therefore, it is essential to define the connotation of ICT literacy of STEM teachers and effectively improve the professional literacy of STEM teachers. The current problem is that schools overemphasize the importance of equipping STEAM learning facilities (3D printer, maker space etc.) and require teachers to acquire abundant technological knowledge (TK) instead of encouraging the holistic development of TPACK knowledge among teachers. According to the existing literature, the TPACK model

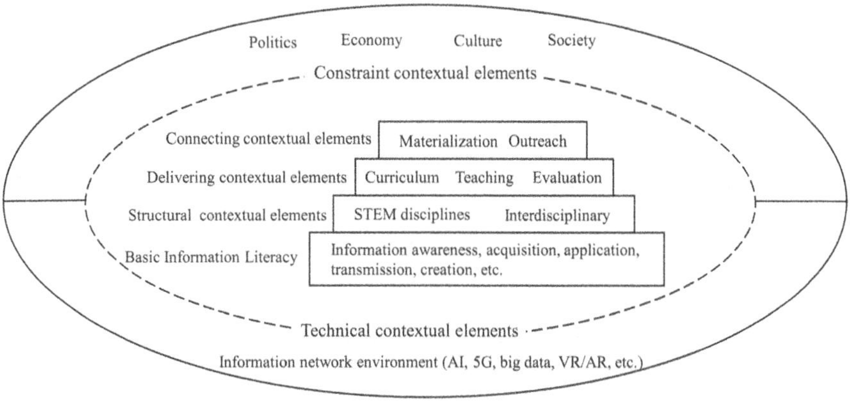

Fig. 3.2 Hierarchical elements of information literacy for STEM teachers, adapted from Dong et al. (2020a, p. 74)

has proven to be effective on teacher professional development (Chai et al., 2013), Lin and colleagues (2018) hence proposed the TPACK-based STEAM teacher development model to assist teachers to design more relevant, high-quality and efficient STEAM curriculum and activities. This modified model consists of three optimized paths: (1) TK optimization by guiding teachers to learn STEM educational technologies and related products, instead of learning general technological products, (2) content knowledge (CK) optimization by guiding teachers to implement interdisciplinary learning guiding by the real-world problems, and (3) pedagogical knowledge (PK) optimization, by facilitating teachers acquiring design-based STEM innovative teaching approaches.

Dong and colleagues (Dong et al., 2020a) explored elements of STEM teachers' ICT literacy by analyzing the development of their ICT literacy and its concepts, and categorized STEAM teachers' ICT literacy into two dimensions (contextual elements and extensional environment elements) at six levels, as shown in Fig. 3.2.

The external circle shows the technological and restrictive context elements. With the rapid development of technologies such as artificial intelligence, 5G, big data, virtual reality, and augmented reality, STEAM teachers need to continue learning and keep up with the development of these new technologies, in order to establish a good foundation for information literacy improvement. Meanwhile the social and cultural context play the role of restrictive conditions, which may include impact factors from politics, social-economic, cultural changes, which require STEAM teachers to create STEAM courses and teaching materials that are in line with the current social-cultural happenings when carrying out activities, and meet the requirements of the real context, teachers should be able to adopt appropriate materials to successfully carry out the STEAM learning activities.

The inner circle showed the four levels of information literacy teachers should acquire: (1) the basic information literacy, which includes the awareness, obtaining, applying, delivering and creating of the information, (2) information literacy to support disciplinary and interdisciplinary teaching, (3) information literacy for curriculum and instructional and evaluation, and (4) information literacy for materializing and publishing the learning and teaching results. To achieve the goal of improving teachers' information literacy, three strategies were proposed: (1) improving the basic ICT literacy of STEM teachers on a large scale through targeted teacher training; (2) promoting the contextual ICT literacy of the STEM teachers by providing opportunities for teachers to observe and participate; and (3) promoting the integration of the STEM teachers' information literacy elements by their personal practice.

Furthermore, with online learning becoming the new norm in the post-COVID era, to design and deliver STEAM courses online or in the virtual learning environment also become an essential skill for teachers. Hu and colleagues (2021) designed an online training model of STEM Teachers' teaching design ability by combining the backward design model (Wiggins & McTighe, 1998), virtual internship (Shaffer, 2006), and collaborative instructional design, as shown in Fig. 3.3. The innermost circle is the STEM instructional design guided by the reverse design process, which

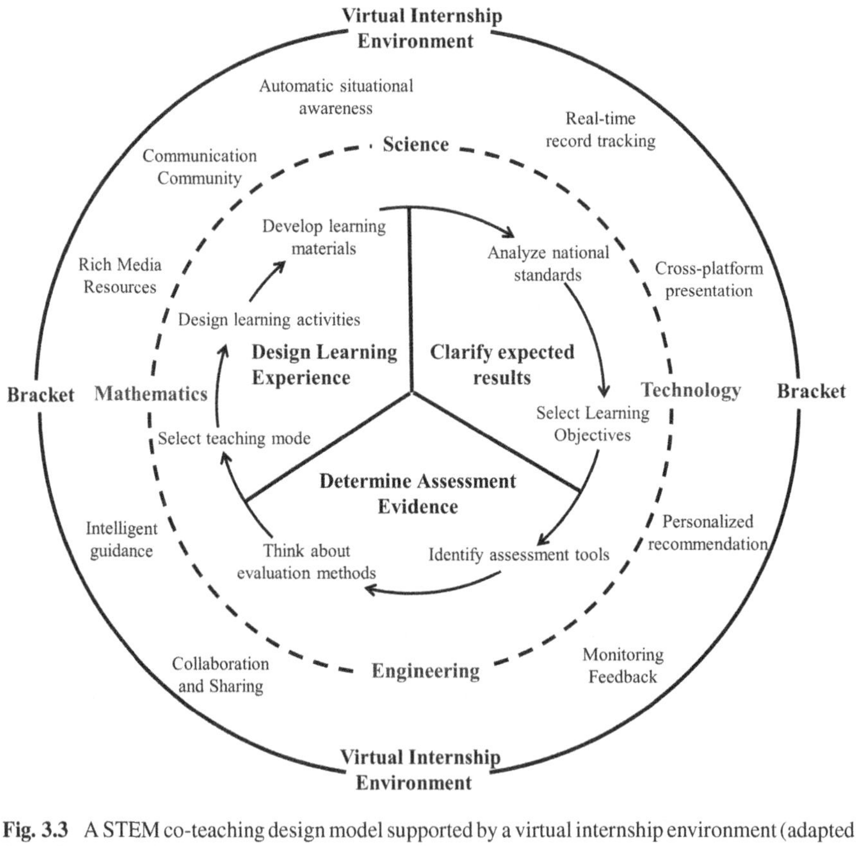

Fig. 3.3 A STEM co-teaching design model supported by a virtual internship environment (adapted from Hu et al., 2021, p. 34)

includes three stages of clarifying expected results, determining evaluation evidence, and designing learning experience, which are further subdivided into seven steps. The middle circle contains the core feature of STEM-interdisciplinary integration, and the way in which teachers of different disciplines collaboratively design lesson plans; the outermost circle is the virtual practice environment and support design that supports STEM collaborative teaching design, including the personalized functions provided by the technical environment and the guidance and intervention of virtual tutors.

Besides developing the ICT literacy and online instructional design abilities, scholars also pointed out the potential lack of CK of engineering. Zhan et al. (2021) conducted a mixed methods research on science teachers in Shanghai, and found that despite having good understanding of engineering and engineering education, science teachers failed to differentiate the engineering practices and scientific practices, the integration level of science and engineering, the cognitive level designed by the teachers for their students were low, and there was a lack of high-quality integrated STEM teacher professional development programs.

Conclusion and Future Directions

In this chapter, we provide an overview of the development of STEM education in mainland China from four aspects: policy formation and promulgation of STEM education, learner characteristics in STEM education, research on STEM learning process, research and practice on teaching approaches and teacher professional development. At the national policy level, education informatization is the key word leading the development of the education reform and STEM education development, STEAM education as an interdisciplinary educational paradigm has been integrated with the CVV and the K-12 curriculum, both formal and informal educational institutes have been actively participating in the STEM practices. Despite the fact that the research on the learning process of STEAM is still in its emerging phase, researchers in mainland China have been exploring the biological learning behavioral data, and with the lens of cognitive science to unwrap the insights of collaborative learning. For teacher professional development, we observed that multiple models have been developed, for both offline and online teaching in the smart learning environment. With engineering as a relatively new discipline in STEM education, teachers should be given more support to facilitate their understanding of engineering subject-knowledge, as well as relevant pedagogy. There is a need for teacher training and collecting empirical evidence to support the implementation of STEM in K-12 education. With STEM being integrated into the course format of CPAC, it is projected that more practical experience and empirical research evidence reflecting the development of STEM education in mainland China will be accumulated.

Acknowledgements This work was funded by the International Joint Research Project of Huiyan International College, Faculty of Education, Beijing Normal University (ICER201902).

References

Bybee, R. W., Taylor, J. A., Gardner, A., Van Scotter, P., Powell, J. C., Westbrook, A., & Landes, N. (2006). *The BSCS 5E instructional model: Origins and effectiveness*. Colorado Springs, Co: BSCS. https://media.bscs.org/bscsmw/5es/bscs_5e_full_report.pdf.

Chai, C. S., Chin, C. K., Koh, J. H. L., & Tan, C. L. (2013). Exploring Singaporean Chinese language teachers' technological pedagogical content knowledge and its relationship to the teachers' pedagogical beliefs. *The Asia-Pacific Education Researcher, 22*(4), 657–666. https://doi.org/10.1007/s40299-013-0071-3.

Department of Education of Zhejiang Province. (2017). *Zhèjiāng shěng jiàoyù tīng bàngōngshì guānyú gōngbù zhōng xiǎoxué STEM jiàoyù xiàngmù zhǒngzǐ xuéxiào hé péiyù xuéxiào míngdān de tōngzhī* [Notice of the Office of the Department of Education of Zhejiang Province on publishing the list of seed schools and breeding schools for STEM education projects in primary and secondary schools]. Retrieved from http://jyt.zj.gov.cn/art/2017/12/18/art_1532973_27485255.html.

Dong, Y., & Sun, W. (2019). Research on the Production-based Learning (DoPBL) Model for promoting interdisciplinary learning: Based on the integrated perspective of Problem-based Learning and Project-based Learning. *Journal of Distance Education, 2019*(2), 81–89. https://doi.org/10.15881/j.cnki.cn33-1304/g4.2019.02.010 (in Chinese).

Dong, Y., Sun, W., & Xu, C. (2019). Research on interdisciplinary learning under integration of information technology. *E-education Research, 2019*(11), 70–77. https://doi.org/10.13811/j.cnki.eer.2019.11.010 (in Chinese).

Dong, Y., He, J. Y., Xu, C., & Zheng, Y. F. (2020a). Situated analysis and development strategies of STEM teachers' information literacy. *China Educational Technology, 2020*(8), 70–77 (in Chinese).

Dong, Y., Wang, J., Yang, Y., & Kurup, P. M. (2020b). Understanding intrinsic challenges to STEM instructional practices for Chinese teachers based on their beliefs and knowledge base. *International Journal of STEM Education, 7*(1), 47. https://doi.org/10.1186/s40594-020-00245-0.

Du, Y. P. (2020). *Xiǎoxué zònghé shíjiàn huódòng kèchéng róngrù STEAM jiàoyù lǐniàn de lùjìng yánjiū* [A study on the path of integrating STEAM education concept into integrated practical activity courses in elementary school]. *Shanghai Research on Education, 2020*(4), 62–67 & 57. https://doi.org/10.16194/j.cnki.31-1059/g4.2020.04.013.

English, L. D., & Mousoulides, N. G. (2015). Bridging STEM in a real-world problem. *Mathematics Teaching in the Middle School (MTMS), 20*(9), 532–539. https://doi.org/10.5951/mathteacmiddscho.20.9.0532.

Gao, Y. R., Dong, Y., & Zhao, L. (2020). Research on STEM ecosystem construction and service function evaluation in China. *China Educational Technology, 2020*(11), 80–86 (in Chinese).

Han, S., Capraro, R., & Capraro, M. M. (2015). How science, technology, engineering, and mathematics (STEM) project-based learning (PBL) affects high, middle, and low achievers differently: The impact of student factors on achievement. *International Journal of Science and Mathematics Education, 13*(5), 1089–1113.

Hu, P., Jiang, J. B., & Chen, Z. C. (2016). The realistic problems and path choice of STEAM education in Chinese elementary and secondary schools. *Modern Educational Technology, 26*(8), 22–27 (in Chinese).

Hu, Y. L., Peng, X. L., & Wu, B. (2021). STEM jiàoshī jiàoxué shèjì nénglì zàixiàn péiyǎng móshì yánjiū [An online model for developing instructional design skills of STEM teachers]. *Distance Education in China, 2021*(8), 31–39. https://doi.org/10.13541/j.cnki.chinade.2021.08.004.

Li, K. D., & Li, Y. (2019). 5EX Design Model for interdisciplinary learning activity in STEM education. *E-education Research, 2019*(4), 5–13. https://doi.org/10.13811/j.cnki.eer.2019.04.001 (in Chinese).

Li, Y. Y., Dong, X. N., Li, X., & Zhang, Y. (2020). Construction of quality evaluation indicator system for STEM education based on the Delphi method and analytic hierarchy process. *Modern Distance Education Research, 32*(2), 48–55, 72 (in Chinese).

Lin, X. F., Hu, Q. T., & Liang, Z. M. (2018). Optimize design research on promoting developments of STEM education based on TPACK. *China Educational Technology, 2018*(9), 24–30 (in Chinese).

Liu, J. Z. (2020). *2019 Nián zhōngguó STEAM jiàoyù hángyè gàilǎn* [China STEAM Educational Industry Overview in 2019]. LeadLeo Research Institute. http://qccdata.qichacha.com/ReportData/PDF/d016352fb8459ad483ce710b5d37811b.pdf.

Lu, K. L. (2020). *The interpersonal neural substrate underlying group creativity* [Doctoral dissertation, East China Normal University, Shanghai]. China National Knowledge Infrastructure (CNKI) (in Chinese).

Ministry of Education, P. R. of China. (2016). *Jiàoyù bù guānyú yìnfā "jiàoyù xìnxī huà "shísānwǔ" guīhuà" de tōngzhī* [The 13th five-year plan for educational informatization]. Retrieved from http://www.moe.gov.cn/srcsite/A16/s3342/201606/t20160622_269367.html.

Ministry of Education, P. R. of China. (2017a). *Yìwù jiàoyù xiǎoxué kēxué kèchéng biāozhǔn* [Science curriculum standards for primary school]. Retrieved from http://www.moe.edu.cn/srcsite/A26/s8001/2017a02/W020170215542129302110.pdf.

Ministry of Education, P. R. of China. (2017b). *2017b Pǔtōng gāozhōng shēngwù xué kèchéng biāozhǔn* [Ordinary high school biology curriculum standard]. Retrieved from https://wenku.baidu.com/view/b8be582acbaedd3383c4bb4cf7ec4afe04a1b1d0.html.

National Institute of Education Sciences. (2017). *China STEM education white paper.*

National Research Council. (1996). *National science education standards*. National Academy Press. https://doi.org/10.17226/4962.

National Science Teaching Association. (2012). *Position statement. STEM education teaching and learning.* https://www.nsta.org/nstas-official-positions/stem-education-teaching-and-learning.

Saavedra, A. R., Rapaport, A., Morgan, K. L., Garland, M. W., Liu, Y., Hu, A., Hoepfner, D., & Haderlein, S. K. (2021). Project-based learning in AP classrooms: Lessons from research. *Phi Delta Kappan, 103*(3), 34–38. https://doi.org/10.1177/00317217211058522.

Sanders, M. (2009). STEM, STEM education, STEMmania. *The Technology Techer, 68*(4), 20–27.

Shaffer, D. W. (2006). *How computer games help children learn.* Palgrave Macmillan.

The State Council, P. R. of China. (2016). *Guówùyuàn bàngōng tīng guānyú yìnfā quánmín kēxué sùzhì xíngdòng jìhuà gāngyào shíshī fāng'àn (2016–2020 nián) de tōngzhī* [Implementation Plan of the Action Plan for the Quality of Science for All (2016–2020)]. Retrieved from http://www.gov.cn/zhengce/content/2016-03/14/content_5053247.htm.

The State Council, P. R. of China. (2017). *Guówùyuàn guānyú yìnfā xīnyídài réngōng zhìnéng fāzhǎn guīhuà de tōngzhī* [Notice of the State Council on the Issuance of the Development Plan for a New Generation of Artificial Intelligence]. Retrieved from http://www.gov.cn/zhengce/content/2017-07/20/content_5211996.htm.

Wei, X., Lin, L., Meng, N., Tan, W., & Kong, S. C. (2021). The effectiveness of partial pair programming on elementary school students' computational thinking skills and self-efficacy. *Computers & Education, 160*, 104023.

Wiggins, G., & McTighe, J. (1998). *Understanding by design.* Association for Supervision and Curriculum Development.

Wu, H. (2018). Huádōng shīdà chūbǎnshè xiéshǒu lègāo jiàoyù tuīchū STEAM jīchǔ kèchéng [East China Normal University Press and the Lego Group Developed the Lego Basic Course of STEAM Jointly]. *Modern Teaching, 2018*(23), 76.

Wu, Y. H., Chang, X. Y., Wang, J. W., & Guo, S. C. (2019). Research on STEAM curriculum design under the guidance of "learning-research-career development". *China Educational Technology, 2019*(2), 51–56, 125 (in Chinese).

Xie, H., Howell, A., Schreier, M., Sheau, K. E., Manchanda, M. K., Ayub, R., Glover, G., Jung, M., Reiss, A. L., & Saggar, M. (2019). Finding the neural correlates of collaboration using a three-person fMRI hyperscanning paradigm. *bioRxiv, 117*(37). https://doi.org/10.1101/782870.

Yakman, G., & Lee, H. (2012). Exploring the exemplary STEAM education in the US as a practical educational framework for Korea. *Journal of the Korean Association for Science Education, 32*(6), 1072–1086.
Yan, X. M., Pei, X. N., & Zheng Y. H. (2018). Problems and suggestions in the development of science education in China. *Science and Society, 8*(3), 13–21. https://doi.org/10.19524/j.cnki.10-1009/g3.2018.03.013 (in Chinese).
Yao, J. X., & Guo, Y. Y. (2017). The curriculum development and practical challenge of science education in primary school. *Curriculum, Teaching Material and Method, 37*(9), 98–102. https://doi.org/10.19877/j.cnki.kcjcjf.2017.09.016 (in Chinese).
Yu, S. Q., & Hu, X. (2015). STEM education and its model for interdisciplinary integration. *Open Education Research, 21*(4),13–22. https://doi.org/10.13966/j.cnki.kfjyyj.2015.04.002 (in Chinese).
Yu, S. Q., & Wu, L. (2019). Research on evidence - Oriented STEM instructional model. *Modern Distance Education Research, 31*(5), 20–31, 84 (in Chinese).
Zhai, X. S., & Shu, Y. H. (2018). The connotation, measurement and cultivation of creativity: An interview with Brad Hokanson. *Open Education Research, 24*(5), 4–10. https://doi.org/10.13966/j.cnki.kfjyyj.2018.05.001 (in Chinese).
Zhan, X., Sun, D., Wan, Z. H., Hua, Y., & Xu, R. (2021). Investigating teacher perceptions of integrating engineering into science education in mainland China. *International Journal of Science and Mathematics Education, 19*(7), 1397–1420. https://doi.org/10.1007/s10763-020-10117-2.
Zhang, Y., Xu, Q., Lao, J., & Shen, Y. (2021). Reliability and validity of a Chinese version of the STEM Attitude Scale for primary and secondary school students. *Sustainability, 13*(22), 12661. https://doi.org/10.3390/su132212661.
Zhu, Z. T., & Lei, Y. (2018). An analysis of the national policies and practical models for STEM education. *E-education Research, 2018*(1), 75–85. https://doi.org/10.13811/j.cnki.eer.2018.01.011 (in Chinese).

Nanxi Meng received her Ph.D. in Learning Technologies, and has been an educator for over a decade. As a researcher, she has focused primarily on STEM education, project-based learning pedagogy, teacher professional development, and literacy education. In 2019, she was certified by the National Institute of STEM Education of the United States as a STEM teacher and teacher trainer, and has led multiple K-12 STEM curriculum design projects worldwide. She is currently a senior lecturer at University of North Texas.

Yang Yang is currently an Assistant Professor in Center for Educational Science and Technology, Institute of Advanced Studies in Humanities and Social Sciences, Beijing Normal University, China. Research area includes curriculum design and teacher professional development in STEM education, measurement & evaluation, and quantitative research methods.

Xinyu Zhou is currently a senior undergraduate student in the School of Educational Information Technology at South China Normal University and has been recommended for admission to Beijing Normal University for postgraduate study in Science and Technology Education, with research interests in STEM education, game-based learning and artificial intelligence education.

Yan Dong received her Ph.D. in School of Psychology from University of Beijing Normal University in June 2007, majoring in Education Technology (Beijing, China). She was a visiting scholar in the Institute of Learning and Instruction from University of Sydney (Sydney, Australia) from 2004.7 to 2004.11. By 2015 she has been a professor working in Faculty of Education, Beijing Normal University. She also serves as vice-Dean of the School of Education Technology & the Research Institute of Science Education in Beijing Normal University. By 2021 she has also been a part-time professor working in Faculty of Education, Zhejiang University. Her research interests include teachers' TPACK knowledge, 21st-century learning, technology-enhanced learning and STEM project-based learning.

Chapter 4
Status of STEM/STEAM Learning in Japan: International Perspectives on Preparedness for Society 5.0

Yoshisuke Kumano

Abstract The twenty-first century is already one-fifth over. This chapter focuses on the contexts of STEM/STEAM learning in Japan. Japan is currently the third largest economy in the world, and major learning changes are underway. This chapter is organized into three sections. First, the broad context is briefly explained for STEM/STEAM area learning in relation to the National Defence Education Act of 1960 in the United States. Many changes and developments have recently occurred in STEM/STEAM learning, especially since 2016. These changes are described with reference to several drivers, including the drive to promote 21st-century skills or competencies. The second section focuses on changes at the governmental level, including the Ministry of Economy, Trade and Industry (METI) and the Ministry of Education, Culture, Sports, Science and Technology (MEXT). The third section focuses on how the Japanese prefectural governments and independent city governments have engaged in local actions connected to STEM/STEAM learning. In this section, the Shizuoka STEM Academy which adopts action research is described. The chapter concludes with a prediction that there will be country-wide development of systemic reform in Japan. However, communication of leaders and team members at each school, institution, university, and company, as well as local and national policymakers, will be essential to this development. In addition, systemic funding reform is needed to move towards a 'Society 5.0', defined by the Cabinet Office of Japan in 2021 as 'A human-centred society that balances economic advancement with the resolution of social problems by a system that highly integrates cyberspace and physical space'.

Keywords STEM education · Japanese educational policy towards STEM · Status study

Y. Kumano (✉)
Faculty of Education, Shizuoka University, 836 Ohya, Suruga-Ku, Shizuoka-Shi 422-8529, Japan
e-mail: kumano.yoshisuke@shizuoka.ac.jp

M. M. H. Cheng et al. (eds.), *Concepts and Practices of STEM Education in Asia*,
https://doi.org/10.1007/978-981-19-2596-2_4

Introduction

This chapter describes the development of STEM/STEAM area learning in Japan from 1960–2021 and its impact on Japanese society and other countries. In addition, as part of the exemplary STEM/STEAM learning model, central and local actions are explained. Furthermore, approaches for supporting high-quality STEM/STEAM learning are discussed.

Method

This chapter adopts a descriptive approach to study the status of Japanese STEM/STEAM area learning from 1960 to 2021. It includes reporting on an exemplary STEM model for grades 5–9 in the Shizuoka STEM Academy funded by Japan Science and Technology Agency (JST) from 2018 to 2021. This model was connected with a series of investigations into STEM/STEAM education.

STEM/STEAM Learning in Japan: A Brief Background

To describe the context of STEM/STEAM area learning in Japan, we can go back to the 1960s. The year 1960 marked 15 years after the termination of World War II. All Japanese people were concentrating on the success of the Tokyo Olympics in 1964. It was the beginning of identifying the importance of science and technology (STEM-related) education throughout the world. In 1962, the first Kousen (Technology College) opened (which we can now identify as one of the STEM model schools in Japan); at present, more than 50,000 students are studying science and technology at 51 Kousens throughout the regions of Japan. In the 1960s, in the United States, the National Defense Education Act of 1960 was enacted. High-quality science education curricula were developed, including work by the PSSC (Physical Science Study Committee), BSCS (Biological Science Curriculum Study), CHEMS (Chemical Education Material Study), ESCP (Earth Science Curriculum Program), and SAPA (Science—a Process Approach), supported by the National Science Foundation. These curricula were adopted and adapted by many researchers in science education in different countries, who developed or translated textbooks, adopted them into their contexts, and conducted science teacher training.

The K-12 National Course of Study of Japan for science was influenced by the high-quality science curriculum from the United States. The science lessons focused on the importance of inquiry lessons and understanding basic laws and principles in the pure sciences.

The Environmental Education Act of 1970 in the United States brought about the Back to Basics Movement (the importance of reading, writing and arithmetic).

Subsequently, it was the so-called Japan Shock or German Shock when Japanese or German products became best-selling products because of their high quality. As a result, the National Science Foundation (NSF) started to provide more funding in the field of science and technology education starting from around 1985 and Science for All Americans (Rutherford & Ahlgren, 1989) was published in 1989. The Japanese translation was published in 2005. In Japan, the Zest for Life educational policy came into force in 2002, highlighting the importance of all subjects. This educational policy decreased the amount of time spent on science, technology, and mathematics lessons in a year.

In January 1996, the National Science Education Standards were developed in the United States because of many discussions and research papers that provided evidence of the need for new state science standards for general scientific literacy. Additionally, the OECD's PISA (Programme for International Student Assessment) results had revealed issues related to student learning in many countries. Japan was positioned in the top group. However, there were many discussions about issues such as attitudes towards science, technology and mathematics, and comprehensive reading skills. Ministry of Education, Culture, Sports, Science and Technology (MEXT, 2008) made major changes to the Course of Study, placing more emphasis on STEM area learning starting from 2008. These changes included: increased science and mathematics lesson hours, recommendation of inquiry learning or problem-solving learning, encouragement of authentic assessment, and support for 'maker education'. The importance of authentic assessment in all subjects was identified as part of the 2008 Course of Study in Japan.

By 2000, the so-called twenty-first century skills or competencies were being actively discussed in the United States, the European Union, and other countries, including Japan. In the case of the United States, many researchers argued that the National Science Education Standards (NSES) had not put sufficient emphasis on the development of skills or competencies needed for students to be scientifically literate citizens. Additionally, advances in science and innovative technologies were creating new settings and contexts.

Work in relation to 21st-century skills, such as the Battelle for Kids (n.d.) and Rotherham and Willingham (2009), has had widespread influence. Many curriculum specialists and researchers in each of the school learning subjects have investigated 21st Century skills in their subjects. In science education, a summary report was developed by the National Research Council (NRC, 2010) in the US. The committee identified five important 21st Century skills for science education: adaptability, complex communication/social skills, non-routine problem solving, self-management/self-development, and systems thinking. Subsequently, A Framework for K–12 Science Education: Practices, Crosscutting Concepts and Core Ideas (NRC, 2011) was published, followed by the Next Generation Science Standards (NRC, 2013).

MEXT in Japan started investigating 21st Century skills following the American developments (National Institute for Education Policy Research [NIER], 2013). The Central Council for Education defined the 21st Century skills as proactive, interactive, and deep learning for all subjects (MEXT, 2013, 2014). Proactive learning includes learning with intrinsic motivation, or subjective or agentic learning. Interactive learning includes collaborative learning or higher order communication. Deep learning focuses on understanding the nature of science, mathematics, and so on. In addition, MEXT (2021) explained that the skills involve ‘fostering three types of strengths, namely, “knowledge and skills” for living and working in real society, “abilities to think, make judgement and express themselves” for responding to unprecedented circumstances, and “motivation to learn, and humanity”’ (p. 8).

Developments in STEM/STEAM Learning in Japan

Before 2016, Japanese society did not widely understand the importance of STEM/STEAM education as the new movement. There were several reasons for this. First, the meaning of technology and the meaning of engineering are considered to be quite similar in Japan. In a broad sense, engineering was an academic area of research at the university level, whereas technology education was the area of ‘making education’. Technical High Schools made up 7.5% of all high schools in Japan in 2020. In contrast, in 1972, they represented 18% of all high schools. More than 550,000 high school students were able to secure a job after graduation at that time. Now, we have fewer than 220,000 high school students going into employment in 2021. Many more students go to universities or specialized technical schools. In addition, almost 50,000 students graduating from the Kousens will become leaders in all types of industry in Japan. Additionally, MEXT established Super Science High Schools in 2002, and we have 217 high schools where science/technology-oriented curricula and project-based learning (PBL) are strongly emphasized.

In 2016, the Science and Technology Basic Plan Phase 5 declared the advent of Society 5.0, stating that all Japanese policy needs to change to prepare for this stage. According to the Cabinet Office of Japan (2016), Society 5.0 can be defined as ‘a human-centred society that balances economic advancement with the resolution of social problems by a system that highly integrates cyberspace and physical space’. Of course, this was influenced by the East Japan Giga Earthquake of 2011 and new educational policies of other countries. From this time, every governmental policy shifted towards Society 5.0. This was the beginning of systemic reform in Japan.

Figure 4.1 shows changes in the number of Japanese research projects receiving funding for STEM/STEAM education over time. As can be seen, few researchers in Japan worked in the area of STEM or STEAM education in the early 2000s. The push towards STEM/STEAM research started around 2011, when the K-12 Science Education Framework was developed in the United States. In January 2016, MEXT declared the Science and Technology Basic Law, followed with the Science and Technology Basic Plan, which was a turning point in Japanese educational policy.

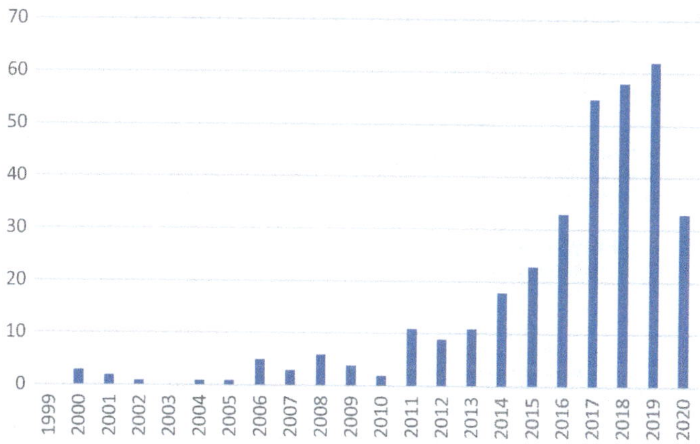

Fig. 4.1 Number of studies funded by the Japan Society for the Promotion of Science (JSPS) on the Kaken Database, 1999–2020

What Is Happening in STEM/STEAM in Japan?

After the Scientific and Technology Basic Planning was established in Japan, many new initiatives were implemented. First, almost all elementary and middle schools had received one computer for every student by the end of March 2021. This important action by the government aimed to move schools towards Society 5.0. Importantly, the Chromebooks or iPads could be used for on-demand classes at home through the lockdowns and school closures resulting from COVID-19. Many teachers in compulsory education worried about whether they could manage lessons via Chromebook or iPad; however, local boards of education started designing support for teachers to develop their practices. This was the first time that every student had received a computer throughout Japan, representing a historical event in Japanese educational policy.

Prior to this, METI and MEXT had declared the Future Classroom as a new educational pilot project to be planned in 2017. The planning was conducted by four strategic committees in 2018 and continued into 2021. As shown in Fig. 4.2, to establish Society 5.0, the Future Classroom consists of three important strategies: (1) STEAMifying learning; (2) intrinsic learning or individually optimized learning, taking into account individual students' motivations, interests, or aspects of cognition; and (3) development of the learning environments, such as focusing on information and communication technology (ICT) or the educational system, and introducing school business process reengineering (BPR). The committee identified three issues: (a) there was a lack of STEAM learning programmes, lesson plan models, and assessment strategies; (b) schools did not have enough space and time to implement PBL; and (c) there are many students who cannot properly communicate with others due to mental problems such as the Attention-Deficit Hyperactivity

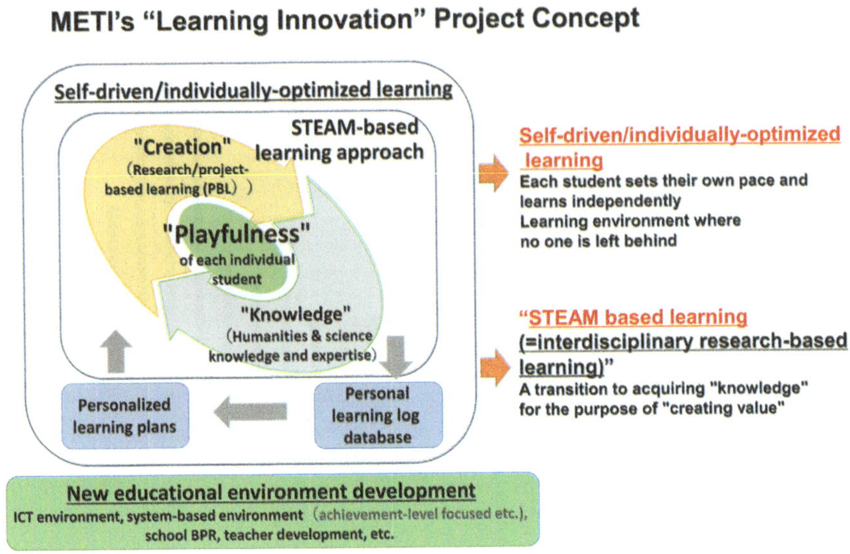

Fig. 4.2 Goals of future classrooms (adopted from METI, 2019, p. 4, modified and translated by Kumano)

Disorder. As a result, the committee developed four action plans (described below) to resolve these issues.

Development of a STEAM Library

The STEAM library is the online shared library with piloted STEAM models which teachers can access, many of which depend on specific subjects or integrating studies, such as the SDGs (Sustainable Development Goals). A new school can find models that might be helpful for the school in the library. In Japan, as part of the Future Classrooms, we already have more than 180 schools that are conducting STEAM schools or model STEAM lessons in 2021, supported by around 81 education-related companies via METI grants.

The EdTech (Combination of Education and Technology) Library

The EdTech library is a shared library where teachers can access piloted models for using IT in relation to specific subjects or integrating studies, such as the SDGs. A new school can find models that might be helpful for the school in the library. The provision of the Chrome Books and iPads for each student at the 180 STEAM schools motivated the schools to adopt the model STEAM lessons in as many different ways as possible.

School Business Process Reengineering

School BPR comprises collections of school management models for STEAM schools, showcasing exemplary actions among the 180 STEAM schools. A new school can find a certain model that might fit the school.

Newsletters for the Classrooms of the Future

Newsletters are shared among the 180 STEAM schools and other interested schools, allowing them to exchange ideas or success stories showcasing what they have done in their own contexts. Through these newsletters, teachers are able to share advantages, challenges, and experiences of implementing STEM/STEAM, and might be able to see new possibilities or move to new ideas.

Expansion of STEM/STEAM Learning in Japan

Since the new Society 5.0 government policy was initiated in 2016, all educational policies have moved towards the same goals of developing innovations in STEAM areas, including humanities, science and technologies, and engineering. For example, in the category of Super Science High Schools, 217 schools can be identified as STEAM schools.

These Super Science High Schools began to redevelop their planning, inviting STEAM model ideas developed by Future Classrooms, METI. In addition, the Global Science Campus, comprising 10 universities, provides a systematic support programme for high school students, encouraging them to conduct research with university researchers (JST, 2020). In addition, the Next-Generation Scientists project under Fostering Next Generation Human Resources has initiated a project called Fostering Junior Doctor Scientists for fifth to ninth grade students in 30 institutions, including universities, Kousens, and non-profit organizations, in many locations of Japan.

On 4th April, 2021, the Cabinet Office officially declared the 6th Science, Technology, and Innovation Basic Plan developed by the Council for Science, Technology and Innovation. According to this plan, US$150 billion for five years will be used to support the Society 5.0 policy. The greatest expansion will be seen in the next five years.

An Exemplary STEM Project: Shizuoka STEM Academy

The year 2021 represents the fourth year that Shizuoka STEM Academy has been operating. It was established when JST started a competitive funding system, named the Fostering Next-Generation Scientists Program, which included the Fostering Junior Doctor Scientists project and the Shizuoka STEM Academy.

Structure of the Shizuoka STEM Academy

The Shizuoka STEM Academy is a one-year program that is open to the public. It is advertised by sending a two-page announcement flyer to students in grades 5–9 at elementary and middle school in Shizuoka Prefecture (Fig. 4.3). We also included the flyer on the homepage of the Kumano Laboratory at Shizuoka University.

Each year, we receive permission for the Shizuoka STEM Academy program from the Boards of Education of Shizuoka Prefecture, Shizuoka City, Hamamatsu City, Omaezaki City, Makinohara City, Fujieda City, Yaizu City, Shizuoka City, Mishima City, and Numazu City. For the application, students need to write a 400-word essay on their research conducted in past years and their future research interests in the area of STEAM. The Shizuoka STEM Academy consists of three stages: Stage 1.0, Stage 1.5, and Stage 2.0. Stage 1.0 includes two activities—STEM learning model activities in the morning and advising lessons for individual or group STEM area research in the afternoon. STEM learning was developed through communication and investigation of the STEM lesson model from the NGSS. In addition, Prof. Gillian Roehrig and her team from the Minnesota STEM Education Center have been invited periodically to Shizuoka University for STEM lesson training for our staff. In addition, the team of

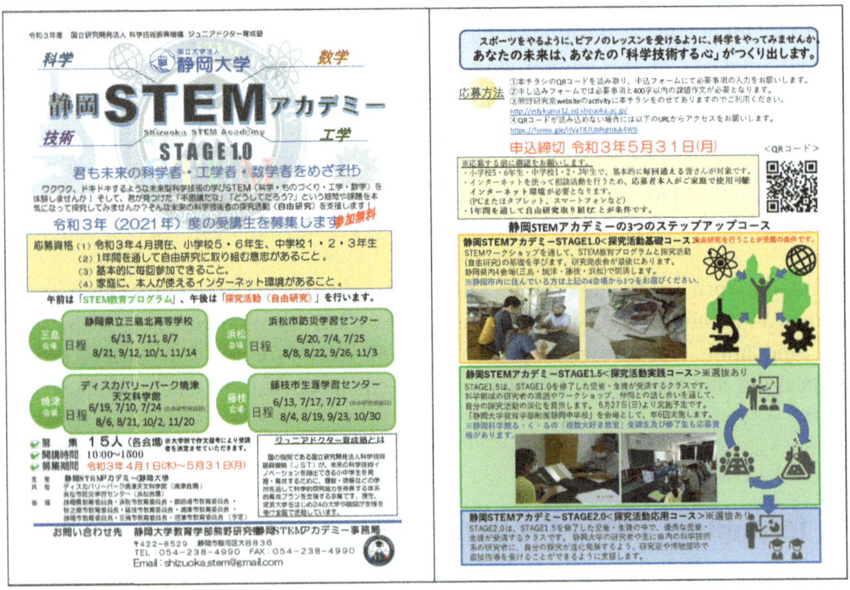

Fig. 4.3 Two-page announcement flyer targeting students in Shizuoka prefecture

Shizuoka STEM Academy visited the Minnesota STEM Education Center and their affiliated STEM schools in order to understand the management of schools supported by JSPS (Kumano, 2020, 2021).

Minimum Elements of STEM Lessons

Our team at the Shizuoka STEM Academy agreed that STEM lessons should at least include a number of elements (minimum elements) in the context of Japan, namely the intrinsic learning model, conducting practices or inquiries, collecting data, interpretations using mathematical processes, finding solutions or results, presenting findings to others, or engaging in communication.

From an Intrinsic Learning Model or Agentic Learning to Conducting Practices or Inquiries

To conduct STEM learning, it is important to share the questions, problems, or issues that students raise. For Stage 1.0, we carefully find exemplary questions, problems, or issues that are understandable for the students. To solve these questions, problems, or issues, students first need to develop certain designs. Then, it is important to discuss the designs with others. As a group, they will find many models or inferences for establishing solutions. Group activities are highly recommended in this context. The groups will discuss the kinds of scientific experiments or engineering practices that can be applied, and decide on the variables for collecting data.

Collecting Data, Analysis, and Interpretations Using Mathematical Processes

For the data collection stage, students are highly encouraged to collect data with handmade sensors, such as Micro:bit, Arduino, and Raspberry Pi; data are collected using certain calculating software. Then, students learn how to use computers more effectively. Trial and error is important in that they may redesign practices or inquiries as a group. However, for model STEM learning, it is better to conduct simple activities. Thus, Shizuoka STEM Academy introduced balloon rocket practice as the first activity with a clear objective as one of the sophisticated models in STEM learning.

Finding Solutions or Results, Presenting Findings to Others, Engaging in Communication

If students identify a certain solution or product as a result of the practices or inquiries, then they need to develop a paper and create a PowerPoint presentation to explain it to others. They need to prepare for questions after the presentation from peer students. This activity is part of the reflection of scientific inquiries or engineering practices and the assessment or evaluation of their work. Through this activity, they will find the next targets of their new work.

Adding Characteristics of Stage 1.5

In Stage 1.5, we find interested scientists or engineers who have developed innovations during their careers. Last year, we invited a biologist whose specialty was biodiversity in extreme environments, Emeritus Professor Dr. Tomoaki Masuzawa. Students attended a speech at the Shizuoka University on the nature of biodiversity and research in severe environments, such as Antarctica.

In the last two years, we invited Emeritus Professor Dr. Shinji Tsuyumu, College of Agriculture, Shizuoka University, who developed a special yeast called "Shizuoka Koubo" for Japanese rice wine. He explained DNA development and innovation for our students at Shizuoka University. It is important for students to meet scientists or engineers who have discovered new things or invented new things or ideas for the public.

Receiving Good Advice from Mentors

In our informal STEM learning setting, we have been conducting a mentoring system in every stage. Here, we allow time for each student to talk about their research with their mentors in the afternoon. We have hired two special mentors—a retired elementary school principal (Mr. Aoki Yoshiaki) and a retired junior high school principal (Mr. Masuda Toshihiko), both former directors of Shizuoka City Children's Museum. They have experience of encouraging students to conduct interesting research. They conducted lectures in preservice science teacher training several times last year at Shizuoka University. There are theories and practices for STEM mentoring just like sports coaches (National Academies of Science, Engineering, Medicine, 2019).

For students at Stage 2.0, we find scientists or engineers whose research is close to the students' research. The Stage 2.0 students and the mentors visit these scientists or engineers at their institutions and receive more specific advice.

STEM Camp for Stages 1.5 and 2.0

STEM camps are stimulating STEM learning activities for students at Stage 1.5 and Stage 2.0 considering time and space concerns. Students are stimulated by their fellow students, university staff, and mentors. We used to prepare longer team-oriented STEM activities, and students can practice their presentation skills. In addition, we used to ask one special engineer or scientist to give a special lecture with activities.

Unfortunately, the STEM camps for 2020 and 2021 were cancelled because of the Covid-19 pandemic.

Submitting to Local or National Science Competitions

From October to December, students at Stages 1.0, 1.5, and 2.0 submit their research to competition committees at the local and national levels, including the Japan Student Science Award and Natural Science Observation Competition. Last year, one Stage 2.0 student achieved first place in Japan at the Natural Science Observation Competition. We have a total of 29 students who have received prizes at the prefecture or national level out of 86 students who carried out research at the Shizuoka STEM Academy in 2020.

Presentation Day When All the Students Receive Peer Reviews and Questions from Staff

The final day of the Shizuoka STEM Academy closes with students' presentation of their research. They assess others' presentations with rubrics and take questions from peer students and staff. Students must prepare PowerPoint presentations. Top students exchange their presentation at the STEM school connected with the STEM Education Center at the University of Minnesota or another university in the United States.

Discussion and Implementation of STEM/STEAM Learning in Japan

In the context of Japan, what the NGSS (Next Generation Science Standards) proposed may not work as it is. Clear definitions of technology and engineering may not be necessary in Japan. This is because the sixth Science, Technology, and Innovation Basic Plan was declared on 24th March, 2021, without clear definitions of technology and engineering. We have been using the term Science and Technology, which comprises science, engineering, and technology, for over 150 years. The NGSS recommends 'practices' instead of 'inquiry', and the ideas of 'practices' are quite logical and well supported in many recent papers such as the historical analysis of 'inquiry' (Barrow, 2006); however, it is almost impossible to accept the ways of thinking for 'practices' because 'inquiry' is used everywhere in the contexts of Japan, even in the new Course of Study developed by MEXT. Japan may need another decade to adopt 'practices' as the higher quality 'inquiry' for STEAM learning.

A number of studies have been conducted in Japan to inform the systemic reform towards Society 5.0. The implementation of Japanese model for STEM Education in Indonesia was introduced and teachers' perceptions of STEM integration into curriculum was discussed (Suwarma & Kumano, 2019). Researchers have identified authentic learning and assessment as one of the important characteristics of STEM lessons (Anwari et al., 2015). In a study conducted by Kumano and Goto (2016), scientific processes and their relation to the NGSS were examined. Also, research engaging young children in STEM learning was conducted for many pre-school children (Sakata & Kumano, 2018). One study involving an exemplary trial for Bio-STEM learning at a high school was reported (Okumura & Kumano, 2016). Saito et al. (2016) examined the STEM integrated learning environment using a historical approach. More efforts in research related to student learning were reported as Mutakinati et al. (2018) looked into the development of critical thinking skills among students in Project Based Learning. In addition, we conducted research on changes in US high school biology textbooks, and made comparison with Japanese textbooks, focusing on the impact of STEM (Kosaka & Kumano, 2021). For university STEM learning, an action research project was conducted for the undergraduate students of the College of Science at Ehime University (Kuroda & Kumano, 2018). In addition, we conducted a study on STEM teacher training for preservice teachers in Japan and Indonesia (Putra & Kumano, 2018). Recently, exemplary STEM education focusing on the geology and culture of Niijima Island in Japan was developed (Takebayashi & Kumano, 2020). Furthermore, exemplary in-service teacher training for the development of STEM teachers was conducted in Hamamatsu, Shizuoka, Japan (Takemoto et al., 2020). Last year, the research in STEM education through the Engineering Design Process are conducted at the Shizuoka University Attached Junior High School (Nurul, et al., 2021). Drawing on the findings, these STEM/STEAM studies have become the evidence and rationale supporting the development of the systemic reform towards Society 5.0 in Japan.

It is important to consider the concept of '21st-century skills' in the United States or '21st-century competencies' in the European Union, but we do not really know how we should re-define many skills or competencies in view of new developments or innovations of science and technologies. Understanding fundamental principles and theories will still be important until brain science develops in such a way that we can input knowledge into our brains automatically. There is no doubt that intrinsic learning or agentic learning is becoming more important, and this is connected to high-quality STEAM learning. Thus, we need more practices to identify new questions, issues, or problems from everyday school life. Questions and problems are the starting moment of real learning. An individual or a group of students will develop the design of projects or research, conduct experiments, make observations, or make instruments for the projects. In these processes, they might need to predict or to infer results; this is deductive thinking.

Trying things out and making mistakes are most welcome in situations of inductive thinking. Students can use mathematical thinking to collect data, develop graphs, and interpret the results. Through these practices, students learn the nature of science and engineering. Science and engineering processes should not be memorized without practice. A lesson or teaching unit supporting STEM/STEAM learning needs about at least 3–5 hours; then, we need a well-structured curriculum that is managed throughout the school year. Comparatively, it is much easier to manage the STEM/STEAM practices via informal science education, such as that offered by Shizuoka STEM Academy.

Systemic Reform to Conduct STEM/STEAM Learning

Japan has the capacity to develop systemic STEM/STEAM learning innovation. What is missing is the STEM/STEAM leaders, mentors, and structured budgets. To advance STEM education, we should cover both formal and informal education as much as we can. In Japan it is not common to donate private funds to schools, as is done in Western countries. This habit of mind needs to change in order to support STEM/STEAM learning in schools. In 2021, at least 81 companies are helping to develop STEAM schools in many locations of Japan with government funding. They are developing good role models for systemic reform in the context of Japan. Systemic changes for STEM/STEAM are critical for the advancement to Society 5.0.

Conclusion

In this chapter, we found that systemic innovation for STEAM learning started in 2016 with the Society 5.0 government policy development by the Cabinet Office of Japan. This was a historic moment for the country. However, the speed of change in the nation has been slow compared with the systemic reform in the United States or other countries. In the Japanese context, the Science and Technology Basic Law and Plan of 2016 first described Society 5.0. Starting in 2017, around 180 schools and companies funded by METI were engaged in endeavours contributing to the evidence of STEAMifying. Data on STEAM learning have been collected, and such evidence has developed the STEAM Library supporting schools that would like to be STEAM schools. The sixth Science, Technology, and Innovation Basic Plan was developed as of 26th March, 2021, and the Cabinet of Japan is planning real systemic reforms as a country.

It is highly possible that students at all elementary schools, junior high schools, and high schools in Japan will receive one Chromebook or iPad with a high-speed Wi-Fi system through MEXT's Global and Innovation Gateway for All (GIGA) school project by 2023. Thus, in the context of Japan, interesting advancements in terms of STEAM learning will be seen in the next three to five years; however, we know that the advancement of education is not easy and takes a long time. Shizuoka STEM Academy could be an interesting model for supporting STEAM informal learning among students (Kumano, 2020, 2021).

References

Anwari, I., Yamada, S., Unno, M., Saito, T., Suwarma, I. R., Mutakinati, L., & Kumano, Y. (2015). Implementation of authentic learning and assessment through STEM education to improve students' metacognitive skills. *K-12 STEM Education, 1*(3), 123–136.

Barrow, L. H. (2006). A brief history of inquiry: From Dewey to standards. *Journal of Science Teacher Education, 17*(3), 265–278. https://doi.org/10.1007/s10972-006-9008-5.

Battelle for Kids. (n.d.). *Partnership for 21st century learning: A network of Battelle for kids.* Retrieved August 20, 2021, from https://www.battelleforkids.org/networks/p21.

Cabinet Office of Japan. (2016). *Kagakugijyutsu Kihonkeikaku* [The science, technology basic plan]. Government of Japan. https://www8.cao.go.jp/cstp/kihonkeikaku/5honbun.pdf.

Cabinet Office of Japan. (2021). *Kagakugijyutsu Innovation Kihon Keikaku* [Science, technology, and innovation basic plan]. Government of Japan. https://www8.cao.go.jp/cstp/kihonkeikaku/6gaiyo.pdf.

Japan Science and Technology Agency. (2020). *JST-GSC: Global Science Campus.* https://www.jst.go.jp/cpse/gsc/about/images/flyer_aboutgsc2020.pdf.

Kosaka, N., & Kumano, Y. (2021). A study of changes in US high school biology textbooks: Focusing on the impact of the STEM (science, technology, engineering and mathematics) education reform by the Next Generation Science Standards (NGSS). *Japanese Journal of Biological Education, 62*(3), 128–139 (In Japanese).

Kumano, Y. (2020). *Reiwa Gannendo Nenndo Junia Doctor Ikuseijyuku Houkokusho* [2019 Shizuoka STEM Academy annual report, junior doctor fostering Juku, fostering next-generation scientists program]. Japan Science and Technology Agency. http://hdl.handle.net/10297/00027418.

Kumano, Y. (2021). *Reiwa Ninendo Nenndo Junia Doctor Ikuseijyuku Houkokusho* [2020 Shizuoka STEM Academy annual report, junior doctor fostering Juku, fostering next-generation scientists program]. Japan Science and Technology Agency. http://hdl.handle.net/10297/00028222.

Kumano, Y., & Goto, M. (2016). Issues concerning scientific processes in science lessons involving outdoor and indoor activities: A comparative study of scientific processes in Japanese science classes and the chronological development of scientific processes in the US through NGSS. *Bulletin of the Faculty of Education, Shizuoka University, Educational Research Series, 47*, 93–103. https://doi.org/10.14945/00009541.

Kuroda, T., & Kumano, Y. (2018). A study of social skills training program design for STEM students in a global society: Focusing on evaluation and curriculum design. *Journal of Science Education in Japan, 42*(2), 82–99.

Ministry of Economy, Trade and Industry. (2019). *Miraino Kyoushitsu Bijyon, Keizaisangyousyou Mirainokyoushitsu to Edotekku Kenkyuukai* Dainiji *Teigen* ["Future Classrooms" Vision, The 2nd proposal committee on "Future Classrooms" and EdTech]. https://www.meti.go.jp/shingikai/mono_info_service/mirai_kyoshitsu/pdf/20190625_report.pdf.

Ministry of Education, Culture, Sports, Science and Technology. (2008). *Chuugakkou Gakushuushidouyouryou Kaisetsu Rikahen* [New course of study guidelines for science]. https://www.mext.go.jp/component/a_menu/education/micro_detail/__icsFiles/afieldfile/2011/01/05/1234912_006.pdf.

Ministry of Education, Culture, Sports, Science and Technology. (2013). *Atarashii Nouryoku to Gakushuuhyouka no Wakugumi* [New competencies" and framework of learning evaluation]. https://www.mext.go.jp/b_menu/shingi/chousa/shotou/095/shiryo/1330122.htm.

Ministry of Education, Culture, Sports, Science and Technology. (2014). *Ikusei subeki shishitsu/nouryoku wo fumaeta kyouiku mokuhyou/naiyou to hyouka no arikata ni kansuru kentoukai* [Summary of the panel on educational objectives, contents and evaluation based on the quality and competencies to be developed]. https://www.mext.go.jp/b_menu/shingi/chousa/shotou/095/houkoku/1346321.htm.

Ministry of Education, Culture, Sports, Science and Technology. (2021). *Overview of the Ministry of Education, Culture, Sports, Science and Technology—Japan*. https://www.mext.go.jp/en/content/20210325-mxt_kouhou02-200000029_1.pdf.
Mutakinati, L., Anwari, I. K., & Kumano, Y. (2018). Analysis of students' critical thinking skill of middle school through STEM education project-based learning. *Journal Pendidikan IPA Indonesia, 7*(1), 54–65.
National Academies of Sciences, Engineering, Medicine. (2019). *The science of effective mentorship in STEMM*. The National Academes Press. https://www.nap.edu/catalog/25568/the-science-of-effective-mentorship-in-stemm.
National Institute for Education Policy Research. (2013). *Shakai no Henkani Taiousuru Shishitsuya Nouryokuo Ikuseisuru Kyouikukateino Kihongenri* [Basic principles of curriculum design for competencies and skill building towards societal changes]. https://www.nier.go.jp/kaihatsu/pdf/Houkokusho-5.pdf.
National Research Council. (2010). *Exploring the intersection of science education and 21st century skills: A workshop summary*. National Academy Press. https://doi.org/10.17226/12771.
National Research Council. (2011). *A framework for K-12 science education: Practices, crosscutting concepts, and core ideas*. National Academies Press. https://doi.org/10.17226/13165.
NGSS Lead States. (2013). *Next generation science standards: For states, by states*. National Academies Press. www.nextgenscience.org/next-generation-science-standards.
Okumura, J., & Kumano, Y. (2016). Practical research on attitude changes in communication activities in upper secondary school biology classes -Verifying the effects of reflective learning using video data-. *Journal of Research in Science Education, 56*(4), 421–434 (in Japanese).
Putra, P. D. A., & Kumano, Y. (2018). Energy learning progression and STEM conceptualization among pre-service science teachers in Japan and Indonesia. *The New Educational Review, 53*, 153–162. https://doi.org/10.15804/tner.2018.53.3.13.
Rotherham, A. J., & Willingham, D. (2009). 21st century skills: The challenges ahead. *Educational Leadership, 67*(1), 16–21.
Rutherford, F. J., & Ahlgren, A. (1989). *Science for all Americans: American Association for the Advancement of Science, Project 2061*. Oxford University Press.
Saito, T., Anwari, I., Mutakinati, L., & Kumano, Y. (2016). A look at relationships (Part I): Supporting theories of STEM integrated learning environment in a classroom—A historical approach. *K-12 STEM Education, 2*(2), 51–61.
Sakata, S., & Kumano, Y. (2018). Attempting STEM education in informal Japanese educational facilities through the theme of "Sand". *K-12 STEM Education, 4*(4), 401–411.
Sulaeman, N. F., Putra, P. D. A., Mineta, I., Hakamada, H., Takahashi, M., Ide, Y., & Kumano, Y. (2021). Exploring student engagement in STEM education through the engineering design process. *Journal Penelitian dan Pembelajaran IPA, p-ISSN 2477–1422, 7*(1), 1–16.
Suwarma, I. R., & Kumano, Y. (2019). Implementation of STEM education in Indonesia: Teachers' perception of STEM integration into curriculum. *Journal of Physics: Conference Series, 1280*, 052052. https://doi.org/10.1088/1742-6596/1280/5/052052.
Takebayashi, T., & Kumano, Y. (2020). Exemplary STEM education focusing on the geology and culture of Niijima Islands in Japan with cross-cutting concepts. *Southeast Asian Journal of STEM Education, 1*(1), 76–92.
Takemoto, I., Ogawa, H., Horita, T., & Kumano, Y. (2020). Characteristics of the relationship between teachers, researchers and engineers in web-based elementary school STEM class design meetings: Through network analysis using utterance data. *Journal of Science Education in Japan, 44*(4), 338–352 (In Japanese).

Yoshisuke Kumano is Professor Emeritus, Appointed Professor, Vice Director of STEAM Education Institute, Shizuoka University. He is also currently the Director of NPO Shizuoka STEAM Education Development Center and the Director of Shizuoka STEM Academy supported by Japan Science and Technology Agency. He is a past president of East Asian Association for Science

Education (EASE). He has been Professor of Science Education, Shizuoka University for about 30 years after getting Ph.D. in Science Education at the University of Iowa through the Fulbright Program in 1993. His focus area of study is currently STEAM education.

Part II
Implementing STEM Education in Asia

Chapter 5
What Does STEM Education Offer and How Is It Relevant? A Content Analysis of Secondary School Websites in Singapore

Yann Shiou Ong and Yew-Jin Lee

Abstract Science, technology, engineering, and mathematics (STEM) form the basis of many educational programmes around the world. In Singapore, school-based STEM education appears within STEM Applied Learning Programmes (ALP) offered by some primary and secondary schools. In this chapter, we present an in-depth survey of the diverse offerings and benefits of STEM education here; specifically, we examine STEM learning/activities from the websites of 15 secondary schools (Grades 7–10/11). Using a theoretical model of relevance for science education from the literature, we identified the benefits and pathways that STEM education has been reported to afford its participants, that is, how STEM education can be made relevant for students through ALP. Relevance is defined in terms of fulfilment of intrinsic or extrinsic needs in the present or future, and along the three dimensions of individual, societal, and vocational needs in this model. Our main findings indicate that this sample of STEM ALP websites did not sufficiently yield statements that supported the present or future aspects of intrinsic relevance within the societal and vocational dimensions. On the other hand, multiple descriptions in relation to the extrinsic and future aspects across the individual, societal, and vocational dimensions of relevance were provided. Three implications of these findings for STEM education in Singapore are highlighted: (i) greater consideration of student choices, identities, and agency, (ii) greater awareness and discussion of undesirable/negative impacts of STEM solutions on society, and (iii) greater emphasis on the epistemic aspects of STEM.

Keywords Relevance of science · STEM education · Content analysis · Singapore

Y. S. Ong
Natural Sciences & Science Education (NSSE), National Institute of Education, Nanyang Technological University, NIE 7-03-81, 1 Nanyang Walk, Singapore 637616, Singapore
e-mail: yannshiou.ong@nie.edu.sg

Y.-J. Lee (✉)
Natural Sciences & Science Education (NSSE), National Institute of Education, Nanyang Technological University, NIE 7-03-104B, 1 Nanyang Walk, Singapore 637616, Singapore
e-mail: yewjin.lee@nie.edu.sg

M. M. H. Cheng et al. (eds.), *Concepts and Practices of STEM Education in Asia*,
https://doi.org/10.1007/978-981-19-2596-2_5

Introduction

Science, technology, engineering, and mathematics (STEM) form the basis of many school and after-school educational programmes around the world. But what exactly STEM is, and how it is defined and characterized, has remained problematic over the past three decades since this acronym was first coined (e.g. Martín-Páez et al., 2019; McComas & Burgin, 2020). Often, discussions related to STEM education have focused on the content matter of STEM (i.e. a mix of various disciplines/specializations), including associated concepts, procedural, and/or epistemic knowledge, and/or the pedagogies of STEM, that is, providing opportunities to apply knowledge, skills and/or practices in an integrative fashion to solve authentic, complex problems. Beyond these broad categories, little consensus exists regarding which constitutive focal areas in STEM education are foundational, how they are related or sequenced, and how progress in STEM learning should be measured (e.g. Bybee & Gardner, 2006; Kloser et al., 2018).

These difficulties, however, have not prevented educators from focusing on STEM education as a pervasive as well as persuasive curriculum organizer (Millar, 2020). Moreover, STEM learning outcomes are often described in terms of enhancing cognitive and procedural knowledge, in addition to enhancing students' attitudes towards, and interests and identities in these domains (Martín-Páez et al., 2019; National Academy of Engineering [NAE] & National Research Council [NRC], 2014). Its reported benefits are arguably less contentious to accept, and indeed form necessary and sufficient justifications for the widespread promotion of STEM education across diverse regions and contexts. STEM educators appear to have taken some wisdom from continental philosophy to heart where it was claimed that '[t]he question is not: is it true? But: does it work? What new thoughts does it make it possible to think?' (Deleuze & Guattari, 1988, p. xv). In this chapter, we present an in-depth survey of the diverse offerings and benefits of STEM education ('what has worked') as expressed by school-based STEM programmes in Singapore. Specifically, we examine the relevance of STEM education for learners, as advertised on secondary school (Grades 7–10/11) websites. Our aim was to identify some of the many benefits and pathways that STEM education has been reported to afford its participants.

What might be the significance of examining the practical outcomes and relevance of STEM education over basic questions regarding its ontogeny? Is it desirable for schools to bypass difficult issues of what really constitutes STEM education and instead publicize its (positive) outcomes? This situation is understandable if one considers schools to be sites for knowledge transmission compared to academia (and sometimes industry), which are the typical sites of knowledge production. Schools are thus 'downstream' and tend to dwell on more established knowledge, leaving ontological and epistemological discussions about STEM to universities. Schools also deal with matters of practical relevance and benefits in order to engage their clients—their students. Be that as it may, we want to make the case that identifying 'products' of STEM education to students is neither straightforward nor predictable, and even more so when talking about scenarios situated in the future. For example, it is premature to claim to be able to prepare students for desirable jobs when future careers or STEM labour markets are notoriously difficult to predict (e.g. McComas & Burgin, 2020). However, K–12 STEM education has frequently been promoted based on lofty, albeit largely unexamined visions, outcomes, and benefits. It therefore seems important to have a better understanding of what STEM education claims to be able to offer, and the ways in which these benefits are being made relevant to learners.

While hopeful promises of the benefits of STEM education are likely to be found in all K–12 systems, we locate our study in Singapore, which is widely regarded for the quality of its science and mathematics education. In what follows, we briefly introduce the sites of STEM education in this country, followed by details of our sampling procedures and analytic methods before we present our findings.

The STEM Applied Learning Programmes (ALP)

In a bid to make academic learning more relevant and meaningful, the Singapore Ministry of Education (MOE) has recently encouraged every primary and secondary school to initiate an Applied Learning Programme (ALP). These are school-owned, non-examinable opportunities for enrichment to increase students' joy of learning. Depending on the school, these weekly or monthly activities might be within timetabled time, or as an after-school activity, facilitated by school teachers or through the services of external vendors. Activities in ALP are very flexible depending on the grade level, pedagogy choice (e.g. face-to-face or virtual), external partnerships/providers, and of course, the school-specific nature of the ALP themes. Many schools offer a tiered approach to STEM ALPs. That is, all students, typically in lower secondary (grades 7–8), experience the basic STEM ALP program. At the upper secondary levels (grades 9 to 11), selected students are offered more advanced, elective STEM ALP programs. To the best of our knowledge, there is no publicly available information on the stipulated number of curriculum hours required for an ALP. Above all, the ALP aims to help students make connections between academic knowledge and skills and the real world, that is, the 'applied' aspect of academic learning, so as to deepen their appreciation of the relevance and value of their school learning. It is hoped that by making these connections, students will be more motivated to develop the knowledge and skills that can be applied in society and industries (MOE, 2021). By 2023, all primary and secondary schools in Singapore will offer an ALP; schools currently offer ALP programmes in one of the following areas: STEM, languages, humanities, business and entrepreneurship, aesthetics, and interdisciplinary (e.g. thinking skills).

In Singapore, ALP is the only place where STEM takes centre stage; it is mentioned only in passing within the secondary science curriculum in Singapore. We also observed that the official aims of ALP are broadly aligned with the model of relevance in science education by Stuckey et al. (2013), which considers relevance in terms of fulfilment of intrinsic or extrinsic needs in the present or future, along the three dimensions of individual, societal, and vocational needs (details of the model will be explained later). As such, this conceptual model is deemed appropriate for guiding our analysis to answer the following research question: How is STEM education (through the ALP) made relevant for secondary students in Singapore?

Sample

Based on publicly available ALP websites, we identified 66 secondary schools (Grades 7–10/11) in Singapore that offered STEM ALP in 2020 (MOE, 2021). (There are about 136 secondary schools in the country.) We further grouped these, based on information available on the websites, according to five MOE-prescribed STEM themes:

1. Emerging Technologies (including design thinking/engineering)—31 schools;
2. Sustainability—14 schools;
3. Health and Food Science—14 schools;
4. Cities and Urban Technology—6 schools; and
5. Future of Transportation—1 school.

Approximately 20% of schools were then sampled from each of the five STEM themes (refer to Table 5.1 below). Our general criterion for inclusion was whether a school webpage revealed sufficiently detailed, extensive information about its STEM ALP. As there is no fixed ALP template from the MOE, schools crafted their webpages as they saw fit. Some webpages were rather brief, while others relied extensively on promotional video clips, which we excluded from our analysis. As few STEM ALPs fell under the last two themes, one school from each was selected. The content from a total of 15 STEM ALP websites was therefore analysed in this study.

Method of Analysis

Ground-Up and Top-Down Approaches to Content Analysis

Our analysis was initially conducted via a ground-up approach (Hsieh & Shannon, 2005). We first derived broad themes through reviewing the content of the STEM ALP webpages; both authors assessed the websites independently and then proposed possible categories. Through iterative rounds of discussions, we checked on each other's interpretations and reached consensus on three themes associated with the benefits/outcomes from participation in STEM ALP: 'benefits for the individual', 'potential or desired futures and identities', and 'ultimate purpose of being in STEM ALP'. We then reinterpreted these three tentative themes through the model of relevance (described next), taking a more top-down, theoretically-driven approach to the analysis. Hsieh and Shannon (2005) would classify our analytic process here as

Table 5.1 The websites of the 15 selected schools, their STEM ALP themes, URL links, and ALP name

Theme	School name/STEM ALP URL	STEM ALP name
(A) Emerging technologies	**1. Clementi Town Secondary School** https://clementitownsec.moe.edu.sg/key-programmes/applied-learning-programme-alp	Coding to discover and empower (CODE)
	2. Montfort Secondary School https://montfortsec.moe.edu.sg/curriculum/distinctive-programme/applied-learning-programme	Made in Montfort
	3. Hai Sing Catholic School https://haisingcatholic.moe.edu.sg/hai-programmes/direct-school-admission-robotics-n-engineering/applied-learning-programme	Robotics & engineering
	4. Woodlands Ring Secondary School https://woodlandsringsec.moe.edu.sg/applied-learning-programme/	The robotics education & enterprise (TREE)
	5. Loyang View Secondary School https://www.loyangviewsec.moe.edu.sg/signature-programmes/applied-learning-programme-alp	Design and engineering
	6. Admiralty Secondary School https://admiraltysec.moe.edu.sg/academic-curriculum-n-applied-learning-program/applied-learning-program	Design thinking through innovation and technology
	7. Regent Secondary https://regentsec.moe.edu.sg/departments/aesthetics-and-innovation/alp/	Emerging technology
(B) Sustainability	**8. Queensway Secondary School** https://queenswaysec.moe.edu.sg/programmes/promoting-a-sustainable-environment-alp	Promoting a sustainable environment
	9. Bukit View Secondary School https://bukitviewsec.moe.edu.sg/joules-programme/	Clean energy and environmental technology
	10. Fajar Secondary School https://fajarsec.moe.edu.sg/signature-programmes/applied-learning-programme-alp	Sustainability through 21st century applied critical and inventive thinking skills (ACIT)

(continued)

Table 5.1 (continued)

Theme	School name/STEM ALP URL	STEM ALP name
(C) Health and food science	**11. Chung Cheng High School (Yishun)** https://chungchenghighyishun-moe-edu-sg-admin.cwp.sg/our-curriculum/academic-development/department/science	Food science and technology
	12. Dunearn Secondary School https://dunearnsec.moe.edu.sg/school-programs/food-science-and-technology-applied-learning-programme-alp	Food science and technology
	13. Bendemeer Secondary School https://bendemeersec.moe.edu.sg/alp/	Making health science alive through authentic problem-based learning
(D) Cities and urban technology	**14. Manjusri Secondary School** https://manjusrisec.moe.edu.sg/passionate-learners/applied-learning-programme	Smart city and assistive technologies
(E) Future of transportation	**15. Serangoon Secondary School** https://serangoonsec.moe.edu.sg/distinctive-programmes/stem-applied-learning-programme-alp	Future of transportation

first adopting conventional content analysis followed by directed content analysis. We found that the topics within the original three themes aligned well with the three dimensions and present-future/intrinsic-extrinsic spectrum in Stuckey et al. (2013) model of relevance discussed earlier. Notably, it was found that while the webpage content reflected mainly extrinsic needs, some appeals to students' intrinsic needs were also articulated.

Relevance in Science/STEM Education

Based on a wide-ranging review of the literature, Stuckey et al. (2013) proposed a model of relevance in science education, which is an expansive concept that overlaps with interest, meaningfulness, and worth with respect to learning the subject. Because this model is not tied to either the content or pedagogy of science education, we argue that it is similarly applicable to understanding what STEM education offers and how it is relevant to learners. Accordingly, (STEM) learning is considered relevant education if learning has positive consequences for/in students' lives. The model includes: (1) fulfilment of *present* needs relevant to students' current interests or educational needs that students are currently aware of, and (2) anticipation of *future* needs that students may not necessarily be aware of. Relevance can also fulfil a range of *intrinsic* and *extrinsic* needs. For example, the former includes students' personal interests and motivations while the latter include ethically justified expectations of one's actions and behaviours by virtue of membership in one's personal environment, communities, and society. In addition, Stuckey et al. (2013) showed how relevance can be clustered into three main dimensions—individual, societal, and vocational—that are now described below using STEM education as an exemplar:

- The *individual* dimension. STEM education is relevant for individual students if it matches the student's curiosity and interests, and provides students with necessary intellectual knowledge and skills for coping with their everyday lives in the immediate and distant future. We interpret such knowledge and skills to include domain-specific conceptual, procedural, and/or epistemic knowledge akin to the scientific knowledge elements articulated in the PISA science framework (Organisation for Economic Co-operation and Development [OECD], 2017), as well as more generic ways of thinking, such as those associated with critical thinking, problem-solving and creative thinking, often called twenty-first century competencies (World Economic Forum [WEF], 2016).
- The *societal* dimension. STEM education is relevant from the societal standpoint if it prepares students for self-determination and a responsibly-led life in society through understanding the interdependence and interactions of STEM and society, and developing skills for societal participation. For example, this includes twenty-first century competencies such as communication and collaboration (WEF, 2016), and other competencies for contributing to society's sustainable development.
- The *vocational* dimension. STEM education is relevant in the vocational dimension if it offers orientation for future (STEM) careers, preparation for further (STEM) academic or vocational training, and opening up (STEM) career opportunities (e.g. by having sufficient academic and non-academic achievements to qualify for a particular higher education programme of study).

Each of the three dimensions comprises a spectrum of present and future needs coupled with a range of intrinsic and extrinsic needs. Two characteristics of the model of relevance are worth noting according to the authors. Firstly, the three dimensions could be interrelated and overlap. For instance, orientation for future STEM careers (vocational dimension) can fulfil individual students' interests (individual dimension-present-intrinsic) and/or satisfy the workforce demand for more engineers (vocational-future-extrinsic). Secondly, different dimensions and needs could be emphasized differently across school levels. For example, the individual dimension might be the most important pole at the primary level, whereas the vocational dimension might be more salient at upper levels. Even within a dimension, emphasis on needs could shift according to school level. Thus, within the vocational dimension, while the present-extrinsic need of passing qualifying examinations at every level is important, future needs associated with careers (future-intrinsic/extrinsic) gain prominence near the end of compulsory education. Given the comprehensive explanatory reach of this model of relevance, we employed it to understand what STEM ALP websites were offering in terms of relevance and products/benefits of STEM education to secondary students in Singapore.

Findings

We now present the ways in which STEM has been promoted and made relevant for secondary students from our sample of STEM ALP webpages. Tables 5.2, 5.3, and 5.4 in the following pages are adapted from Stuckey et al., (2013, p.19) and provide illustrative examples of the relevance of STEM education from STEM ALP webpages for the individual, societal, and vocational dimensions.

Table 5.2 The individual dimension of STEM programme relevance, showing the present/future and extrinsic/intrinsic spectrum

Individual-present-extrinsic *(Doing well in school)*	**Individual-future-extrinsic** *(Acting responsibly in life)*
• Help students learn advanced content in STEM subjects and other school subjects (i.e. conceptual knowledge)	• twenty-first century competencies associated with epistemic aspects/ways of thinking (e.g. critical and creative thinking), literacies (e.g. scientific and digital literacies), methodologies (e.g. scientific and engineering methods), and dispositions • Applied knowledge (e.g. associated with Science, Technology, Society and the Environment [STSE] issues; using health science knowledge to lead healthy lifestyles)
Individual-present-intrinsic *(Satisfying curiosity & interest)*	**Individual-future-intrinsic** *(Skills for coping with future personal life)*
• Programs chosen by students with particular interests	• Life skills (e.g. self-directed learning, confidence, communication)

Table 5.3 The societal dimension of STEM programme relevance, showing the present/future and extrinsic/intrinsic spectrum

Societal-present-extrinsic *(Learn how to behave and act in society)*	**Societal-future-extrinsic** *(Behave as responsible citizens and contribute towards society's sustainable development)*
• Social and emotional skills (e.g. collaborative skills, teamwork, leadership, confidence, communication, graciousness)	• Responsible global citizens • Care for others • Participate in decisions and problem-solving about issues that affect humankind and Earth • Advocate for causes that benefit society
Societal-present-intrinsic *(Find one's place in society)*	**Societal-future-intrinsic** *(Promote own interests in societal discourse)*
No examples	No examples

Table 5.4 The vocational dimension of STEM programme relevance, showing the present/future and extrinsic/intrinsic spectrum

Vocational-present-extrinsic *(Qualify for coming education)*	**Vocational-future-extrinsic** *(Contribute to economic growth)*
• Selected students could participate in modules offered by institutes of higher learning	• STEM-related careers e.g. scientists, engineers, digital workforce
Vocational-present-intrinsic *(Orientation towards potential careers)*	**Vocational-future-intrinsic** *(Get a desired job)*
No examples	No examples

Relevance for Individuals

Guided by the overarching vision for ALP, that is, students experiencing applied learning, school webpages that we included in our study contained a diverse spectrum of individual benefits (see Table 5.2). Relevance for **individual-present-extrinsic** needs involved acquiring more conventional 'final form' conceptual knowledge of school subjects. For example, Montfort school spoke of infusing different school subjects into their design, coding, and maker ALP thereby helping their students learn 'advanced content in related subjects'. This approach was perceived to be a 'win–win for both the subject and the ALP'.

Aligned with MOE's aim of the ALP helping students to make connections between academic knowledge and skills with the real world, a number of schools targeted knowledge outcomes relevant for acting responsibly in life (rather than doing well in school subjects), that is, applied knowledge and 21st century competencies were identified as being benefits of the STEM ALP. We categorized this as reflecting **individual-future-extrinsic** needs. Examples of applied knowledge include knowledge associated with science, technology, society, and the environment (STSE) related issues such as 'knowledge in environmental issues' (Bukit View) and the 'importance of using renewable energy' and using the acquired knowledge to 'liv[e] out sustainability in their own lives' (Fajar). Another example was learning 'concepts and principles related to health science' and using such knowledge to lead healthy lifestyles by 'tak[ing] greater responsibility for their own health as well as the health of their family members' (Bendemeer). While we note that this latter example could also be relevant as a present need, we chose to maintain the distinction of individual-present-extrinsic needs as needs associated with 'doing well in school' versus individual-future-extrinsic needs as needs associated with 'acting responsibly in life', as reflected in Stuckey et al. (2013).

Developing '21st century competencies' as an outcome of engaging in the STEM ALP is a catch-all phrase of highly-desired intellectual and affective competencies that encompass criticality, problem-solving, student agency, communication, collaboration, self-regulation, adaptability, information literacy, and (re-)learning, among others (Sinnema & Aitkin, 2013). Like 'relevance', the term '21st century skills/competencies' has been understood very loosely in the literature (NRC, 2011). It was no surprise that many of the webpages in our sample made reference to 21st century competencies: Admiralty, Bendemeer, Clementi Town, Queensway, Serangoon, Fajar, Chung Cheng High, Dunearn, and Woodlands Ring. This attests to the enormous rhetorical power and popularity of using such slogans when describing the value of educational programmes. Here, we highlight examples of 21st century competencies associated with epistemic thinking, that is, ways of thinking. We consider that these are relevant to **individual-future-extrinsic** needs. While these ways of thinking could definitely be relevant to present needs—to the extent that they assist student achievement in school and are often engaging for students—they seem more appropriate to future utility as suggested by the popular narrative of developing students' 21st century competencies to help them meet the needs of the future.

We observed that many STEM ALPs aimed to inculcate a multitude of ways of thinking, although there was a general lack of elaboration or definition by the schools on these various 'thinkings'. This ambiguous scenario reminded us of a cognitive thesaurus whereby a diversity of allied words can be used to describe a particular mental state (Wiggins & Potter, 2017). One prime example is, in fact, the word 'thinking', which has nuances of meanings such as believing, feeling, dreaming, contemplating, remembering, considering, judging, realizing, reasoning, and so forth. Using each of these different words would connote different meanings among readers, which are usually positive and desirable in this context. Hence, among the sampled STEM ALP, many referenced learning domain-general or generic ways of thinking, such as critical and creative thinking, scientific thinking, problem-solving, decision-making skills, reasoning, analytical thinking, adaptive thinking, logical thinking, and inventive/innovative thinking. All these forms of 'thinking' are generally regarded as highly desirable and valued skills across many educational systems. Four illustrative examples of how these kinds of thinking (in **bold**) were often grouped together are provided below:

> Develop… confident and creative young scientists, who possess **critical thinking skills**, **creativity** and **inventive thinking**, and communication and ICT skills (Dunearn).

> The emphasis is on application of **thinking skills, integrating knowledge** across subject disciplines, **stretching the imagination** and **applying** these in real-world settings in society and industries (Woodlands Ring).

> Students will be exposed **to collaborative learning and thinking**, **critical thinking**…This encourages self-directed learning, and builds their **critical thinking**, **problem-solving** and **decision-making** skills (Regent).

> Through the participation of students in competitions, we aim to provide students with additional opportunities to further develop **critical thinking**, **problem-solving abilities** and interpersonal skills to work in collaborative environments (Loyang View).

In another example, Montfort explicitly stated that their ALP would enable students to strengthen four kinds of thinking processes, namely critical, creative, logical, and systems thinking. Other STEM ALP websites reported learning ways of thinking in disciplinary domains that were closely tied to their espoused STEM activities, such as design thinking (e.g. Admiralty, Bukit View, Clementi Town, Fajar, Loyang View, Manjusri) and computational thinking (Clementi Town, Fajar).

Another cognitive thesaurus associated with epistemic thinking is the myriad of STEM disciplinary-related literacies and methodologies mentioned on the websites. Schools reported benefits/outcomes related to scientific literacy and/or scientific inquiry (e.g. Bendemeer, Chung Cheng High, Fajar), (the) scientific method (e.g. Bendemeer, Fajar), the engineering (design) method (e.g. Fajar), as well as more generic literacies including new media/digital literacy (e.g. Bendemeer, Chung Cheng High), coding/ICT literacy (e.g. Fajar, Regent), and civic literacy (e.g. Hai Sing Catholic). Chung Cheng High defined scientific literacy as 'the ability to think, reason and analyse scientifically, using scientific knowledge and processes'. More frequently, these ideas about literacy were not defined, and thus assumed that readers understood these loaded terms. Indeed, due to the lack of definitions of 'literacies' and 'methods' as used in the abovementioned websites, it was not possible to ascertain if the schools had intended them to be synonymous with the various 'thinkings' or beyond. However, we do not interpret 'thinkings' and 'literacy'/'methods' as synonymous as the latter goes beyond just 'thinkings', and thus, we have reported examples of each separately.

If we regard dispositions as akin to habits of mind, then a few schools wanted their students to achieve these ways of thinking too. Montfort and Queensway spoke of 'STEM dispositions' where the latter reported that these dispositions are manifested through a sense of wonder, competency, and collaboration. On the other hand, Dunearn stated that their experiential food science program would succeed in developing the 'dispositions associated with scientific literacy'.

In contrast, appeals to individual-intrinsic needs either in the present or future have far less presence in the school webpages for the basic STEM ALP experienced by all students. Only in elective programs offered to selected students were **individual-present-intrinsic** needs mentioned. Examples of appeals to students' individual-intrinsic needs are (in **bold**) as follows. Bukit View's STEM ALP (JOULES programme) for selected student leaders provided them with 'the opportunity to pursue **their interest** in design thinking, programming, the environment and sustainable energy'. Loyang provides '[s]tudents **with higher interest and ability** in STEM fields' with 'opportunities to further **develop their passion and potential** in STEM and design thinking'. Some schools also highlighted their intent to promote students' interest in STEM/joy of learning (rather than catering to students' interests) through the STEM ALP. For example, Montfort hopes that its STEM ALP will 'spark [students'] interest and surface inert talents' in areas of technology, coding and computers. Bendemeer wanted their students to 'learn that science is not frightening or boring', and experiencing fun is an aim of Hai Sing Catholic's STEM ALP.

As for relevance for **individual-future-intrinsic** needs, examples of the STEM ALPs include the development of life skills, such as developing self-directed learning/learners (e.g. Admiralty, Fajar, and Hai Sing Catholic) which would enable students to meet their self-motivated future learning needs, independence (e.g. Hai Sing Catholic), and communication skills (e.g. Dunearn, Hai Sing Catholic). These dispositions and skills also have relevance for the societal dimension.

Relevance for Society

In terms of the relevance of STEM ALPs for the societal dimension, only relevance for extrinsic but not intrinsic needs was identified from the sampled school webpages (see Table 5.3). While present extrinsic needs that were identified also seem relevant as future extrinsic needs, we distinguished between the two categories based on the extent of contribution to society rather than temporal separation (i.e. present versus future). That is, meeting societal-present-extrinsic needs helps students learn how to behave or function well in society, while meeting societal-future-extrinsic needs helps students contribute to society's development as responsible citizens. Nonetheless, this does not imply that students can only achieve the latter as adults.

In terms of meeting **societal-present-extrinsic** needs, some STEM ALPs spoke about developing learners' social intelligence, collaborative skills (e.g. Bendemeer, Queenstown), communication skills (e.g. Dunearn, Hai Sing Catholic), cross-cultural skills (e.g. Hai Sing Catholic), and teamwork (e.g. Clementi Town, Fajar). Others aimed to develop students' leadership qualities (e.g. Hai Sing Catholic, Manjusri), confidence (e.g. Admiralty, Hai Sing Catholic, Queensway), imagination (Dunearn), and independence and graciousness (e.g. Hai Sing Catholic). Collectively, these areas can be broadly construed as social and emotional skills (WEF, 2016), required to effectively interact with others at school/work or in daily living.

As for the STEM ALPs addressing **societal-future-extrinsic** needs, numerous schools aspire to have their students become responsible citizens and make contributions to others, the community, society and nation, as well as to humanity at large. For example, Fajar wants their students to become 'responsible global citizens of the Earth' through learning concepts of sustainability and 'applying and living out sustainability in their own lives' after participating in their sustainability-themed STEM ALP. In terms of making contributions, some schools spoke of students helping unspecified others, for example, students who have a 'caring heart with a desire to empathise with others and to help others' and 'ready to serve the community' (Admiralty). Many schools were explicit in pinpointing the wider community or society in general as the ultimate beneficiaries. For example, family and underprivileged children in the community were mentioned by Hai Sing Catholic. This school neatly summarized how it regarded the chain of beneficiaries on its website: 'School:community:nation'. Manjusri aimed to inspire their students to '[build] a sustainable, safe and secure living space for Singaporeans … for the elderly and less-abled [sic]'. The nation-building aspect of ALPs was also featured through tie-ins with efforts to improve life and work for Singapore. This nation-building aspect was particularly prominent in the following quotations:

> 'keen interest in nation building in creating a more vibrant, exciting and advanced society for generations to come.' (Manjusri)
>
> 'Help Singapore become a smart nation.' (Montfort)

Some schools were even more ambitious or forward-looking for their STEM ALP. For example, Chung Cheng High wants their students to be able to 'participate in decisions and problem solving that affect mankind'. Additionally, several schools wanted their students to contribute towards a sustainable world, as illustrated by the following quotations:

> 'effective workers and persons who positively contribute to a sustainable 21st Century World.' (Fajar)
>
> 'pursue STEM courses in institutes of higher education, especially in the area of environment and sustainable energy and thereby contribute to Singapore and the world.' (Bukit View)

In particular, two of the three schools with sustainability-themed STEM ALPs mentioned wanting students to become advocates for causes that benefit society. For example, Fajar wanted their students to become 'environmental advocates', while Bukit View hoped to nurture 'ethical advocates who will pursue causes that benefit society ensuring their values guide them as they make a difference beyond their school years'.

Overall, we observed that the abovementioned goals for students are often linked to the values espoused by the particular school hosting the STEM ALP. For example, Manjusri's goals to develop students' leadership qualities and have students contribute towards nation building, including creating a better living space for the 'elderly and less abled [sic]', are aligned with their school value of 'developing passionate learners and compassionate leaders who are future-ready and anchored in values'.

Relevance for STEM Vocations

Similar to the relevance of the STEM ALPs in relation to the societal dimension, relevance in relation to the vocational dimension focused on extrinsic and not intrinsic needs (see Table 5.4). Specifically, relevance was mainly targeted at supporting and motivating students to qualify for and select STEM areas of post-secondary study that would put them on the path to STEM-related careers. It should be noted that we redefined the description for vocational-future-intrinsic needs as 'get a desired job' from Stuckey et al. (2013) original description of 'getting a good and well paid job' (p. 19). We argue that the former better describes students' future intrinsic needs in the vocational dimension, as a desired job need not be one that is well paid, and what counts as a 'good' job should be up to the students' interpretation.

Schools in our sample mainly positioned the relevancy of the STEM ALPs for students' **vocational-present-extrinsic** needs by providing opportunities for selected students to enrol in elective modules offered by polytechnics, that is, post-secondary educational institutions in Singapore. For example, as part of their 'talent development' for selected secondary three students (ninth graders), Manjusri students could enrol in an Advanced Elective Module related to smart gadgets design offered by a polytechnic. Similarly, Bendeemer made explicit that the skills their students develop through their health science STEM ALP 'will prepare students for Advanced Elective Modules…and post-secondary options at Institutes of Higher Learning (IHL)'. Two names were given for the IHLs, which 'a large proportion of [Bendeemer] students go to after their secondary school education'. Furthermore, a list of modules and courses offered at the IHLs relevant to Bendemeer's STEM ALP were provided on their webpage. Hence, Bendemeer signalled a targeted approach to align their STEM ALP to students' vocational-present-extrinsic needs, ensuring that they would qualify for identified post-secondary education after completing their STEM ALP. The STEM ALP websites of Bukit View and Dunearn also referenced Advanced Elective Modules, but did not provide further information on these modules.

Relevancy of the STEM ALPs in relation to students' **vocational-future-extrinsic** was extensively described. Among the eight schools that made some reference to preparing students for their future career, only one took the approach of equipping students with the necessary skills to pursue their future career without explicitly encouraging them to aim for a STEM career. Specifically, Regent mentioned that Singapore is moving towards becoming a smart nation and that they wanted to equip their students with relevant skills to 'give them a head start in their future career', with no explicit encouragement for students to aim for a STEM career. Two schools explicitly stated that the aim of the STEM ALPs was to ignite students' passion or aspirations to 'pursue STEM related courses and careers' (Fajar and Manjusri). Other schools were more specific in the types of STEM careers they wanted their students to pursue. Serangoon wanted students to consider careers in 'Aerospace Engineering, Transportation and Infrastructure, Entertainment and Media, Telecommunications, Health Sciences and Health Care, Architecture and others'. Many of these listed careers go beyond their STEM ALP theme of future of transportation. Other schools wanted students to pursue STEM careers more closely related to their STEM ALP theme. For example, Clementi Town (with a STEM ALP in emerging technologies, with an emphasis on coding) hoped to ignite students' interest 'to pursue [computer science] CS-related courses in Institutes of Higher Learning and a career in CS-related industries', while Admiralty (also with emerging technologies theme) wanted to produce 'caring and future-ready engineer[s]'.

Two schools also made explicit the link between how their STEM ALP targets STEM careers for students that meet nation building agendas, including economic needs. Montfort's STEM ALP on the emerging technologies theme aims to '[support] Singapore's vision of becoming the first Smart Nation in the world'. Through their STEM ALP, the school hopes that 'more students [would] make relevant Computer Science and STEM related courses their choice in their tertiary education' so as to 'alleviate the problem of a shrinking pool of engineers and technical personnel currently faced by the nation'. Similarly, through their health and food science themed STEM ALP, Dunearn wished to raise the profile of learning food science and technology as it was not a popular subject among Singapore school students. Dunearn felt that 'the potential for [their] students to further their interest through courses at ITE, polytechnic and university are vast and sustainable'. In an interview featured on their school webpage, a Dunearn representative states that their rationale for specializing in food science and technology for their STEM ALP considered that 'the nation needs good scientists to provide healthy food and also to raise awareness on what we can eat and what we should cut down on'. Bendemeer highlighted career opportunities in the healthcare and health science industries in view of needs of 'an aging population in Singapore and many developed countries'. Though Bendemeer did not explicitly mention their intent for students to take on careers in healthcare and health science industries on their webpage, the strong connection between their health science STEM ALP and their targeted preparation for students to qualify for health science related studies in IHLs suggests that this is a likely intent.

What is obvious from some of the examples cited above is that many schools positioned their STEM ALP as an attempt to appeal to students' vocational-present-extrinsic needs, that is, to inspire students to be interested and motivated in STEM-related post-secondary studies, and subsequently to pursue a STEM-related career. In addition to the school-based STEM activities, as aptly summarized by Clementi Town, 'to expose students to STEM careers at an early age, Industrial Visits and Learning Journeys were conducted to organizations and STEM-related events'. Indeed, many of the schools appeared to have worked hard to get secondary students interested in STEM careers and courses through early exposure to these areas of study and careers. However, it is not known to what extent such attempts are aligned with students' vocational-present-intrinsic needs, that is, students' present, intrinsic interest and orientation towards potential careers. Moreover, schools did not make known the impact of their efforts on orienting students towards a future (STEM) career on their webpages, if any.

Discussion and Critique

The purpose of this content analytic study was to examine the outcomes and relevancy of STEM education as presented on secondary school STEM ALP websites in Singapore. This narrow and somewhat unusual sample was because these are the only places within the local school curriculum that features STEM education; it is absent in local textbooks and is only mentioned in passing in the official science curriculum. On the other hand, STEM ALPs are an MOE-sanctioned programme that we found comprised five STEM themes, and that schools use to develop their own distinctive STEM-related niches. In Singapore, all school leaders have been constantly urged by the MOE to be innovative and to create niche programmes of excellence that can set their institutions apart from others (Shanmugaratnam, 2006). We thus see these official inducements spilling over into the process of planning for STEM ALP in local schools too. In order to make sense of the huge amount of text from the 15 selected websites, we adapted the model of relevance from Stuckey et al. (2013) as an organizing framework to categorize the benefits, outcomes, and relevance arising from STEM education.

Even though we did not quantify the number of statements along the present/future and extrinsic/intrinsic spectrum for each dimension in the model of relevance, certain patterns clearly emerged. One immediate observation was that the STEM ALP websites in Singapore did not yield any examples of statements that supported the present or future aspects of intrinsic relevance in the societal and vocational dimensions. This meant that no examples existed from our sample that talked about helping learners appreciate how to find their current place in society (i.e. societal-present-intrinsic) or to promote their own interest in the societal context as adults (societal-future-intrinsic). In addition, these websites did not appeal to or foster students' current interests in relation to potential careers (vocational-present-intrinsic), nor did they make connections between the STEM ALPs and preparing students to pursue a career that appeals to them (vocational-future-intrinsic). Within the individual dimension, the intrinsic aspects of the present were limited, and catered to a select group of talented or motivated students from the ALP.

On the other hand, the STEM ALP websites provided multiple descriptions in relation to extrinsic and future aspects across all three dimensions of relevance, although we are unable to ascertain whether the 'present' or the 'future' predominated in the websites. To review our findings briefly, we found that the STEM ALP programmes spoke much about doing well in school (individual-present-intrinsic) because being involved in STEM ALP could support learning in science as well as other school subjects. Moreover, the websites reported that there were many kinds of 21st century competencies (e.g. various 'thinkings' and literacies) and application of knowledge that enabled one to act with responsibility in later life (individual-future-extrinsic) in the individual dimension. This same pattern was observed for the societal dimension: the websites identified many social and emotional skills as short-term outcomes of STEM education (societal-present-extrinsic), as well as having caring and altruistic attributes as responsible citizens in the community and nation (societal-future-extrinsic). We were heartened to see these websites mentioning how students in STEM could aspire to a future role in service to others around them, and even to mankind in general. Finally, for the vocational dimension, STEM ALPs were identified as being relevant because they could help students qualify for post-secondary education, with some STEM ALPs focusing on more promising or talented students (vocational-present-extrinsic). Over time, participation in STEM ALP would contribute to the economy as people took on STEM-related careers that are vital to Singapore and the larger world (vocational-future-intrinsic).

When considering the findings overall, we identified three main implications. Firstly, sampled STEM ALPs seemed to make limited reference to **student choices, identities, and agency**. Just as some doors are opened through participation in STEM ALPs, other career or study options are not presented. For example, environmental advocacy roles of the likes of Greta Thunberg were absent; nor did we see any mention of pursuing careers in science communication regarding which the recent Pulitzer Prize winner Ed Yong comes to mind. Students did not have many opportunities to gain awareness of intrinsic relevance—present or future—in the individual dimension. Schools generally assumed that participation in STEM ALP activities would spark students' interest in the school's STEM areas/themes. While more talented students or those interested in the STEM ALP themes offered by a school had opportunities to further their interest, it was not clear how the intrinsic needs of students with interests in other aspects of STEM or those not interested in STEM would be addressed. Rather, the majority of the statements in the websites were located at the extrinsic pole, be it in relation to either the present or the future. For example, some STEM ALP websites seemed to overtly serve current/future national agendas in terms of recommending certain STEM-related careers (e.g. scientists, engineers) and study choices (e.g. healthcare, food science and technology, computer science). We argue that these should be exploratory moves rather than anything more concrete for young students; it seems unwise to restrict career choices so early. We also raise the well-known argument that future jobs and labour markets are difficult to anticipate or predict, which should cause us to pause in our crystal-ball gazing efforts. Furthermore, while 21st century competencies have notions of increased learner agency, the competencies were not well defined in the ALP websites. This study did not investigate whether the espoused competencies are actually developed through the STEM ALPs.

Secondly, STEM applications and knowledge (often taking the form of products or solutions) appear to be presented as positive, unproblematic, trustworthy, and able to solve complex problems. The field of Science Studies has already shown that STEM knowledge can be complexly problematic, uncertain, and not value-neutral. Thus, when teachers fail to bring to learners' attention the negative aspects or by-products of STEM in the real world, students' holistic grasp of scientific literacy is diminished (e.g. Bruna & Vann, 2007; Hoeg et al., 2015). While we do not have sufficient information about the STEM ALPs to determine whether dilemmas or issues of STEM applications are being discussed with students, we argue that this is an aspect schools should incorporate into their programmes, if this is not currently done. Examples of issues or potential biases worthy of discussion include: use of unmanned vehicles, such as driverless cars for transportation or unmanned drones for delivery and military purposes and their potential social, moral, and ethical issues, in addition to their benefits. Acknowledging that STEM does not hold all the answers to the problems facing society would arguably make STEM ALPs more interesting and authentic for learners. However, we are again heartened that there is explicit mention of empathy by some schools, such as on Admiralty and Queensway's STEM ALP websites.

Thirdly, there is a need for greater emphasis on the epistemic aspects of STEM. Akin to the call for science education to harmonize conceptual, epistemic, and social goals (Duschl, 2008), we argue that STEM education needs to pay attention to epistemic goals. In the construction of STEM products or solutions—common among the STEM ALPs we reviewed—there is a need for students to take on the dual role of constructors and critiquers of claims (Ford, 2008) in order to gain a deeper understanding of the nature of the STEM enterprise and to participate as members of the enterprise. While sampled websites mentioned developing a multitude of 'thinkings' and inquiry through the STEM ALPs, these are presented with little detail. As per our second implication, we urge teachers to attend to epistemic aspects of how STEM products or solutions are generated. Students should be given opportunities and shared epistemic authority (with the teacher) to negotiate the criteria for evaluating the 'goodness' of a STEM product or solution (including those generated by themselves or their peers), as well as to engage in peer critiquing to evaluate the product or solution (e.g. by weighing its trade-offs) and/or to determine the 'best' product or solution. Through engaging in such epistemic practices, students learn how knowledge claims (embedded in STEM products and solutions) are proposed, communicated, assessed, and legitimized within a community, which is as important as the final form of knowledge itself (Ford, 2010; Kelly, 2016).

While we applaud the push by many local secondary schools to promote STEM education in Singapore, we realize that there is some way to go on a number of fronts regarding the curriculum and pedagogy. These issues that we have raised are sometimes subtle, especially when the more exciting or fun aspects of STEM education take the limelight. Nonetheless, we argue that STEM education deserves more vigorous critique and debate for its improvement, precisely because STEM education is worth the effort.

References

Bruna, K. R., & Vann, R. (2007). On pigs and packers: Radically contextualizing a practice of science with Mexican immigrant students. *Cultural Studies of Science Education, 2*, 19–59. https://doi.org/10.1007/s11422-006-9041-x.

Bybee, R. W., & Gardner, A. L. (2006). High school biology and the physics first movement. *The American Biology Teacher, 68*(3), 134–138.

Deleuze, G., & Guattari, F. (1988). *A thousand plateaus: Capitalism and schizophrenia*. The Athlone Press.

Duschl, R. (2008). Science education in three-part harmony: Balancing conceptual, epistemic, and social learning goals. *Review of Research in Education, 32*(1), 268–291.

Ford, M. J. (2008). 'Grasp of practice' as a reasoning resource for inquiry and nature of science understanding. *Science & Education, 17*(2–3), 147–177.

Ford, M. J. (2010). Critique in academic disciplines and active learning of academic content. *Cambridge Journal of Education, 40*(3), 265–280.

Hoeg, D., Lemelin, N., & Bencze, J. L. (2015). Sociopolitical development of private school children mobilising for disadvantaged others. *Cultural Studies of Science Education, 10*, 1155–1174. https://doi.org/10.1007/s11422-014-9611-2.

Hsieh, H. F., & Shannon, S. E. (2005). Three approaches to qualitative content analysis. *Qualitative Health Research, 15*(9), 1277–1288. https://doi.org/10.1177/1049732305276687.

Kelly, G. J. (2016). Methodological considerations for the study of epistemic cognition in practice. In J. A. Greene, W. A. Sandoval, & I. Bråten (Eds.), *Handbook of epistemic cognition* (pp. 393–408). Routledge.

Kloser, M., Wilsey, M., Twohy, K. E., Immonen, A. D., & Navotas, A. C. (2018). "We do STEM": Unsettled conceptions of STEM education in middle school S. T. E. M. classrooms. *School Science and Mathematics, 118*, 335–347. https://doi.org/10.1111/ssm.12304.

Martín-Páez, T., Aguilera, D., Perales-Palacios, F. J., & Vílchez-González, J. M. (2019). What are we talking about when we talk about STEM education? A review of literature. *Science Education, 103*(4), 799–822. https://doi.org/10.1002/sce.21522.

McComas, W. F., & Burgin, S. R. (2020). A critique of "STEM" education. *Science & Education, 29*(4), 805–829. https://doi.org/10.1007/s11191-020-00138-2.

Millar, V. (2020). Trends, issues and possibilities for an interdisciplinary STEM curriculum. *Science and Education, 29*, 929–948.

Ministry of Education. (2021). *Applied learning programme (ALP)*. Retrieved May 23, 2021, from https://www.moe.gov.sg/secondary/courses/express/electives/?term=Applied%20Learning%20Programme%20(ALP).

National Academy of Engineering & National Research Council. (2014). *STEM integration in K-12 education: Status, prospects, and an agenda for research*. National Academics Press. https://doi.org/10.17226/18612.

National Research Council. (2011). *Assessing 21st century skills: Summary of a workshop (2011)*. The National Academies Press. https://doi.org/10.17226/13215.
Organisation for Economic Co-operation and Development. (2017). *PISA 2015 assessment and analytical framework: Science, reading, mathematic, financial literacy and collaborative problem solving (revised edition)*. PISA, OECD Publishing. https://doi.org/10.1787/9789264281820-en.
Shanmugaratnam, T. (2006, Sep 29). *Speech at MOE workplan seminar 2006*. https://www.nas.gov.sg/archivesonline/data/pdfdoc/20060929994.htm.
Sinnema, C., & Aitkin, G. (2013). Emerging international trends in curriculum. In M. Priestly & G. Biesta (Eds.), *Reinventing the curriculum: New trends in curriculum policy and practice* (pp. 141–163). Bloomsbury Academic.
Stuckey, M., Hofstein, A., Mamlok-Naaman, R., & Eilks, I. (2013). The meaning of "relevance" in science education and its implications for the science curriculum. *Studies in Science Education, 49*(1), 1–34. https://doi.org/10.1080/03057267.2013.802463.
Wiggins, S., & Potter, J. (2017). Discursive psychology. In C. Willig & W. Stainton-Rogers (Eds.), *The SAGE handbook of qualitative research in psychology* (pp. 73–90). Sage.
World Economic Forum. (2016). *New vision for education: Fostering social and emotional learning through technology*. http://www3.weforum.org/docs/WEF_New_Vision_for_Education.pdf.

Yann Shiou Ong is an Assistant Professor at the National Institute of Education (NIE), Nanyang Technological University, Singapore and a core team member of the meriSTEM@NIE Singapore, a multi-centric education, research and industry centre. Her research interests include students' epistemic practices of argumentation during science inquiry and STEM problem solving. She designs learning environments to foster students' epistemic practices using the Productive Disciplinary Engagement framework. Her recent work involves construction of a scientific practices survey instrument for secondary students and an integrated STEM classroom observation protocol. She is currently an associate editor of Pedagogies: An International Journal (NIE publication).

Yew-Jin Lee is Assistant Head of Primary Science education at the National Institute of Education (NIE), Nanyang Technological University, Singapore and his research interests include curriculum studies, classroom assessment, epistemic knowing and questions of knowledge, and learning in formal/informal environments across learners of all ages. He was a past co-editor of Pedagogies (Routledge) and serves on the editorial boards of Research in Science Education (till 2018), Studies in Science Education, and Asia-Pacific Science Education (till 2019). In 2008/9, he received a Fulbright Academic Exchange award to study urban science education with the City University of New York. He is currently involved in the Science Expert Group planning for the OECD PISA 2025.

Chapter 6
Fostering STEM Education for Early Childhood in Thailand

Tepkanya Promkatkeaw, Navara Seetee, and Chanyah Dahsah

Abstract STEM education is a national agenda for educational policy in many countries including Thailand. The Thai policy aims to encourage youth to take an interest in STEM learning and to be able to use STEM knowledge and skills in their everyday life and future careers. Much literature (e.g., Campbell et al., 2018; McClure, 2017; The National Research Council [NRC], 2011) suggests that the earlier we start in students' educational journeys, the more effectively we can enhance our students' learning in STEM. Thus, early childhood education (ECE) is the right place to start learning STEM. In Thailand, STEM education became a national agenda in 2015 but it began in Early Childhood Education since 2006. The ECE policy aims to lay fundamental skills and promote learning experiences related to STEM learning for young children. There are several professional development programs and STEM conferences for ECE teachers and school administrators hosted by government agencies and the private sector, delivered via face to face and distance training. In addition, various STEM learning activities and resources have been created to promote young children's hands-on and minds-on learning experiences. This chapter will provide a brief history of STEM education in ECE in Thailand, an overview of teachers' professional development, examples of STEM activities in formal and informal learning, and an overview of research studies.

Keywords Early childhood · Kindergarten · STEM curriculum · Professional development · Curriculum development

T. Promkatkeaw · N. Seetee · C. Dahsah (✉)
Science Education Center, Faculty of Science, Srinakharinwirot University, 114 Sukhumvit 23, 10110 Bangkok, Thailand
e-mail: chanyah@g.swu.ac.th

T. Promkatkeaw
e-mail: tepkanya@g.swu.ac.th

N. Seetee
e-mail: suwapid@g.swu.ac.th

M. M. H. Cheng et al. (eds.), *Concepts and Practices of STEM Education in Asia*,
https://doi.org/10.1007/978-981-19-2596-2_6

Introduction

STEM education in Thailand is generally recognized as an integrated learning approach of science, technology, engineering, and mathematics to support learners to apply these knowledge and skills for solving real-life problems or to develop new processes or products through an engineering design process. The purpose is to make STEM relevant to students' everyday life experiences and their future careers (The Institute for the Promotion of Teaching Science and Technology [IPST], 2014a).

STEM education has been a national agenda since 2014, with the Thai government emphasizing the building of an innovative society through the promotion of teaching and learning systems that integrate science, mathematics, and technology. In 2015, the National Legislative Assembly of Thailand (NLA) in cooperation with the IPST, a government agency under the Ministry of Education (MOE), initiated a proactive policy for STEM education for developing youth and human capital in STEM. This policy proposed guidelines for implementing STEM education, such as embedding it into the national long-term strategic plan; promoting change in teaching, learning, and assessment; raising public awareness; and setting up a committee and operational system among government organizations to cooperatively drive STEM education (NLA, 2015). In 2017, STEM education was included in the Thai National Education Plan B.E. 2560–2579 (A.D. 2017–2036). In order to drive the STEM education strategy into practice, various national education institutes have responsibility for promoting STEM education at all educational levels (Office of the Education Council, 2017).

Early childhood education (ECE), as part of basic education in Thailand, needs to promote STEM education according to the national agenda. However, ECE has its own uniqueness. Many questions may come to teachers' and educators' minds: How do young children learn STEM? What learning content and competencies in STEM should be promoted? How should STEM be taught? How can ECE teachers' understanding and ability of integrating STEM be promoted?

This chapter will share how STEM education has been fostered in Thai ECE. The practices and experiences in the Thai context can provide insights into these questions and may be helpful for other countries looking for ways to promote STEM education in ECE.

STEM in the Thai National ECE Curriculum

The Early Childhood Curriculum

The United Nations Educational, Scientific and Culture Organization (UNESCO, 2019) and many countries define early childhood as the period from birth to 8 years old. However, the Early Childhood Development Act B.E. 2562 (A.D. 2019)

(Office of the Council of State, 2019) defined early childhood as children under 6 years old. ECE in Thailand has been promoted since the early 1900s specific to the policies related to the provision of quality early childhood care and education in the National Education Plan. Since 2002, ECE has been part of the basic education. It can be provided for all children in various settings, both inside and outside schools, and through collaboration between the government sector, the private sector, communities, and families (Phatdunyadunrut, 2019). In order to provide quality standards for ECE, the MOE has provided the National Early Childhood Curriculum as the guideline for providing experiences and learning activities for children under the age of 6 years old since 1997. The recent Early Childhood Curriculum B.E. 2560 (A.D. 2017) was established by revising the previous Early Childhood Curriculum B.E. 2546 (A.D. 2003) to reflect changes in the economy and society and developments in information technology, and to align with national laws and policies.

The recent early childhood curriculum emphasizes the holistic development of children from birth to age 6 through caring for and promoting children's learning process in order to allow them to reach their full potential according to the stages of child development (MOE Thailand, 2017). According to this curriculum, ECE is divided into two stages. The first stage is child care and development for children under 3 years old, which may be provided by parents or caregivers at home, nursery or pre-kindergarten. The second stage is child care and education for children from 3 to 6 years old, which may be provided by kindergarten, preschool, or child development centres. In formal education, ECE is divided into the three grade levels of Kindergarten 1 (3–4 years old), Kindergarten 2 (4–5 years old), and Kindergarten 3 (5–6 years old).

The early childhood curriculum for children aged 3–6 years old has four goals for learning and development: physical growth and development; mental health, aesthetic values, and moral values; life skills according to the philosophy of sufficiency economy, self-discipline, and other social values; and thinking skills, communication skills, and inquiry. There are 12 'desirable characteristics' standards with indicators and expected competencies under each standard that children in each age group should be able to demonstrate after receiving the provision of services from early childhood development settings. This curriculum also provides learning content as a medium for providing learning experience, consisting of four key experiences to enhance physical, emotional, social, and cognitive development; and four content areas, namely self, people and places, the natural environment, and things around children. The learning content is flexible and can be adjusted to respond to children's age, needs, interests, and real-life experiences and environment (MOE Thailand, 2017).

The early childhood curriculum emphasizes that the provision of experiences for children 'should not be organized into subject matter, but rather integrated into activities through play' (MOE Thailand, 2017, p. 47). The educational setting can therefore use different learning activities that are relevant to the context and that meet the needs and interests of children on a daily basis. However, these everyday activities should cover all aspects of development, including large and small muscles; emotions, cultivating morality, and righteousness; social character, thinking, and languages; and imagination and creative thinking (MOE Thailand, 2017).

Endeavours Related to the Integration of Science, Mathematics, Technology, and STEM Education in the Early Childhood Curriculum

The promotion of STEM education in ECE in Thailand can be traced back to around 2006 when the IPST started the project for enhancing science learning in ECE. In December 2006, 40 early childhood educators and experts from various institutes joined together with science educators in a brainstorming meeting on the development of science curriculum standards for early childhood (IPST, 2006). As a result of the discussion, it was agreed that science learning was important to young children and should be provided for laying fundamental learning skills and promoting young children's cognitive development. Science learning was embedded into the Early Childhood Curriculum B.E. 2546 (A.D. 2003) both in key experiences and content areas, but most early childhood teachers lacked science content knowledge and pedagogical knowledge. Thus, although they already delivered science learning experiences in their everyday activities—such as observing and classifying things, and doing or demonstrating simple science experiments—they could not effectively teach fundamental scientific concepts, science process skills, or inquiry abilities. There was thus a need for the development of science learning standards for ECE to guide teachers to effectively integrate science learning into early childhood activities with an emphasis on the process of inquiry.

The science learning standards for preschool level (3–6 years old) aligned with the Early Childhood Curriculum B.E. 2546 were first published in 2008 (IPST, 2008) and were introduced at a seminar hosted by the IPST in January 2009, which attracted around 400 early childhood education personnel (ThaiPR.net, 2009). The seminar was followed by workshops for around 670 early childhood lead teachers and teacher supervisors across the country during April–May 2009. These workshops focused on enhancing participants' understanding of inquiry-based learning in ECE, and demonstrating how inquiry-based science learning can be integrated into everyday activities (IPST, 2009). In 2010, the science learning standards were revised to align with the Basic Education Core Curriculum B.E. 2551 (A.D. 2009) (IPST, 2010a).

For enhancing mathematics learning in ECE, it was found that mathematics learning was commonly integrated into early childhood everyday activities (MOE Thailand, 2004) because it was more clearly stated than science in the learning content of the Early Childhood Curriculum B.E. 2546 (A.D. 2003). However, many fundamental mathematics concepts and skills that should be promoted in early childhood were missing. The learning scope and sequences were also not clearly indicated in the curriculum. Thus, to help early childhood teachers to effectively design and provide mathematics learning experiences, the mathematics learning standards for preschool level (3–5 years old) were developed by the IPST in 2010 by adopting a similar process as the development of the science learning standards.

In 2011, sets of standards frameworks for science and mathematics learning for ECE (3–6 years old), aligned with the Early Childhood Curriculum B.E. 2546 (A. D. 2003), were published (IPST, 2011a, 2011b, 2011c). These documents are guidelines for what and how to teach science and mathematics to young children (3–6 years old). These standards were introduced at a seminar hosted in March 2011 by the IPST (ThaiPR.net, 2011) and were followed by workshops for around 600 early childhood lead teachers and teacher supervisors across the country during April–May 2011 (IPST, 2011d).

For enhancing technology learning in ECE, technology was embedded within the science learning standards framework published in 2011 under a learning standard, 'the nature of science and technology'. According to this standard, technology was defined as the application of knowledge, skills, and resources to build objects or develop procedures through the problem-solving process, responding to desire, or enhancing human ability to work. Children learn how to use simple and everyday equipment and tools properly and safely and to be aware of the advantages and disadvantages of using scientific processes, equipment, and tools in everyday life (IPST, 2011c). This technology learning is based on the idea of design and technology education with the use of the technological process including identifying problems; gathering information to develop possible solutions; selecting the best solution; designing, making, testing, and improving models; and assessing solutions. Technology learning could be easily integrated into early childhood activities because in the early childhood curriculum, children are required to learn how to use simple tools and equipment, such as scissors and glue; to learn about technologies in daily life, such as electric devices and vehicles; and to learn to make arts and crafts. The IPST has also provided examples of design and technology learning activities with the integration of science and mathematics, and has trained early childhood lead teachers and teacher supervisors on relevant topics through workshops since 2010 (IPST, 2010b).

The recent Early Childhood Curriculum B.E.2560 (A.D. 2017) has embraced science, mathematics, and technology (SMT) in ECE, with the expectation that SMT learning will promote the development of children's thinking skills, communication skills, and inquiry knowledge. There are three of twelve standards for child's attributes that are related to SMT learning outcomes, including standard 10—children develop thinking skills for basic learning; standard 11—children develop their imaginations and creative thinking; and standard 12—children develop a positive attitude towards learning and searching for new knowledge. Young children's SMT concepts and skills would be developed through the key experience of cognitive development and the content areas of the Natural Environment and Things around Children. Details of these key experiences and content areas are shown in Fig. 6.1.

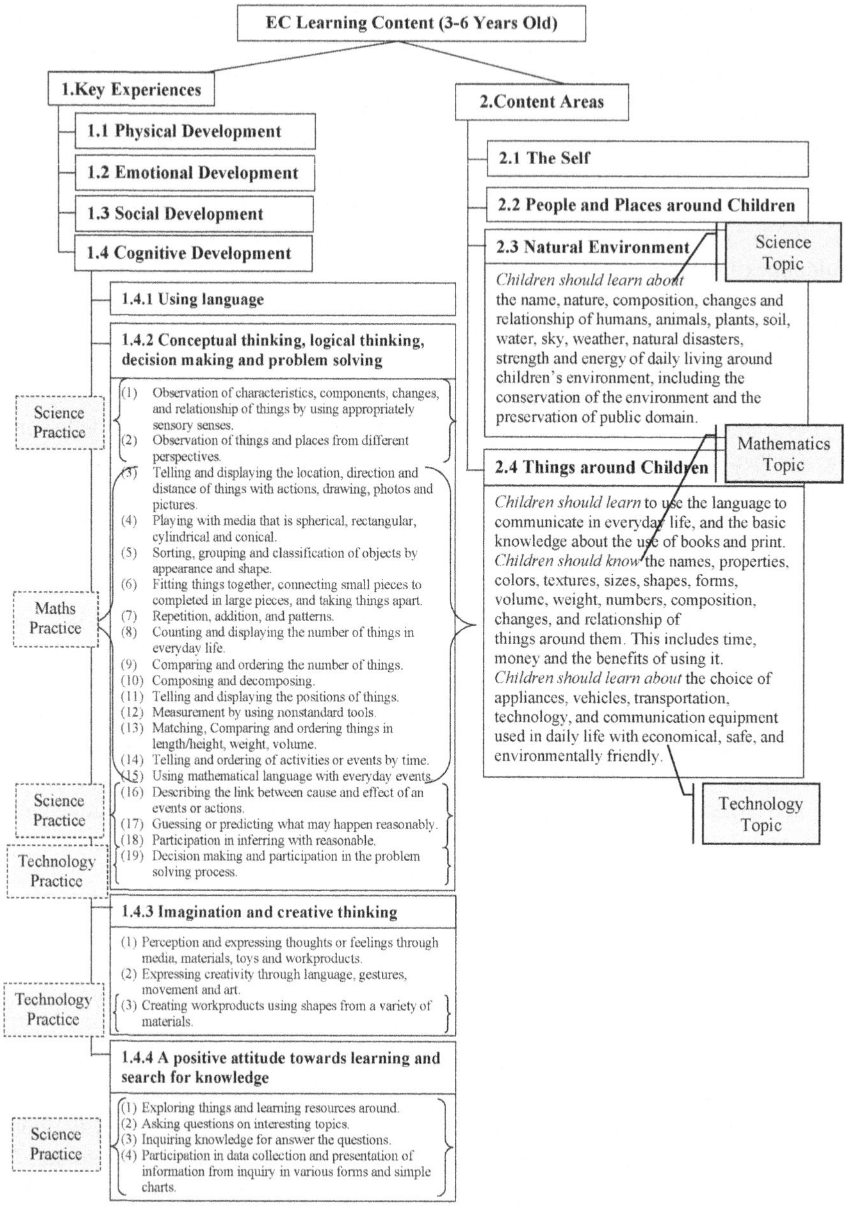

Fig. 6.1 Key experiences and content areas related to science, mathematics, and technology in the early childhood curriculum B.E.2560

In order to help teachers design learning experiences of SMT and STEM for early childhood, in 2020 the IPST published a learning framework and guidelines for integration of SMT in early childhood learning experiences aligned with the Early Childhood Curriculum B.E. 2560 (A.D. 2017) (IPST, 2020). This learning framework provides the expected learning outcomes and learning content for SMT for children (3–6 years old) and guidelines, together with basic information for teachers to design STEM learning activities by integrating SMT with an engineering design process and computational thinking into early childhood learning experiences.

Professional Development in STEM Education for Early Childhood Teachers

To foster STEM education in ECE, one major strategy for enhancing STEM teaching and learning is to provide professional development (PD) for early childhood teachers. There are various training programs for teachers, hosted by government agencies and the private sector using both face to face and distance training modes. This chapter introduces two national PD programmes that have had the most significant impact on early childhood STEM education in Thailand: the project of the IPST, and the Little Scientists' House (LSH) project.

During 2018–2019, the IPST was granted a budget of about 1.8 million USD for conducting STEM workshops for early childhood teachers and school administrators under the MOE. There were four continuing 2-year workshops on the topics of integrated science, mathematics, and technology; emphasizing computational thinking; and STEM learning. The workshops aimed to develop early childhood teachers' understanding of the integration of STEM learning in early childhood learning experiences and to develop school administrators' understanding of how to support and promote STEM education in ECE. The workshops were implemented through distance training by broadcasting via the Educational Television Channel and live through a YouTube channel to 200 training centres across the country, followed by web-based online training courses. Initially, the training programmes began with the IPST providing face-to-face workshops for mentor teams. Then, the mentor teams organized training in their centres, along with the IPST broadcasting workshop activities, and supported teachers and school administrators participating in the training during and after the workshops. After the workshops, participant teachers implemented learning activities and materials received from the IPST, and school administrators coached and mentored the teachers. About 1,200 mentors, 17,000 early childhood teachers, and 3,000 school administrators from 3,700 schools participated in the programme.

The LSH project is the most impactful long-term project that has promoted STEM learning in ECE. This project was launched in 2006 by Stiftung Haus der Kleinen Forscher, a non-profit organization in Germany, and has been implemented in Thailand since 2010 to develop positive attitudes towards STEM among Thai children aged 3–6 years old. The project is organized by the Foundation of Her Royal Highness Princess Maha Chakri Sirindhorn and eight private and public organizations. The project began in 2010 in 221 schools under eight local networks. In 2021, the number of local networks and schools increased to 232 and 29,000 respectively. The LSH project provides ongoing PD opportunities for early childhood teachers. Twice a year, the teachers attend a workshop in which they receive updated information about the project, gain new teaching ideas, and have access to new teaching materials developed by local trainers. The local trainers are trained and supervised by core trainers who have a solid background in STEM and have been trained by the trainers from the original project in Germany. The project also provides certificates to certify participating schools. Schools which aim to apply for a project certificate in a particular year need to implement 20 activities and one inquiry-based project in each academic year. The certificate is valid for the following two years. At present, approximately 5,000 schools apply for the certificate each year (LSH Project Thailand, 2019).

In addition to training programs, government agencies and the private sector have hosted several early childhood STEM conferences. For example, the IPST has organized the Early Childhood and Primary Science Mathematics and Technology Education conference (ECaP) every 2 years since 2012. There are activities which give participants new knowledge, showcase best practices, share experiences and promote discussions on early childhood science, mathematics, technology, and STEM teaching and learning. Around 300–400 early childhood personnel attend the conference each year (IPST, 2016).

STEM Activities and Learning Resources

Various STEM activities and learning resources have been created by public and private organizations both in formal and informal settings. Again, two national projects, by the IPST and LSH, are used as examples of learning activities in a formal setting. Examples of activities developed by teachers are also identified. In relation to informal settings, most of the activities are developed by the museums, both public and private.

The IPST developed STEM learning units with lesson plans and sets of learning materials as a prototype for teachers to use with 5-year-old students and distributed these to schools participating in the IPST PD programmes. The units were designed for developing SMT fundamental concepts and basic skills by enhancing children's experiences of solving problems through an engineering design process. For example, in the first lesson of the 'Fun with Sound' learning unit, children explore different sounds. The aim is for them to develop basic scientific ideas about sources of sounds. In the second lesson, they listen and try to identify unknown boxes according to the sounds made by the objects inside, which helps them develop the basic skills of observation, inferencing, sorting, and matching. In the third lesson, they make sounds and patterns of sounds using parts of their body, which is an integration of the mathematics concept of pattern. In the fourth lesson, children make sounds from different materials, which helps them explore the processes of science and technology. In the last lesson, a teacher introduces some percussion instruments and asks children to design and make percussion instruments from surrounding materials and to use their creativity to create music using their instruments. Thus, in this final lesson, children apply knowledge which they have gained from previous lessons into an engineering design process, present their percussion instruments, and perform their own music (IPST, 2019).

The LSH project provides STEM learning activity cards created by the LSH Foundation in Germany, translated into the Thai language and adapted to suit Thai contexts. Over 100 science, mathematics, design technology, computer science, and STEM activities are provided on the website (www.littlescientistshouse.com), which can be downloaded for free. In each activity, children explore STEM experiences through simple experiments and explorations related to the children's daily lives, and learn to explore ideas or solve problems of their interests through inquiry-based learning, called an inquiry cycle. For example, in the topic of water and technology, children explore objects sinking and floating. Then, they investigate more about factors that make objects sink or float. After the children have developed basic understanding, they investigate how to make a plasticine boat that can carry heavy objects and still sturdily float on water through an inquiry cycle (LSH Project Thailand, 2021).

Besides STEM learning activities developed by public and private organizations, early childhood teachers have also created STEM learning activities themselves. The Pre-school Education Association of Thailand (2017) has collected and shared via their website (https://preschool.or.th/content/documents/The-full-report-of-the-study.pdf) 40 STEM lessons created by teachers for children aged 4–6 years old on various topics related to science, mathematics, or technology learning experiences. The lesson plans are designed based on problem-based STEM learning using an engineering design process. Examples include making juice, making toys from used materials, and designing equipment for protection from sunlight. For example, in designing the sunlight protection equipment activity, children explore the advantages and disadvantages of sunlight; observe properties, shapes, and sizes of objects and tools; inquire about data from various sources; and use the engineering design process to design and create their own sunlight protection equipment from appropriate materials.

In 2014, the IPST hosted the second ECaP conference with the topic of STEM education in early childhood, at which 16 STEM project-based learning activities created by early childhood teachers were selected to be presented. These learning activities have been published and shared on the conference website. In each of the STEM projects, children are encouraged to observe their surroundings and pose questions or problems of interest to them, and then investigate further through scientific inquiry and/or an engineering design process. Examples in the units include finding ways to filter dirty water; to create watercolours from plants; or to create new toys, models, or equipment from natural or waste materials. For example, in the water project, students learn to work together to investigate, design, and create their water filter using simple materials, testing and improving their prototype, and creating an exhibition to share their ideas. The teachers indicated that children developed a positive attitude towards nature and environmental conservation, and improved their self-esteem and responsibility (IPST, 2014b).

In informal settings, STEM education for early childhood has been promoted in many public and private museums. For example, the National Science Museum (2021) has many hands-on and minds-on STEM activities for young children, such as a science circus, science lab, little scientists house, and a maker space. In the Children's Discovery Museum, there are exhibition zones that provide hands-on exhibitions and activities for enhancing young children's STEM learning, such as creative science, Dino detective, build our city, and the Inventor's club (Museum Siam, 2016). In KidZania Bangkok (2021), children encounter the roles STEM careers play in different contexts, such as an auto repair shop, bank, construction site, Crime Scene Investigation (CSI), convenience food store, dental care, drinking water research centre, eye care centre, design studio, fire station, flight stimulation, hospital, and vet clinic.

Overview of Research in STEM Education for Early Childhood

Many research studies related to STEM education in ECE in Thailand have been conducted by early childhood educators, university instructors, and graduate students since the 2010s. Many of them have focused on the impact of STEM learning on young children's cognitive development. The findings have indicated that STEM learning activities can improve young children's problem-solving skills (Hemra & Samahito, 2016; Kaittawee et al., 2019; Karsuwan & Samahito, 2015; Meesilpa, 2019), children's scientific process skills (Butda, 2021; Chemae, 2020; Kaewwichian et al., 2017; Kijkrajang & Kongruang, 2019; Sangpet et al., 2020; Thanopajai & Lehmongkol, 2016; Thongkam, 2021), analytical thinking (Saguansri, 2020; Tanonnok et al., 2019), critical thinking (Charoenket & Chamnankit, 2014), creativity skills and creative thinking (Kijkrajang & Kongruang, 2019; Loha & Jongkonklang, 2019; Sethirunkul & Samahito, 2019), and communication skills (Iaolek et al., 2019; Phonsutham & Samahito, 2017). Children were also very positive about their STEM learning, and enthusiastic and actively engaged in asking and answering questions (Kaewwichian et al., 2017).

Other studies have suggested that the use of mathematical games (Chalermphajong & Chamnankit, 2014; Pongkhamsing et al., 2010; Ratsirijan et al., 2020), cooking activities (Somsri, 2015), and mathematical tales (Phanomtung et al., 2012) can promote children's basic mathematical skills (5–6 years old). In addition, it has been shown that STEM learning can be enhanced by using local resources, local wisdom, and participation of the community (Rodkumnerd & Samahito, 2016; Wimuttipanya, 2019). For example, Rodkumnerd and Samahito (2016) studied the effects of a project approach based on STEM education using local resources in Samut Songkhram province on preschool children's understanding of science concepts (5–6 years old). The local resources included a salt field, mackerel fish market, and coconut farm. The results indicated that children developed better understanding of several science concepts, such as sinking and floating, forces, changes of substances, structure of organisms, and growth of living things.

Conclusion

This chapter's overview of STEM education in ECE in Thailand may provide some insights into how to promote STEM education in ECE in other educational contexts. It has shown that in order to foster STEM education in ECE, many educational agencies—both public and private—should be involved over the long term, with the support of government policy and funding. Further, children should have opportunities to experience STEM learning in both formal and informal education. In relation to formal education, STEM learning should be embedded in the national early childhood curriculum. Teachers should be provided with a learning framework and guidelines for the integration of science, mathematics, technology, and STEM learning in early childhood learning experiences for enhancing their ability to design and implement effective STEM education. Examples of STEM learning activities, lesson plans, learning materials and resources, and various forms of PD should be provided for early childhood teachers.

For the future development of STEM education in ECE in Thailand, the IPST continues to develop more STEM learning activities and materials with an emphasis on integrating computational thinking and unplugged coding in early childhood learning activities. The LSH project in Thailand also emphasizes the integration of the United Nations' sustainable development goals (SDGs) into STEM learning. In addition to the STEM activities and resources provided by organizations such as the IPST and LSH, Thai early childhood teachers are encouraged to develop their own learning activities and resources to promote STEM learning related to their school contexts and children's interests. They can share their learning activities and resources with others through websites, conferences, and professional learning communities.

In relation to future research, to increase the quality of STEM education in ECE, more research studies should be conducted by early childhood teachers. Thus, the future PD for early childhood teachers should focus on enhancing early childhood teachers' research knowledge and skills. In addition, most of the PD programmes about STEM education in early childhood are done with in-service teachers, so we should place more emphasis on the development of pre-service teachers, embedding STEM education into pre-service early childhood teachers' courses.

Hopefully, the efforts to foster STEM education in ECE in Thailand will contribute to the improvement in quality of STEM education and will increase the future STEM workforce in Thailand and in the rest of Asia.

References

Butda, N. (2021). A set of experience-enhancing activities to improve science process skills "by using objects fun coupled learning" for preschool children kindergarten year 2. *Panya, 28*(1), 38–46. https://so06.tci-thaijo.org/index.php/panya-thjo/article/view/247730/168290 (in Thai).

Campbell, C., Speldewinde, C., Howitt, C., & MacDonald, A. (2018). STEM practice in the early years. *Creative Education, 9*, 11–25. https://doi.org/10.4236/ce.2018.91002.

Chalermphajong, S., & Chamnankit, B. (2014). The effects of educational experience by using mathematical games on the basic mathematics ability of early childhood. *Journal of Graduate Studies in Northern Rajabhat Universities, 4*(5), 129–141. https://so02.tci-thaijo.org/index.php/JGNRU/article/view/29365 (in Thai).

Charoenket, S., & Chamnankit, B. (2014). The effect of scientific experience arrangement on critical thinking ability of early childhood. *Journal of Graduate Studies in Northern Rajabhat Universities, 4*(6), 117–128. https://so02.tci-thaijo.org/index.php/JGNRU/article/view/29364 (in Thai).

Chemae, S. (2020). The learning experiences by using a set of science experiment activities to develop scientific process skills for preschool children in 3rd kindergarten. *Journal of Educational Technology and Communications (hosted by Faculty of Education Mahasarakham University), 3*(8), 41–54. https://so02.tci-thaijo.org/index.php/etcedumsujournal/article/view/244465/165713 (in Thai).

Hemra, K., & Samahito, C. (2016). STEM education activity provision on good products in Roi Et to develop problem-solving ability in young children. *Kasetsart Educational Review, 31*(3), 159–160. https://kukr2.lib.ku.ac.th/kukr_es/index.php?/BKN/search_detail/result/363544 (in Thai).

Iaolek, C., Butkatanyoo, O., & Samahito, C. (2019). Effects of science learning experience provision by using inquiry and engineering design process on young children's science communications. *Journal of Research Unit on Science, Technology and Environment for Learning, 10*(1), 79–93. http://ejournals.swu.ac.th/index.php/JSTEL/article/view/11346/9286 (in Thai).

Kaewwichian, S., Khamkhad, U, & PichaiKum, J. (2017). Development of scientific process skill for early childhood through STEM learning method. In *Proceedings of the 12th RSU National Graduate Research Conference, Rangsit University, Pathum Thani* (Vol. 12, pp. 794–804). https://rsujournals.rsu.ac.th/index.php/rgrc/article/view/711/479 (in Thai).

Kaittawee, S., Tongthaworn, R., & Upanno, P. (2019). Learning arrangement from the King's Projects through STEM education to develop early childhood children's problem-solving skills and engineering. *Journal of Education Studies (hosted by Chulalongkorn University), 47*(1), 508–525. https://so02.tci-thaijo.org/index.php/EDUCU/article/view/179875/127726 (in Thai).

Karsuwan, P., & Samahito, C. (2015). The effects of STEM education activity provision for early childhood on problem solving thinking ability. *Kasetsart Educational Review, 30*(3), 175–185. https://so04.tci-thaijo.org/index.php/eduku/article/view/78530 (in Thai).

KidZania Bangkok. (2021). *Activities*. https://bangkok.kidzania.com/en-th/activities.

Kijkrajang, S, & Kongruang, S. (2019). Results of learning experiences emphasizing on engineering design towards science process skills and creativity skills of pre-school children. *Journal of Management Science Review, 21*(2), 155–167. https://so03.tci-thaijo.org/index.php/msaru/article/view/248262/167128 (in Thai).

Little Scientists House Project Thailand. (2019). *Rāyngān kār dảnein ngān khorngkār b̄ān nạk withyāṣ̄āstr̒ n̂xy pratheṣ̄thịy pracả pīkārṣ̄ụks̄'ā 2562* [Annual Report of Little Scientists House Project Thailand Academic year 2019]. https://www.obec.go.th/wp-content/uploads/2020/06/รายงานสรุปโครงการ-ปี-2562-นักวิทย์น้อย.pdf.

Little Scientists House Project Thailand. (2021, May 8). *Bı kickrrm reụ̄xng n̂ả lǽa thekhnoloyī* [Water and technology activity card]. https://www.littlescientistshouse.com/download.

Loha, P., & Jongkonklang, S. (2019). The development of assertive behavior and creative thinking of early childhood students by STEM education. *Ratchaphruek Journal, 17*(1), 121–128. https://so05.tci-thaijo.org/index.php/Ratchaphruekjournal/article/view/175226/125361 (in Thai).

McClure, E. (2017). More than a foundation: Young children are capable STEM learners. *Young Children, 72*(5), 83–89. https://www.jstor.org/stable/90015862.

Meesilpa, R. (2019). Inquiry method provision of learning experiences to the problem-solving ability of kindergarten. *NEU Academic and Research Journal, 9*(2), 97–106. https://so04.tci-thaijo.org/index.php/neuarj/article/view/213341/148341 (in Thai).

Ministry of Education Thailand. (2004). *Khū̂mụ̄x h̄lạks̄ūtr kār ṣ̄ụks̄'ā pt̄hmwạy phuthṭhṣ̄ạkrāch 2546 (s̄ảh̄rạb dĕk xāyu 3–5 pī)* [Manual of early childhood curriculum A.D. 2003 (for 3–5 years old)]. Bureau of Academic Affairs and Educational Standards, Office of the Basic Education Commission. https://docs.google.com/viewer?a=v&pid=sites&srcid=ZGVmYXVsdGRvbWFpbnxyYWJlaXlieHB0aHxneDozNDM2MDgxMDE5ZWE2MTU.

Ministry of Education Thailand. (2017). *Early childhood curriculum B.E. 2560 (A.D. 2017)* [English version]. Bureau of Academic Affairs and Educational Standards, Office of the Basic Education Commission. http://academic.obec.go.th/images/document/1572317446_d_1.pdf.

Museum Siam. (2016). *Children's discovery museum Bangkok 1 (Chatuchak)*. Museum Thailand. https://www.museumthailand.com/en/museum/Childrens-Discovery-Museum-Bangkok-1-Chatuchak?fbclid=IwAR38GF0OacH88SGO-Mx0OEq-Ke3L96OL3Ow6bRt-WCM69zo68Z54QBrCzEE.

National Science Museum. (2021). *National Science Museum*. https://www.nsm.or.th/en/.

Office of the Council of State. (2019). *Phrarāchbạyỵạti kār phạt'hnā dĕk pt̄hmwạy ph.Ṣ. 2562* [Early Childhood Development Act of A.D. 2019]. https://www.krisdika.go.th/data/ebook/thailaw/10-2562.pdf.

Office of the Education Council. (2017). *Phœnkār ṣ̄ụks̄'ā h̄æ̀ng chāti ph.Ṣ. 2560–2579* [Thai national education plan A.D. 2017–2036]. Office of the Education Council. https://planipolis.iiep.unesco.org/sites/default/files/ressources/thailand_educationplan.pdf.

Phanomtung, N., Erawan, P., & Boonchai P. (2012). A development of mathematics skills by using the mathematics tales for early children at Anuban Wisit Amnuaysin school, Bueng Kan: A time cycle process action research. *Journal of Educational Measurement (hosted by Faculty of Educaiton, Mahasarakham University), 17*(1), 201–209. https://so02.tci-thaijo.org/index.php/jemmsu/article/view/149730/109890 (in Thai).

Phatdunyadunrut, K. (2019). Early childhood education management: Policy and Buddhist integrated management. *Mahachula Academic Journal, 6*, 414–435. https://so04.tci-thaijo.org/index.php/JMA/article/view/187456/133241 (in Thai).

Phonsutham, J., & Samahito, C. (2017). STEM education learning experience provision to support communication skill of young children. *Kasetsart Educational Review, 32*(1), 117–123. https://kukr.lib.ku.ac.th/journal/index.php?/KERJ/search_detail/result/371851 (in Thai).

Pongkhamsing, S., Ekgatus, P., & Suikraduang, A. (2010). Development of learning experience plans with natural material games for basic mathematics skills preparation of early childhood students. *Rajabhat Maha Sarakham University Journal, 4*(2), 185–197. https://so05.tci-thaijo.org/index.php/rmuj/article/view/29605/25495 (in Thai).

Pre-school Education Association of Thailand. (2017). *Xeks̄ār rwbrwm p̄hl ngān khorngkār phạt'hnā p̄hū̂ brih̄ār lǽa khrū kæn nả pt̄hmwạy s̄a tĕm ṣ̄ụks̄'ā* [Collection of teachers' works from the project on the development of preschool administrators and core teachers for STEM education]. https://preschool.or.th/content/documents/The-full-report-of-the-study.pdf.

Ratsirijan, P., Wongsan, W., & Somboon, T. (2020). Experiential learning management using educational game for basic mathematical skill development of pre-school pupils of Anongnart kindergarten school. *Journal of Educational Review (hosted by Faculty of Education, Mahachulalongkornrajavidyalaya University), 7*(3), 118–129. https://so02.tci-thaijo.org/index.php/EDMCU/article/view/243831/166458 (in Thai).

Rodkumnerd, L., & Samahito, C. (2016). The effects of project approach experience provision based on STEAM education approach with using learning local resources in Samut Songkhram on preschool children's understanding concepts of science. *Kasetsart Educational Review, 31* (1), 93–100. https://kukr.lib.ku.ac.th/journal/index.php?/KERJ/search_detail/result/363572 (in Thai).

Saguansri, C. (2020). The development of thinking potentiality of the early childhood children using the integration of the mosques scientific models. *Journal of Lawasri (hosted by Thepsatri Rajabhat University), 3*(2), 63–69. https://so04.tci-thaijo.org/index.php/lawasrijo/article/view/246192/167232 (in Thai).

Sangpet, S., Thamsirikhwan, P., & Kulnaphadol, A. (2020). The experience organization with knowledge-based resources for encouraging the science basic skills of the early childhood pupils of Bankhokpanomdee school. In *The 11th National and International Research Conference: Global Goals, Local Actions: Looking Back and Moving Forward*, Suan Sanandha Rajabhat University. http://58.181.147.25/ojsjournal/index.php/8thconference/article/view/2200/2049 (in Thai).

Sethirunkul, T., & Samahito, C. (2019). Provision of STEM learning experiences on Nonthaburi local culture to develop creative thinking of preschool children. *STOU Education Journal, 12* (1), 99–109. https://so05.tci-thaijo.org/index.php/edjour_stou/article/view/191755/140498 (in Thai).

Somsri, W. (2015). A comparison mathematics basic skills and problem-solving thinking abilities of early childhood between cooking activities and educational games. *Santapol College Academic Journal, 1*(1), 32–41. https://so05.tci-thaijo.org/index.php/scaj/article/view/208268/144414.

Tanonnok, P., Tongthaworn, R., & Upanno, P. (2019). Science, technology, and society learning process arrangement to develop early childhood children's analytical thinking. *Journal of Education Studies, 47*(3), 269–289. https://so02.tci-thaijo.org/index.php/EDUCU/article/view/218859/151565 (in Thai).

ThaiPR.net. (2009, February 23). *Dĕk xnubāl VS kār reīyn rū̂ withyāṣ̄āstr̒? Thảmị t̂xng s̄xn withyāṣ̄āstr̒ tậngtæ̀ pṭhmwạy* [Kindergarten children VS learning science? Why we need to teach science in early childhood]. https://www.ryt9.com/s/prg/527052.

ThaiPR.net. (2011, March 28). *Withy̒-khṇit pṭhmwạy s̄ảkhạỵ xỳāngrị t̀x xnākht k̄hxng chāti?* [How science-mathematics are important to the future of the nation?] https://www.ryt9.com/s/prg/1117875.

Thanopajai, S., & Lehmongkol, P. (2016). The utilization of practical activity packages to develop scientific skills of preschool children. *Kasetsart Educational Review, 31*(3), 159–164. http://www.kuer.edu.ku.ac.th/file_upload/abstract/12/12-243.pdf (in Thai).

The Institute for the Promotion of Teaching Science and Technology. (2006). *Rāyngān k̄h̀x s̄rup cāk kār prachum radm khwām khid reụ̄̀xng kār cạd thả mātrṭ̄hān h̄lạks̄ūtr withyāṣ̄āstr̒ s̄ảh̄rạb dĕk pṭhmwạy* [Conclusions Report from the brainstorming meeting on the development of early childhood science curriculum standards]. The Institute for the Promotion of Teaching Science and Technology.

The Institute for the Promotion of Teaching Science and Technology. (2008). *Næwthāng kār cạdkār reīyn rū̂ withyāṣ̄āstr̒ pṭhmwạy tām h̄lạks̄ūtr kār ṣ̄ụks̄'ā pṭhmwạy 2546* [Early childhood science learning guideline aligned with early childhood curriculum A.D. 2003]. http://164.115.23.205/web/index.php/news-and-announcements/49-2010-10-27-03-54-42/86-2010-10-27-04-00-22.

The Institute for the Promotion of Teaching Science and Technology. (2009). *Khū̂mụ̄x h̄lạks̄ūtr kār xb rm cheing pḍibạti kār kār cạd kic kr rm būrṇ ā kār withyāṣ̄āstr̒ pṭhmwạy: Xeks̄ār s̄ảh̄rạb p̄hū̂ h̄ıkār xbrm* [Manual of workshop program on integrated science learning activities in early

childhood for the trainers]. The Institute for the Promotion of Teaching Science and Technology.
The Institute for the Promotion of Teaching Science and Technology. (2010a). *Nœwthāng kār cạdkār reīyn rū̂ withyāṣ̄āstr̒ pṭhmwạy tām h̄lạks̄ūtr kār ṣ̄ụks̄'ā pṭhmwạy 2546 (c̄hbạb prạbprung)* [Early childhood science learning guideline aligned with early childhood curriculum A.D. 2003 (Rev. ed.)]. The Institute for the Promotion of Teaching Science and Technology.
The Institute for the Promotion of Teaching Science and Technology. (2010b). *Khū̂mụ̄x kār cạdkār xbrm cheing pṭibạtikār pheìmphūn ṣ̄ạkyphāph ṣ̄ụks̄'ānitheṣ̄k̒ læa khrū pṭhmwạy: Kār cạd kic kr rm būrṇ ā kār withyāṣ̄āstr̒ khṇitṣ̄āstr̒ læa thekhnoloyī pṭhmwạy* [Workshop manual for teacher supervisors and kindergarten teachers: the integration of science, mathematics, and technology in early childhood learning activities]. The Institute for the Promotion of Teaching Science and Technology.
The Institute for the Promotion of Teaching Science and Technology. (2011a). *Krxb mātrṭhān kār reīyn rū̂ khṇitṣ̄āstr̒ pṭhmwạy tām h̄lạks̄ūtr kār ṣ̄ụks̄'ā pṭhmwạy phuthṭhṣ̄ạkrāch 2546* [Early childhood mathematics learning standards aligned with early childhood curriculum A.D. 2003]. The Institute for the Promotion of Teaching Science and Technology. http://earlychildhood.ipst.ac.th/wp-content/uploads/sites/25/2014/09/Math-framework-for-ECE.pdf.
The Institute for the Promotion of Teaching Science and Technology. (2011b). *Khū̂mụ̄x krxb mātrṭhān kār reīyn rū̂ khṇitṣ̄āstr̒ pṭhmwạy tām h̄lạks̄ūtr kār ṣ̄ụks̄'ā pṭhmwạy phuthṭhṣ̄ạkrāch 2546* [Manual for early childhood mathematics learning standards aligned with early childhood curriculum A.D. 2003]. The Institute for the Promotion of Teaching Science and Technology. http://earlychildhood.ipst.ac.th/wp-content/uploads/sites/25/2018/08/math_m.pdf.
The Institute for the Promotion of Teaching Science and Technology. (2011c). *Krxb mātrṭhān læa khū̂mụ̄x kār cạdkār reīyn rū̂ withyāṣ̄āstr̒ pṭhmwạy tām h̄lạks̄ūtr kār ṣ̄ụks̄'ā pṭhmwạy phuthṭhṣ̄ạkrāch 2546* [Standards framework and manual for early childhood science learning aligned with early childhood curriculum A.D. 2003]. The Institute for the Promotion of Teaching Science and Technology. http://earlychildhood.ipst.ac.th/wp-content/uploads/sites/25/2014/09/Science-Framework-for-ECE.pdf.
The Institute for the Promotion of Teaching Science and Technology. (2011d). *Khū̂mụ̄x kār cạdkār xbrm cheing pṭibạtikār phạt'hnā withyākr kæn nả withyāṣ̄āstr̒ khṇitṣ̄āstr̒ læa thekhnoloyī pṭhmwạy: Kār s̄ngs̄erim kār reīyn rū̂ khṇitṣ̄āstr̒ pṭhmwạy* [Workshop manual for development of science, mathematics, and technology leader instructors: the promotion of mathematics learning in early childhood]. The Institute for the Promotion of Teaching Science and Technology.
The Institute for the Promotion of Teaching Science and Technology. (2014a). *Science Technology Engineering and Mathematics Education (STEM Education)*. http://www.stemedthailand.org/wp-content/uploads/2015/03/Intro-to-STEM.pdf (in Thai).
The Institute for the Promotion of Teaching Science and Technology. (2014b). *Ngān prachum wichākār withyāṣ̄āstr̒ khṇitṣ̄āstr̒ læa thekhnoloyī radạb pṭhmwạy læa prathm ṣ̄ụks̄'ā khrậng thī̀ 2 "s̄a tĕm ṣ̄ụks̄'ā phạt'hnā dĕk thịy"* [The 2nd early childhood and primary science mathematics and technology education conference, 'STEM education enhances Thai children']. https://ecap.ipst.ac.th/ecap-2nd/.
The Institute for the Promotion of Teaching Science and Technology. (2016). *Ngān prachum wichākār withyāṣ̄āstr̒ khṇitṣ̄āstr̒ læa thekhnoloyī radạb pṭhmwạy læa prathm ṣ̄ụks̄'ā khrậng thī̀ 3* [The 3rd early childhood and primary science, mathematics and technology education conference]. https://ecap.ipst.ac.th.
The Institute for the Promotion of Teaching Science and Technology. (2019). *Nœwthāng kār cạd pras̄bkārṇ̒ h̄ıwy kār reīyn rū̂ reụ̄xng 's̄eīyng phā phelin' radạb pṭhmwạy* [Learning experience guideline, 'Fun with Sound' early childhood learning unit]. Bangkok: BOWT Ladprao Printing House.
The Institute for the Promotion of Teaching Science and Technology. (2020). *Krxb kār reīyn rū̂ læa nœwthāng kār cạd pras̄bkārṇ̒ kār reīyn rū̂ būrṇ ā kār withyāṣ̄āstr̒ thekhnoloyī læa khṇitṣ̄āstr̒ nı radạb pṭhmwạy tām h̄lạks̄ūtr kār ṣ̄ụks̄'ā pṭhmwạy phuthṭhṣ̄ạkrāch 2560* [Learning framework

and guideline for integrated science, technology, and mathematics learning in early childhood aligned with early childhood curriculum A.D. 2017]. https://www.scimath.org/ebook-science/item/11322-2020-03-05-01-33-35.

The National Legislative Assembly. (2015). *Rāyngān kār phicārṇā ṣ̄ụks̄'ā k̄hx s̄enx cheing nyobāy s̄a tĕm ṣ̄ụks̄'ā (STEM Education) nyobāy cheing ruk pheụ̄x phạt'hnā yeāwchn lœa kảlạng khn d̄ān withyāṣ̄āstr̒ thekhnoloyī wiṣ̄wkrrmṣ̄āstr̒ lœa khṇitṣ̄āstr̒* [STEM education policy proposal review report, proactive policy for developing youth and manpower in science, technology, engineering, and mathematics]. The Secretariat of the Senate. http://61.19.241.96/w3c/senate/pictures/comm/1547/%E0%B8%82%E0%B9%89%E0%B8%AD%E0%B9%80%E0%B8%AA%E0%B8%99%E0%B8%AD%E0%B9%80%E0%B8%8A%E0%B8%B4%E0%B8%87%E0%B8%99%E0%B9%82%E0%B8%A2%E0%B8%9A%E0%B8%B2%E0%B8%A2STEM.pdf.

The National Research Council. (2011). *Successful K-12 STEM education: Identifying effective approaches in science, technology, engineering, and mathematics*. The National Academies Press. https://doi.org/10.17226/13158.

Thongkam, B. (2021). A set of activities to enhance learning experience to promote cognitive development by using scientific processes for preschool children, kindergarten, year 1. *Panya, 28*(1), 29–37. https://so06.tci-thaijo.org/index.php/panya-thjo/article/view/247667/168289 (in Thai).

United Nations Educational, Scientific and Culture Organization. (2019). *Early childhood care and education*. https://en.unesco.org/themes/early-childhood-care-and-education.

Wimuttipanya, J. (2019). The development of early childhood with STEM education integrated with learning resources and community participation. *Integrated Social Science Journal (hosted by Faculty of Social Sciences and Humanities, Mahidol University), 6*(1), 3–23. https://so02.tci-thaijo.org/index.php/isshmu/article/view/192858/134420 (in Thai).

Tepkanya Promkatkeaw is an instructor at the Science Education Center, Faculty of Science, Srinakharinwirot University, Thailand. Her current research interests include the Nature of Science, STEM Education, and science, mathematics, and technology integration in early childhood education focusing on curriculum development and teacher professional development. She is a former director of compulsory science department at the Institute for the Promotion of Teaching Science and Technology (IPST) and she was a member of working groups for revising the science learning standards in the Thai Basic Education Core Curriculum B.E. 2551 (A.D. 2009) and for developing the Thai National Early Childhood Curriculum B.E. 2560 (A.D. 2017).

Navara Seetee is an instructor at the Science Education Center, Faculty of Science, Srinakharinwirot University, Thailand. She graduated with an Ed.D. in science education from Srinakharinwirot University, Thailand. Her research interests include science teaching and learning at all educational levels, STEM education, assessment in science learning, and early childhood teacher professional development. She was a visiting scholar at National Taiwan University of Science and Technology in 2019. She is also a core trainer in the Little Scientists' House Project, Thailand.

Chanyah Dahsah is an assistant professor of Science Education and director of Science Education Center at the Faculty of Science, Srinakharinwirot University, Thailand. She received a Ph.D. in Science Education from Kasetsart University, Thailand in collaboration with the University of Waikato in 2007. In 2013–2015, she was a researcher scholar at CREATE for STEM Institute, Michigan State University. Her research interests include K-12 STEM learning, assessment, and professional development. Her current projects involve the Little Scientists' House Project, STEM-Entrepreneur, and STEM2 integrated curriculum development and implementation.

Chapter 7
Innovative STEM Curriculum to Enhance Students' Engineering Design Skills and Attitudes Towards STEM

Meng-Fei Cheng and Yu-Heng Lo

Abstract Although STEM education has been advocated internationally, the integration of interdisciplinary learning into STEM education and the gender disparity in the STEM field are challenging. Our research team in Taiwan developed a female-friendly and innovative STEM curriculum with flat (rather than bulky) speakers to enhance male and female students' creativity in developing new technology and to foster their interdisciplinary thinking. Participating year 10 students were encouraged in the 3-hour course to integrate science knowledge into their engineering design processes in order to better develop, evaluate, and revise their technology products. In this study, we examined this STEM curriculum to show the progression of male and female students' engineering designs and their attitudes towards STEM. Through the systematic guidance of the STEM curriculum, students' engineering designs improved, regardless of gender. There were no significant differences between male and female students' performance in engineering design in each stage of the STEM curriculum, However, in terms of the improvement in engineering design ability, female students did not improve whereas their male counterparts did in some activities. Participating in the STEM curriculum developed by this study increased the positive attitudes of both male and female students towards STEM and STEM learning. It also reduced the attitude gap between the two genders seen before the course in the technology dimension. The study findings can contribute to the development of better ways of integrating interdisciplinary learning and teaching and enhancing male and female students' engineering designs and attitudes towards STEM.

Keywords Physics · Engineering design · Attitudes towards STEM · Curriculum design · Assessment of STEM learning

---

M.-F. Cheng (✉) · Y.-H. Lo
Department of Physics, National Changhua University of Education, No.1, Jin-De Road, Changhua City, Taiwan
e-mail: mcheng2@cc.ncue.edu.tw

M. M. H. Cheng et al. (eds.), *Concepts and Practices of STEM Education in Asia*,
https://doi.org/10.1007/978-981-19-2596-2_7

Introduction

Science, technology, engineering, and mathematics (STEM) education has been advocated internationally in order to enhance technology competition at a national level (Freeman et al., 2019). In order to enhance competition in the technology industry in Taiwan, it is also essential to urgently improve the country's industrial structure and expand high-tech industry with cutting-edge innovations. Accordingly, STEM education should be promoted in order to integrate science into engineering learning, enhance technological innovation, and integrate engineering design into the science curriculum to enhance students' interest in learning about science and engineering.

Furthermore, as a vital issue in interdisciplinary teaching in current science education in Taiwan, STEM education must emphasize the priorities of the current curriculum guidelines for K-12 science education issued by the Ministry of Education of Taiwan (2018a). These include the implementation of a literacy-oriented curriculum and the development of learning goals that integrate the STEM subjects into the science and technology curriculum so that students can practice the creative processes of conceptualization and resolution of scientific and technological problems in daily life.

Moreover, in STEM fields, women are generally underrepresented globally (Ceci & Williams, 2007; Freeman et al., 2019), including Taiwan (Ministry of Education of Taiwan, 2019). Therefore, this study aimed to design an innovative female-friendly STEM curriculum that emphasizes interdisciplinary teaching not only to integrate different disciplines and to cultivate inquiry and problem-solving abilities, but also to enhance male and female abilities and interest in learning about STEM subjects. The curriculum designed in this study can be used as a demonstration model and guideline for teachers designing their own STEM courses, thereby facilitating better implementation and promotion of STEM education.

STEM Education

In many countries around the world, the proportion of higher education students who pursue education in engineering and science is limited; this has resulted in a workforce shortage (De Vries, 2018; Han et al., 2015; Raju & Clayson, 2010). For this reason, education scholars from various countries have dedicated themselves to promoting the implementation of STEM curricula in primary and secondary schools to improve students' interest in pursuing STEM-related careers.

Students in Taiwan are more inclined to engage with STEM-related industries than students in other countries. According to statistics published by the Ministry of Education of Taiwan in 2018, Taiwan had the highest proportion of students enrolled in science- and technology-related departments in universities at 40.9% (Ministry of Education of Taiwan, 2018b, see Table 7.1). Consequently, there is no shortage of

Table 7.1 The ratio of university student enrolment by discipline (Ministry of Education of Taiwan, 2018b)

Academic year	Total	Humanities	(%)	Social studies	(%)	Science and technology	(%)
2008	1,006,102	166,953	16.6	380,479	37.8	458,670	45.6
2009	1,010,952	171,656	17.0	384,707	38.1	454,589	45.0
2010	1,021,636	176,439	17.3	393,310	38.5	451,887	44.2
2011	1,032,985	183,051	17.7	403,079	39.0	446,855	43.3
2012	1,038,041	187,950	18.1	412,806	39.8	437,285	42.1
2013	1,035,534	192,151	18.6	416,629	40.2	426,754	41.2
2014	1,037,062	195,292	18.8	421,439	40.6	420,331	40.5
2015	1,035,218	197,237	19.1	423,450	40.9	414,531	40.0
2016	1,015,398	195,602	19.3	414,604	40.8	405,192	39.9
2017	985,927	198,532	20.1	389,494	39.5	397,901	40.4
2018	961,905	194,857	20.3	373,716	38.9	393,332	40.9

science and technology talent in Taiwan. However, we are concerned with improving the quality of this talent in order to promote competition in the technology industry. Hence, there is a need for STEM education in Taiwan that fosters school students who are more capable of innovative, problem-solving, and interdisciplinary thinking, and to enhance the core literacy (systematic thinking and problem solving) promoted by the Ministry of Education of Taiwan in the current education standards for all citizens.

Integrated STEM education often involves teaching science and mathematics subject content through problems related to technology and engineering, thereby facilitating better connections between these different subjects, as well as a more interactive learning process for students (Honey et al., 2014). In other words, integrated STEM education is an interdisciplinary teaching approach. However, integrating multiple subjects into teaching in a meaningful way poses a significant challenge, given that the content and teaching methods of the four contributing subject areas are not always entirely correlated (Bell, 2016; de Vries, 2018; Kertil & Gurel, 2016; Margot & Kettler, 2019; Radloff & Guzey, 2016). In order to address this issue, we designed a teaching method based on the 6E (Engage, Explore, Explain, Engineer, Enrich, Evaluate) teaching model proposed by the International Technology and Engineering Educators Association (Burke, 2014), with the goal of better integrating the different STEM subjects and the practice of scientific and technological inquiry. In our STEM curriculum, we integrated knowledge of scientific models into students' engineering processes and utilized mathematics to collect and analyse data. This supported students into explicitly drawing on scientific and mathematical knowledge and reasoning to better resolve a technological and engineering problem.

STEM Education and Girls

In STEM occupations and departments, males outnumber females, especially in European and American countries (Stephenson et al., 2021). In the hope of addressing this gender inequality, there have been many studies exploring women's interest, confidence, and career choices in STEM fields in recent years (Blackburn, 2017). According to the Department of Statistics of the Ministry of Education of Taiwan, 75% of university students who graduated from STEM-related departments in Taiwan in 2019 were male (Ministry of Education of Taiwan, 2019), demonstrating that the gender gap also needs to be addressed in Taiwan. Understanding how women think and feel about STEM fields and the potential causes of gender inequality will enable researchers to design a female-friendly curriculum that will encourage both men and women to actively participate. In other words, understanding female perspectives can inform the design of interventions to enhance female interest and confidence in STEM fields, thereby alleviating gender inequality in STEM fields in Taiwan. Furthermore, increasing the proportion of women in STEM industries is expected to create emerging and innovative industrial ecosystems through women's ability to think creatively and critically (Gurski & Hammrich, 2017).

First, this paper discusses examples of STEM courses designed to enhance female attitudes towards STEM in Taiwan. Lou and colleagues (2011) implemented a solar energy vehicle-themed STEM course with the goal of participating in the competition of a solar energy vehicle with optimal speed and stability. This course included 40 students in a girls' senior high school, and interviews were conducted afterward to ascertain students' attitudes towards and thoughts on STEM. The students generally responded that the STEM course made them realize the importance of STEM-related knowledge and the fun and practicality of applying it to solve problems. The course also assisted students in exploring future career opportunities and establishing positive attitudes towards STEM learning.

It can be argued on the basis of these research findings that the STEM course positively affected female students' attitudes towards STEM. Nevertheless, in that study, all the research subjects were female, meaning that the difference in attitudes between the two genders before and after the STEM course could not be determined. Therefore, in our study, we developed a female-friendly STEM curriculum for students of both genders. A student attitude survey about STEM and the learning environment was conducted before and after the curriculum in order to explore differences in attitudes between the two genders and whether the STEM course was effective in terms of reducing differences between genders.

Other research has investigated reasons why female learning attitudes and career choices towards STEM differ from those of males. According to studies in the United States (Lindberg et al., 2010; Robinson & Lubienski, 2011), there is no significant difference between boys' and girls' performance in terms of spatial intelligence. However, men's performance gradually outpaces women's after formal education in elementary and secondary school, and this gender difference becomes more pronounced as their educational attainment increases. Wai et al. (2012) also revealed that male students generally perform better than their female counterparts in mathematics and science subjects in elementary school. However, if we look at the performance of each gender in science and the liberal arts, rather than comparing the academic ability of males and females, female performance in science is as good as liberal arts, and their performance in science is not significantly different from their performance in the liberal arts; males, however, perform better in the sciences than in the liberal arts (Wang et al., 2013). Valla and Ceci (2014) attribute the lower proportion of women in STEM careers to the fact that women excel in a wide range of disciplines and therefore invest more time in different fields than do men, who tend to excel in the sciences, resulting in men who focus on the sciences outperforming women in terms of both mathematical and science ability and representation in STEM careers.

On the other hand, research has pointed out that female confidence in mathematics and science is generally lower than that of males at the beginning of the school year, not because of poor academic performance but because females often attribute setbacks in STEM fields to an innate lack of ability rather than to a lack of effort (Chen & Moons, 2015; Ellis et al., 2016; LaCosse et al., 2016). This attribution affects girls' tendency to avoid difficulties when faced with problems in related fields (Dweck, 2007). However, the underlying gender bias that causes females to think this way is related to society's perception that females are generally better at liberal arts than at science. Therefore, it is considered common for females to underperform in the sciences (Jackson et al., 2014).

In terms of career choices, women are generally more inclined towards and interested in careers related to "people" and are therefore less willing to enter STEM fields than are men, who are generally more interested in "things" (Su et al., 2009). Moreover, when choosing their profession, women will often consider whether they can contribute to society and help others; that is, women generally prefer altruism to the pursuit of utilitarianism (Diekman et al., 2015, 2017). As for the choice of work environment, women are generally more likely to choose a workplace in which they can work with others, and are generally more likely to choose a career in which they feel a greater sense of belonging. Thus, women are more likely to choose careers such as teaching and nursing rather than STEM-related careers, which are considered less interactive (Su & Rounds, 2015). A lower sense of belonging and the unfriendly environment can apply to STEM-related classrooms and workplaces: the term "chilly climate," coined by Seaton (2011), refers to the fact that women in STEM-related careers and classes must face gender discrimination, prejudice, and differential treatment. In other words, women may not be treated equally or fairly in this environment, which can be a critical reason for women's reluctance to enter STEM fields.

Finally, gender stereotypes and sexism are among the most significant factors influencing women's choice to enter STEM fields or careers (Moss-Racusin et al., 2015; Ryan, 2014). Although gender discrimination is generally less prevalent than in the past, it is still strongly in evidence, especially in STEM-related fields, where women are exclusively the victims (Jackson et al., 2014). As mentioned above, women tend to attribute their failure or poor performance in science to a lack of ability rather than to a lack of effort. This attribution stems from the common perception in society that when men do not perform well in STEM subjects, they do not try hard enough, whereas it is common for women not to do well in the same subjects (Tiedemann, 2000). Moreover, women who are adept at science are more likely to defy public expectations, which makes women more reluctant to choose STEM-related disciplines and industries because of their fear of going against social expectations (Cheryan et al., 2017; Jones et al., 2013).

Given the various findings presented above, increasing the proportion of women in STEM fields is a significant challenge. Fortunately, many studies have found that science courses with more open-ended discussion in high school can increase girls' willingness to choose STEM fields (Morgan et al., 2013; Redmond-Sanogo et al., 2016). Therefore, it is expected that the proportion of women in STEM fields will increase if the relevant STEM courses can be designed to attract greater female participation.

Dancstep and Sindorf (2018) compiled and identified four pedagogical strategies to develop female-friendly courses through reviewing prior research: enabling social interaction and collaboration, creating a low-pressure setting, providing meaningful connections, and representing females and their interests.

First, enabling social interaction and collaboration recognizes that women generally prefer to cooperate with others to accomplish goals or interact with others during course activities. Conversely, women generally exhibit negative emotions if course activities are competitive. Second, creating a low-pressure setting requires that the course difficulty is moderate. Making the course too difficult will lead to science anxiety and render students less interested and less confident. Setting the course difficulty to moderate, and using open-ended questions and topics will help to reduce the pressure girls feel in the classroom. Third, providing meaningful connections involves ensuring that the topics and products explored in the course are meaningful; that is, course content should relate to solving social problems, improving quality of life, or solving environmental issues. These types of meaningful connections can encourage girls to become more engaged with and interested in the course. Lastly, with regard to girls and their interests, females are generally more interested than males in aesthetics and creativity. Thus, if the course includes these elements, the interest of girls may also increase.

The four strategies for girl-friendly courses align with the findings presented earlier that identified barriers and opportunities for engaging girls in STEM education and future STEM careers. That the course should be cooperative rather than competitive aligns with findings that women generally tend to choose careers that involve getting along with others and working together. That the course should be conducted in a low-stress environment reflects the fact that when girls face science-related setbacks, they often blame their ability rather than a lack of effort, resulting in their unwillingness to accept challenges and setbacks and in their giving up. Thus, setting the appropriate level of difficulty can help increase girls' interest in STEM programmes. The subject of the course should also benefit society and reflect the altruism of women, who tend to prioritize helping others when choosing their career.

Although boys generally out-perform girls in mathematics and science, girls' critical thinking, teamwork, and problem-solving abilities are better than those of boys (Gurski & Hammrich, 2017). Developing STEM courses for women that suit their interests is important to increase their willingness to engage in related fields (Su et al., 2009; Wang et al., 2015) and to balance the gender gap in STEM fields. This will enable Taiwan to leverage different kinds of expertise to solve social problems, as well as creative ideas to develop innovative industries.

The Context of This Study

The STEM curriculum in this study was designed on the basis of the characteristics of female-friendly curriculum design introduced above. First, the focus of the curriculum was the creation of the innovative flat speakers. The goal was to improve traditional stereo speakers, which are bulky and take up much space. The innovative flat speakers would be able to be integrated with artwork, such as oil paintings, and with everyday items such as pillows, helmets, and car surround-sound systems. This project would ensure that the curriculum helps solve problems, improves people's quality of life, and involves artistic elements, in which girls are generally interested, thus increasing their engagement.

Second, this STEM curriculum requires students to work collaboratively in groups. Individuals would discuss their own flat speaker designs with their team members, observe the advantages of each other's designs, and collaborate on the final design of the flat speaker. The team would then work together to produce the speaker, discuss its limitations, and consider methods to improve it. Students would discover that they needed mutual assistance and collaboration to complete the course activity; this is in line with the characteristics of course activities that girls tend to prefer, that is, involving interaction with others rather than competition.

Third, assessment of students' learning progress involves an open-ended engineering structure drawing. The assessment requires students to design and draw a series of structures of a flat speaker that can emit the maximum volume. Students are required to identify the required components, describe the scientific principles of how the components work, and explain why their flat speaker can emit the loudest sound. In order to verify whether their structural drawings meet the desired outcome, students discuss their designs with their peers, analyse the drawings using scientific knowledge, actually make the speaker, and then further revise it on the basis of their findings. In other words, the activity is arranged in such a way that students evaluate their structural drawings and make modifications rather than simply have their output judged as right or wrong. This was intended to reduce the pressure felt by girls to be 'successful' in the learning environment.

Fourth, the enrichment stage of this course involved applying what is learned in the classroom to designing a holiday card with a flat speaker that can emit the maximum volume, allowing students to reflect on and apply what they have learned and to add creative and artistic elements in the creation of holiday cards. The activity combined applicability, aesthetic creativity, and open-ended assessment to reduce pressure in the learning environment and therefore to appeal to girls, increase their engagement, and potentially increase their long-term interest in STEM fields.

Accordingly, this study describes a STEM curriculum designed to integrate interdisciplinary knowledge with consideration of female-friendly topics and teaching strategies in order to enhance both male and female students' engineering design abilities and their learning attitudes. The purpose of the study was to investigate whether this experimental STEM curriculum could enhance male and female students' engineering design abilities and their attitudes towards science, technology, engineering, and the learning environment.

Methodology

This study was aimed at exploring the effects of a STEM curriculum on the engineering design and affective attitudes of male and female students. The curriculum was designed on the basis of the 6E framework. During the curriculum, we conducted a series of assessments to evaluate students' progress in their ability to design engineering products. We also assessed their attitudes before and after the curriculum to shed light on changes in their perceptions of the STEM field and STEM courses.

Participants

The research participants were grade 10 students from seven high schools in central Taiwan. A total of 54 students, including 30 males and 24 females, participated in the study. The participants were recruited to the course through voluntary sign-ups, which they were informed about through an announcement from their schools. This STEM curriculum was run outside of school hours, and the students were all in one class despite attending different schools. Therefore, the participants showed considerable interest in physics, and all studied topics that included electromagnetism and the relationship between electricity and magnetism in middle school.

Learning Activities

The study monitored whether the participants demonstrated improvements in their abilities in engineering design through learning activities in a STEM curriculum. We chose the design of a new and emerging technology—flat speakers—as the focus for the course, and the project involved the application of electromagnetism. The course was designed using the 6E model proposed by Burke (2014). During the 3-hour course (Activities 1~3 took 1 hour; Activities 4~5 took 1 hour; Activities 6~7 took 1 hour), the participants studied physics concepts, designed and modified flat speaker model designs, and built actual models after observing stereo speakers, designing prototypes, conducting laboratory experiments, and discussing their ideas.

Relevant learning was included for each of the underpinning STEM domains. For science, we included theories related to the generation of sound, the form of sound transmitted through conducting wires, and models of electromagnetic induction. For technology, we covered the construction of a flat speaker that features maximized volume and application in daily life, such as pillows, paintings, or cards. For engineering, we delved into creating structure diagrams, constructing flat speakers, and modifying and improving the products. The difference between engineering and technology was that engineering emphasized the process of design, construction, and modification of the products, while technology focused on the products and applications. For mathematics, topics such as measuring decibels and transforming data into graphs were covered. The activities in each phase of the 6E curriculum are listed in Table 7.2. In activities 1, 2, 3, and 7, students drew and recorded their flat speaker structure diagrams individually. However, in activities 4 and 6, students were asked to discuss their ideas with their group members and to create a flat speaker, and then further examined the factors affecting speaker volume in their own groups.

Instruments

Evaluating the impact of the course was divided into two stages: the assessment of the students' engineering design, and the administration of a questionnaire regarding the students' attitudes towards STEM.

Assessment of Students' Engineering Designs. To keep track of the students' learning progress during the engineering design, they were asked to record their structural drawings of volume-maximized flat speakers in their learning journals in four activities: brainstorming on the structure of flat speakers (activity 1), disassembling a real stereo speaker (activity 2), explaining the magnetic effect of an electric current (activity 3), and designing a holiday card with flat speakers playing sound at maximum volume (activity 7). Analysis of these drawings enabled the researchers to investigate changes in the students' design skills as they progressed through the course.

Survey of Students' Attitudes towards STEM and the Learning Environment. The students completed an attitudinal survey before and after the STEM course to investigate any changes in their attitudes with regard to technology, engineering, science, and the STEM learning environment. Since the mathematics part in this STEM course only included drawing graphs of functions, the dimension of mathematics was not included in the attitude survey. Accordingly, survey questions were selected to serve this aim. The 19-question survey, which was adapted from Han and Carpenter (2014) and Unfried et al. (2015), required the participants to rate items on a 5-point Likert scale (*strongly agree, agree, neutral, disagree,* and *strongly disagree*). Each item included a description of a particular attitude, and each participant was instructed to select only one response informed by their own situation. The reliability of each scale ranged from 0.71 to 0.90. Example attitudinal survey items for the engineering dimension are given in Table 7.3.

Table 7.2 Flat speaker design using the 6E teaching model

Teaching activities	6E Model stages	Tasks
Activity 1: Brainstorming about the structure of the flat speaker	Engage	1. A video about innovative technologies involving flat speakers was played to arouse the students' learning motivation 2. Students independently pondered how to create a flat speaker that can produce maximum volume and then visualized and recorded their ideas via structure diagrams
Activity 2: Disassembling a real stereo speaker	Explore	1. The students were asked to reverse engineer real stereo speakers and observe their internal structure 2. Once again, the students were instructed to think about how to create a flat speaker that could produce the maximum volume, and to visualize their ideas in the form of structural drawings
Activity 3: Explaining the magnetic effect of an electric current	Explain	1. The teacher explained the principle of sound generation from vibration, the magnetic effect of an electric current, and the application of these principles to the design of a stereo speaker 2. Students thought about how to create a flat speaker that could produce the maximum volume and visualized their ideas in the form of structure diagrams on the basis of what they had learned about stereo speakers

(continued)

Table 7.2 (continued)

Teaching activities	6E Model stages	Tasks
Activity 4: Creating a flat speaker	Engineer and evaluate	1. Group members shared their individual flat speaker designs and discussed how to construct their speakers on the grounds of the engineering model of a speaker and scientific models of sound and electromagnetic force 2. Limited materials were provided to enable the teams to construct flat speakers on their own 3. Group members examined whether the finished products could operate successfully and evaluated the volume of their flat speakers 4. Each group presented the difficulties and findings that arose during the flat speaker production in front of the entire class, with reflections on the engineering model of a speaker and scientific models underpinning the operation of a speaker
Activity 5: Explaining the factors that affect speaker volume	Explain	1. The teacher led the class in a discussion of the factors and principles that relate to speaker volume, with the scientific model of electromagnetic force as a reference
Activity 6: Discussing and testing the factors affecting speaker volume through experimentation	Engineer and evaluate	1. Each group discussed how they could improve the volume of the flat speakers they had developed in Activity 4 by examining the factors that affect the strength of an electromagnet, with the scientific model of electromagnetic force as a reference 2. During the group discussion, group members designed an experiment using several flat speakers with different designs to explore the relationship between the modified variables and volume 3. The entire class investigated the differences in speaker volume using various variables and compared their findings with predictions they had made based on the scientific model of electromagnetic force

(continued)

Table 7.2 (continued)

Teaching activities	6E Model stages	Tasks
Activity 7: Designing a holiday card with flat speakers playing sound at maximum volume	Enrich and evaluate	1. The students were asked to use what they had learned in the course, along with their creative ideas, to design a flat speaker holiday card that could emit maximum volume 2. During the design process, the students were asked to evaluate whether their design products could operate successfully at maximum volume in accordance with what they had learned about the engineering model of a speaker and the scientific models relating to how a speaker works

Table 7.3 Attitudinal survey for the engineering dimension

Thanks to engineering, there will be greater opportunities for future generations
I like to imagine creating new products
If I learn engineering, then I can improve things that people use every day
I am interested in what makes machines work
Knowing how to use math and science together will allow me to invent useful things

Data Analysis

Assessment of Students' Engineering Designs. The students' engineering designs were classified into four levels (low to high) on the basis of the structure diagrams that they drew in their learning journals (Table 7.4).

Table 7.4 Classification of students' engineering designs into four levels

Level	Description
Level 1	The design does not reflect the major sound-producing components of the speaker and includes only unrelated components or duplicates the appearance of the original
Level 2	The design is informed by the correct selection of components and includes an appropriate description of the sound-producing structure and relative placements
Level 3	The design reflects and illustrates the principles and mechanisms related to how components interact with one another and comprises appropriate explanations
Level 4	The design not only features the components required and their mechanisms but also reflects how components in their diagrams can maximize speaker volume. These enhanced components proposed by students may include the number or position of the magnets or the shape or diameter of the coil

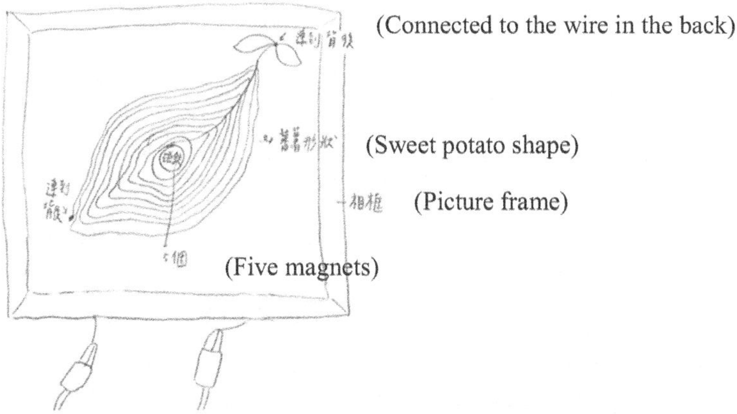

Fig. 7.1 An example of a level 4 flat speaker design

A Level 4 example of an engineering design from one of the students is shown in Fig. 7.1. In this design, the student indicated the main components and mechanisms and reflected on how the components can make the speaker as loud as possible. The coil and five magnets are the main components of the speaker. The principle underlying the design of the coil is that, through the audio source, the coil will produce a magnetic effect of an electric current. The reason for making the speaker as loud as possible is that the circular coils comprise many loops and the inner loop is close to the magnet. The design of the five magnets is based on the principle that magnets will enable the current-carrying coils to vibrate and produce sound waves. The student's experiments show that the speaker will produce the maximum sound with five magnets.

Due to the ordinal scale of the assessment and the small sample size, we performed Mann–Whitney U tests on the four activity stages to compare the levels achieved by the students and to examine how the males and females differed in terms of engineering design abilities before, during, and after the STEM class. We then carried out Wilcoxon signed-rank tests to compare the progression of the male and female students' engineering designs at different stages of the course, and to explore whether the different activities impacted male and female students' design development in a similar or different manner.

Evaluation of Student Attitudes. To investigate changes in the male and female students' attitudes towards technology, engineering, and science as well as classroom performance, we analysed the responses to the attitudinal surveys using Wilcoxon signed-rank tests for the pre- and post-course surveys. We also conducted Mann–Whitney U tests on the students' attitudes in the pre- and post-course surveys to identify whether there were gender disparities.

Results

Gender Gap in Engineering Design

As can be seen from Table 7.5, there were no significant differences between male and female students' performance in engineering design in each stage of the STEM curriculum, and their performance gradually but significantly improved throughout the different stages of the course. Thus, through the systematic guidance of the STEM curriculum, students' abilities to develop, evaluate, and revise their engineering designs improved, regardless of gender.

Through this curriculum, male students' abilities in engineering design progressed from the lower mean score of 1.36 to the higher mean score of 3.61, and female students' abilities in engineering design progressed from the lower mean score of 1.27 to the higher mean score of 3.64 (Table 7.5). Wilcoxon signed-rank tests showed significant improvement in both male students' abilities ($z = -4.50, p < 0.001$) with a large effect size ($r = 0.60$) and female students' abilities ($z = -4.20, p < 0.001$) with a large effect size ($r = 0.63$) in engineering design between activity 1 and activity 7.

In Table 7.5, the Mann–Whitney U tests indicated that there was no significant difference in the engineering design performance of the two genders in each activity. The continuous increase in the mean scores of male and female students' engineering design from Activity 1 to Activity 7 revealed that the engineering design abilities of both genders gradually and continuously improved through the course.

To examine whether male and female students' performance in engineering design progressed differently across the different activities, a Wilcoxon signed-rank test was conducted between different activities (Table 7.6).

According to Table 7.6, Activities 1 to 2 had an impact for both genders (z for males $= -3.70$, $p < 0.01$, $r = 0.49$; z for females $= -4.14$, $p < 0.01$, $r = 0.62$) with a medium and a large effect size respectively, but Activities 2 to 3 and 3 to 7 had an impact only for males. Therefore, Activity 2 is the most impactful activity among all activities, additional impacts being cumulative throughout the course. The

Table 7.5 Mann–whitney U Tests for the differences between male and female performance of engineering design in different activities

Variable	Groups	*N*	*M*	*SD*	*U*	*z*	*p*	*r*
Activity 1	Male	28	1.36	0.87	299.00	−0.252	0.801	0.04
	Female	22	1.27	0.55				
Activity 2	Male	28	2.71	0.98	254.00	−1.113	0.266	0.16
	Female	22	3.00	0.76				
Activity 3	Male	28	3.21	0.79	307.00	−0.021	0.983	0.00
	Female	22	3.23	0.75				
Activity 7	Male	28	3.61	0.74	301.50	−0.170	0.865	0.02
	Female	22	3.64	0.73				

Table 7.6 Wilcoxon signed-rank tests of the male and female students' performance of engineering design in different activities

Stages between the activities	Gender	*N*	*M*	*z*	*p*	*r*
Activity 1 to Activity 2	Male	28	1.36	−3.70***	0	0.49
	Female	22	1.73	−4.14***	0	0.62
Activity 2 to Activity 3	Male	28	0.5	−2.25*	0.025	0.3
	Female	22	0.23	−1.18	0.236	0.18
Activity 3 to Activity 7	Male	28	0.39	−2.31*	0.021	0.31
	Female	22	0.41	−1.83	0.067	0.27
Activity 1 to Activity 7	Male	28	2.25	−4.50***	0	0.6
	Female	22	2.36	−4.20***	0	0.63

Note * $p < 0.05$; ** $p < 0.01$; *** $p < 0.001$

assessment in Activity 2 asked students to revise their original design drawing of a flat speaker after they had disassembled the actual speakers. The assessment in Activity 3 required students to revise their drawing of a flat speaker after they had discussed and clarified the scientific principles underlying the workings of a stereo speaker. Next, in Activities 4 to 6, students constructed their flat speakers and discussed the factors related to speaker volume based on the scientific model of electromagnetic force. Then, they selected and tested the factors that might affect the speaker volume. The assessment in Activity 7 required students to reflect on and employ what they had learned about the engineering model of a speaker and the scientific principles of the workings of a speaker and, taking the model of electromagnetic force as their reference, to create a flat speaker holiday card that could emit the maximum volume.

Hence, Activity 2 (disassembling a real stereo speaker) enabled male students' engineering design to progress from average level 1.36–2.71 and female students' engineering design to progress from average level 1.27–3.00, as shown in Table 7.5. Both female and male students' engineering designs made significant progress from around Level 1 to around Level 3. It appears that Activity 2 (disassembling a real stereo speaker) not only enabled students to start considering the essential sound-producing components of the flat speaker but also encouraged them to start associating their design with relevant scientific models through observation and analysis of the structure of the stereo speaker.

Gender Gaps in Attitude Towards STEM and STEM Learning Environments

In this study, a Wilcoxon signed-rank test was conducted on four major dimensions for mean male and female responses to the attitudinal survey. The results for the male students are shown in Table 7.7, and the results for the female students are shown in Table 7.8.

Table 7.7 Mean scores and wilcoxon signed-rank tests for male students' responses to the attitudinal survey before and after the course

Variable	Pre-test		Post-test		*z*	*p*	*r*
	M	*SD*	*M*	*SD*			
Technology	4.34	0.51	4.54	0.46	−3.16**	0.002	0.42
Engineering	4.28	0.54	4.58	0.51	−3.91***	0.000	0.52
Science	4.14	0.54	4.50	0.49	−4.12***	0.000	0.55
Classroom environment	3.99	0.78	4.54	0.66	−3.35**	0.001	0.45

Note ** $p < 0.01$; *** $p < 0.001$

Table 7.8 Mean scores and wilcoxon signed-rank tests for female students' responses to the attitudinal survey before and after the course

Variable	Pre-test		Post-test		*z*	*p*	*r*
	M	*SD*	*M*	*SD*			
Technology	4.05	0.42	4.42	0.40	−4.02***	0.000	0.61
Engineering	4.35	0.40	4.66	0.40	−3.38**	0.001	0.51
Science	4.05	0.39	4.38	0.52	−2.98**	0.003	0.45
Classroom environment	4.07	0.65	4.77	0.44	−3.41**	0.001	0.51

Note ** $p < 0.01$; *** $p < 0.001$

Tables 7.7 and 7.8 show that the results indicated that the post-course scores of the four major dimensions are significantly higher than the pre-course scores for both genders with the effect sizes between the medium and large ranges. The results therefore indicate that participating in the STEM curriculum developed by this study increased the positive attitudes of both male and female students towards STEM and STEM learning. Although one of the objectives of the course was to develop a female-friendly course, the attitude for male students also improved significantly over the course.

Moreover, this study analysed the responses to the attitudinal surveys before (Table 7.9) and after (Table 7.10) the course for both genders, and conducted a Mann–Whitney U Test for the four major dimensions to compare the differences in male and female students' attitudes towards each dimension.

In the attitudinal survey before the STEM course, male students only had a significantly more positive attitude than female students in relation to the technology dimension (see Table 7.9, $z = -2.31$, $p < 0.05$, $r = 0.33$) with a medium effect size. However, there were no statistically significant differences between males and females after the course (Table 7.10, $z = -1.32$ $p > 0.05$). Tables 7.7 and 7.8 suggest that the STEM course improved the positive attitudes for male and female students (male students' technology dimension in Table 7.7, $z = -3.16$, $p < 0.01$; female students' technology dimension in Table 7.8, $z = -4.02$, $p < 0.001$). Therefore, the

Table 7.9 Mann–whitney U test analysis of the attitudinal survey for the male and female groups before participating in the STEM curriculum

Variable	Groups	*M*	*SD*	*U*	*z*	*p*	*r*
Technology	Male	4.34	0.51	311.50	−2.31*	0.021	0.33
	Female	4.05	0.42				
Engineering	Male	4.28	0.54	449.50	−0.323	0.742	0.05
	Female	4.35	0.40				
Science	Male	4.14	0.54	423.50	−0.70	0.482	0.09
	Female	4.05	0.39				
Classroom performance	Male	3.99	0.78	336.50	−0.15	0.878	0.02
	Female	4.07	0.65				

Note * $p < 0.05$

Table 7.10 Mann–whitney U test analysis of the attitudinal survey for the male and female groups after participating in the STEM curriculum

Variable	Groups	*M*	*SD*	*U*	*z*	*p*	*r*
Technology	Male	4.54	0.46	381.00	−1.32	0.186	0.19
	Female	4.42	0.40				
Engineering	Male	4.58	0.51	451.50	−0.31	0.757	0.04
	Female	4.66	0.40				
Science	Male	4.50	0.49	401.50	−1.03	0.305	0.15
	Female	4.38	0.52				
Classroom performance	Male	4.54	0.66	266.50	−1.55	0.120	0.21
	Female	4.77	0.44				

STEM course not only significantly increased the positive attitudes of both genders in the technology dimension, it also reduced the attitude gap between the two genders seen before the course in the technology dimension.

Discussion

This experimental STEM curriculum emphasizes the teaching methods of continuously guiding students to consider the engineering models of a stereo speaker and the scientific models of sound and electromagnetic force to develop, evaluate, and revise their designs of an innovative flat speaker. This teaching strategy was found to significantly help enhance the engineering designs of both male and female students. The learning process included thinking about the structure of a flat speaker from the students' own experience, disassembling an existing stereo speaker, and using scientific principles first to design a flat speaker and then to revise and depict the design of the speaker on a holiday card. There was no significant difference between the two genders in terms of the quality of the engineering designs.

However, in terms of the improvement in engineering design ability, female students did not improve whereas their male counterparts did, between activities 3 and 7. In these activities, students were encouraged to integrate scientific models in their engineering designs. A possible explanation is that the female students may have already incorporated the scientific principles in their original engineering design, once they had done the reverse engineering of the product. For example, a study by Lou et al. (2011) observed that females have an advantage of integrating engineering and science. Hence, females' abilities in this area are worthy of future research in relation to designing and implementing STEM curricula.

In addition, the study investigated gender differences in attitudes towards STEM. In the attitudinal survey before the course, male students' attitudes in technology were significantly more positive than those of female students. There was no significant difference between the two genders in the science, engineering, and learning environment dimensions. This is consistent with the results of a previous study by American scholars on differences in attitudes of the two genders towards STEM subjects (Chen & Moons, 2015; LaCosse et al., 2016). However, after they participated in the STEM course described in this chapter, the attitudes of male and female students had no significant difference in any of these four dimensions. Additionally, the attitudes of both genders were significantly more positive in all four dimensions. Taken together, the STEM curriculum appeared to effectively reduce the gender gap in female and male attitudes towards technology.

Although many previous studies have investigated the differences in attitudes between males and females in relation to STEM, and have suggested design directions for female-friendly courses (e.g., Dancstep & Sindorf, 2018), few have proposed complete STEM courses and detailed activities that demonstrably address the gap in attitudes between genders. This study proposed detailed STEM course content and teaching methods that reduced the gender gap in relation to students' attitudes towards technology, namely, a gender-friendly STEM course.

In the past, the content of 'female-friendly courses' centred on the activities of women's daily lives, such as kitchen science, hoping to attract female participation. In contrast, the theme of this STEM curriculum was flat speakers, which emphasized the practical application of emerging technologies. This theme was not chosen based on gender stereotypes. However, the course design drew on Dancstep and Sindorf's (2018) strategies for female-friendly courses: (1) enable social interaction and collaboration, (2) create a low-pressure setting, (3) provide a meaningful application to solve daily life issues or improve life quality, and (4) include aesthetic and creative elements. Our course demonstrated a successful application of these strategies in the creation of a STEM course. The resulting course successfully enhanced not only male but also female students' attitudes and engineering design ability.

In summary, this research showed that the design and implementation of a STEM curriculum with gender-friendly and integrated cross-disciplinary learning not only enhanced both male and female students' engineering designs and attitude towards

STEM but also reduced the gender gap between male and female students' STEM learning.

References

Bell, D. (2016). The reality of STEM education, Design and technology teachers' perceptions: A phenomenographic study. *International Journal of Technology and Design Education, 26*(1), 61–79.

Blackburn, H. (2017). The status of women in STEM in higher education: A review of the literature 2007–2017. *Science & Technology Libraries, 36*(3), 235–273.

Burke, B. N. (2014). The ITEEA 6E Learning by design^{Tm} model: Maximizing informed design and inquiry in the integrative STEM classroom. *Technology and Engineering Teacher, 73*(6), 14–19.

Ceci, S. J., & Williams, W. M. (Eds.). (2007). *Why aren't more women in science?* American Psychological Association.

Chen, J. M., & Moons, W. G. (2015). They won't listen to me: Anticipated power and women's disinterest in male-dominated domains. *Group Processes & Intergroup Relations, 18*(1), 116–128.

Cheryan, S., Ziegler, S. A., Montoya, A. K., & Jiang, L. (2017). Why are some STEM fields more gender balanced than others? *Psychological Bulletin, 143*(1), 1–35.

Dancstep, T., & Sindorf, L. (2018). Creating a female-responsive design framework for STEM exhibits. *Curator: The Museum Journal, 61*(3), 469–484.

de Vries, M. J. (2018). Technology education: An international history. In M. J. de Vries (Ed.), *Handbook of Technology Education* (pp. 73–84). Springer International Publishing.

Diekman, A. B., Steinberg, M., Brown, E. R., Belanger, A. L., & Clark, E. K. (2017). A goal congruity model of role entry, engagement, and exit: Understanding communal goal processes in STEM gender gaps. *Personality and Social Psychology Review, 21*(2), 142–175.

Diekman, A. B., Weisgram, E. S., & Belanger, A. L. (2015). New routes to recruiting and retaining women in STEM: Policy implications of a communal goal congruity perspective. *Social Issues and Policy Review, 9*(1), 52–88.

Dweck, C. S. (2007). *Is Math a gift?* American Psychological Association.

Ellis, J., Fosdick, B. K., & Rasmussen, C. (2016). Women 1.5 times more likely to leave STEM pipeline after calculus compared to men: Lack of mathematical confidence a potential culprit. *PloS One, 11*(7). https://doi.org/10.1371/journal.pone.0157447.

Freeman, B., Marginson, S., & Tytler, R. (2019). An international view of STEM education. In A. Sahim & M. J. Mohr-Schroeder (Eds.), *STEM education (2nd ed.): Myths and Truths—What has K-12 STEM education research taught Us?* (pp. 350–363). Brill Sense. https://doi.org/10.1163/9789004405400_019.

Gurski, J. S., & Hammrich, P. L. (2017, June). *Examining the personal nature of the K-14 engineering pipeline for young women.* The 2017 ASEE Annual Conference & Exposition, Columbus, Ohio. https://peer.asee.org/examining-the-personal-nature-of-the-k-14-engineering-pipeline-for-young-women.

Han, S., Capraro, R., & Capraro, M. M. (2015). How science, technology, engineering, and mathematics (STEM) project-based learning (PBL) affects high, middle, and low achievers differently: The impact of student factors on achievement. *International Journal of Science and Mathematics Education, 13*(5), 1089–1113.

Han, S., & Carpenter, D. (2014). Construct validation of student attitude toward science, technology, engineering, and mathematics project-based learning: The case of Korean middle grade students. *Middle Grades Research Journal, 9*(3), 27–41.

Honey, M., Pearson, G., & Schweingruber, H. (Eds.). (2014). *STEM integration in K-12 education: Status, prospects, and an agenda for research* (Vol. 500). National Academies Press.

Jackson, S. M., Hillard, A. L., & Schneider, T. R. (2014). Using implicit bias training to improve attitudes toward women in STEM. *Social Psychology of Education, 17*(3), 419–438.

Jones, B. D., Ruff, C., & Paretti, M. C. (2013). The impact of engineering identification and stereotypes on undergraduate women's achievement and persistence in engineering. *Social Psychology of Education, 16*(3), 471–493.

Kertil, M., & Gurel, C. (2016). Mathematical modeling: A bridge to STEM education. *International Journal of Education in Mathematics, Science and Technology, 4*(1), 44–55.

LaCosse, J., Sekaquaptewa, D., & Bennett, J. (2016). STEM stereotypic attribution bias among women in an unwelcoming science setting. *Psychology of Women Quarterly, 40*(3), 378–397.

Lindberg, S. M., Hyde, J. S., Petersen, J. L., & Linn, M. C. (2010). New trends in gender and mathematics performance: A meta-analysis. *Psychological Bulletin, 136*(6), 1123.

Lou, S. J., Shih, R. C., Diez, C. R., & Tseng, K. H. (2011). The impact of problem-based learning strategies on STEM knowledge integration and attitudes: An exploratory study among female Taiwanese senior high school students. *International Journal of Technology and Design Education, 21*(2), 195–215.

Margot, K. C., & Kettler, T. (2019). Teachers' perception of STEM integration and education: A systematic literature review. *International Journal of STEM Education, 6*(2), 1–16.

Ministry of Education of Taiwan. (2018a). *Shí'èr nián guómín jīběn jiàoyù kèchéng gāngyào—zìrán kē lǐngyù* [Curriculum Guidelines of 12-Year Basic Education- Science Education]. The Ministry of Education of Taiwan.

Ministry of Education of Taiwan. (2018b). *Dàxuéshēng jiùdú lèi kē zhī bǐlǜ* [The ratio of university student enrolment by discipline]. http://stats.moe.gov.tw/files/important/OVERVIEW_U04.pdf.

Ministry of Education of Taiwan. (2019). *Gāoděng jiàoyù zhōng nǚxìng bìyè yú kēxué, jìshù, gōngchéng jí shùxué lǐngyù bǐlì* [Percentage of women graduated from science, technology, engineering and mathematics in higher education]. https://stats.moe.gov.tw/files/gender/106-32.xls.

Morgan, S. L., Gelbgiser, D., & Weeden, K. A. (2013). Feeding the pipeline: Gender, occupational plans, and college major selection. *Social Science Research, 42*(4), 989–1005.

Moss-Racusin, C. A., Molenda, A. K., & Cramer, C. R. (2015). Can evidence impact attitudes? Public reactions to evidence of gender bias in STEM fields. *Psychology of Women Quarterly, 39*(2), 194–209.

Radloff, J., & Guzey, S. (2016). Investigating preservice STEM teacher conceptions of STEM education. *Journal of Science Education & Technology, 25*, 759–774.

Raju, P. K., & Clayson, A. (2010). The future of STEM education: An analysis of two national reports. *Journal of STEM Education, 11*(5&6), 25–28.

Redmond-Sanogo, A., Angle, J., & Davis, E. (2016). Kinks in the STEM pipeline: Tracking STEM graduation rates using science and mathematics performance. *School Science and Mathematics, 116*(7), 378–388.

Robinson, J. P., & Lubienski, S. T. (2011). The development of gender achievement gaps in mathematics and reading during elementary and middle school: Examining direct cognitive assessments and teacher ratings. *American Educational Research Journal, 48*(2), 268–302.

Ryan, M. (2014). *Who is like a scientist? A self-prototype matching approach to women's underrepresentation in STEM fields* (Unpublished doctoral dissertation). University of Washington.

Seaton, G. A. (2011). *Belonging uncertainty and psychological capital: An investigation of antecedents of the leaky pipeline in STEM* (Unpublished doctoral dissertation). Purdue University.

Stephenson, T., Fleer, M., & Fragkiadaki, G. (2021). Increasing girls' STEM engagement in early childhood: Conditions created by the conceptual PlayWorld model. *Research in Science Education*, 1–18.

Su, R., & Rounds, J. (2015). All STEM fields are not created equal: People and things interests explain gender disparities across STEM fields. *Frontiers in Psychology, 6*, Article 189.

Su, R., Rounds, J., & Armstrong, P. I. (2009). Men and things, women and people: A meta-analysis of sex differences in interests. *Psychological Bulletin, 135*(6), 859.

Tiedemann, J. (2000). Gender-related beliefs of teachers in elementary school mathematics. *Educational Studies in Mathematics, 41*(2), 191–207.
Unfried, A., Faber, M., Stanhope, D. S., & Wiebe, E. (2015). The development and validation of a measure of student attitudes toward science, technology, engineering, and math (S-STEM). *Journal of Psychoeducational Assessment, 33*(7), 622–639.
Valla, J. M., & Ceci, S. J. (2014). Breadth-based models of women's underrepresentation in STEM fields: An integrative commentary on Schmidt (2011) and Nye et al. (2012). *Perspectives on Psychological Science, 9*(2), 219–224.
Wai, J., Putallaz, M., & Makel, M. C. (2012). Studying intellectual outliers: Are there sex differences, and are the smart getting smarter? *Current Directions in Psychological Science, 21*(6), 382–390.
Wang, M. T., Degol, J., & Ye, F. (2015). Math achievement is important, but task values are critical, too: Examining the intellectual and motivational factors leading to gender disparities in STEM careers. *Frontiers in Psychology, 6,* Article 36.
Wang, M. T., Eccles, J. S., & Kenny, S. (2013). Not lack of ability but more choice: Individual and gender differences in choice of careers in science, technology, engineering, and mathematics. *Psychological Science, 24*(5), 770–775.

Meng-Fei Cheng is a professor of the Department of Physics at National Changhua University of Education and leads its Physics Education Research Group. She received her Ph.D. degree from the University of Illinois at Urbana-Champaign, USA, in 2012. She has pioneered research in the fields of scientific modeling in physics and STEM education in innovative technology.

Yu-Heng Lo received his Master's degree in physics from National Changhua University of Education, Taiwan, in 2021, and currently teaches physics at Taipei First Girls High School. With funding from the Ministry of Science and Technology, he has developed and implemented a series of STEM curricula as an innovative technology. He also dedicates time to developing STEM teacher training workshops for secondary school science teachers.

Chapter 8
Online Experiments for STEM Education in Hong Kong and Mainland China: Pilot Implementation and Evaluation of a Feasible Approach in Secondary Schools

Yau Yuen Yeung, Parbat Dhungana, and Siew Wei Tho

Abstract In many disciplines of STEM education, online experiments (OEs), commonly called remote-controlled experiments, become an emerging practice in technology-enhanced learning because of their many merits perceived in educational research. We recently formulated an innovative, low-cost, and open-source approach to developing new OEs rapidly. It won a gold medal and a special prize in an international innovation and invention competition in 2018. Using this new approach, teachers and students can easily access OEs with a lightweight web browser using any hardware and any operating system without installing additional software or apps. They can observe and/or control the experiments and collect authentic data (over a very short or long time) for further analysis by plotting graphs and/or conducting calculations to obtain important results. Some well-accepted instructional approaches or pedagogies have been integrated into online science/STEM experiments, which can also be aligned with school-based curricula to provide self-contained courseware for students' online self-regulated learning. In this chapter, we will focus on whether OEs could really and effectively provide the essential elements of experimental learning to 'school students' as expected in traditional laboratory work. From our pilot implementation in four Hong Kong (HK) and Mainland China (CHINA) secondary schools with a total of 172 participants, we explored the OE experiences of different experiments and cognitive performance among students from different regions and genders using the conventional mixed methods research and design-based research. Students' feedback and suggestions for improvement of the OEs were also analysed together with some specific implications for science/STEM education.

Keywords Secondary school · STEM education · Online experiments · Science education · Remote laboratory

Y. Y. Yeung (✉) · P. Dhungana
Department of Science and Environmental Studies, The Education University of Hong Kong, 10 Lo Ping Road, Hong Kong SAR, China
e-mail: yeungy@acm.org

S. W. Tho
Faculty Science and Mathematics, Sultan Idris Education University, Tanjung Malim, Malaysia

M. M. H. Cheng et al. (eds.), *Concepts and Practices of STEM Education in Asia*,
https://doi.org/10.1007/978-981-19-2596-2_8

Introduction

A remote laboratory (also known as remote-controlled, remotely accessed, web-based, or online experiment or laboratory) refers to a learning environment in which students can control, observe, manipulate, and interact with physical experiments from a distance (Abrahams & Reiss, 2012; Crippen et al., 2012; Di Paolo et al., 2003; Tho & Yeung, 2016, 2018). Its main advantage is that it enables students to conduct real-time and real-world experiments anywhere and at any time, extending their laboratory practice beyond the school setting and hours (Scanlon et al., 2004). Other advantages include cost savings and global access to laboratory facilities, space, and technical support and management of resources as well as the elimination of many safety issues for students (Grout, 2017; Lowe et al., 2013; Tho & Yeung, 2016, 2018).

Conducting a Google Scholar search for the phrases 'remote labs' and 'remote experiments', Dintsios et al. (2018) found 2,890 and 3,200 results, respectively, for the years between 2010 and 2018. The figures are nearly triple those for the previous 2 decades combined, showing the growing importance of this field and the rapid rise in the number of research and development projects and publications. The reasons are likely to be the worldwide popularity of the Internet, lower costs for the required facilities, and easier and more diverse methods of system development for applications in various subject disciplines. For example, National Instruments' expensive proprietary software LabVIEW (Ertugrul, 2000) and related hardware have dominated the development of remote laboratories for around three decades. However, in the last decade, a number of open-source or alternative platforms have become available for educational use, such as (i) VISIR (Virtual Instrument Systems in Reality), developed by the Blekinge Institute of Technology for experiments in electric circuits (Tawfik et al., 2013); (ii) the OLAREX (Open Learning Approach with Remote Experiments) project consortium, established to apply ICT-based learning materials and methodologies together with remote laboratories in secondary schools (Dziabenko et al., 2013); (iii) FARLABS (Freely Accessible Remote Labs) for school students to control and interact with equipment located at three Australian universities (Hoxley et al., 2014); and (iv) iLab, which supports Internet-accessible laboratories for sharing among schools and universities (Harward et al., 2008). Many remote laboratories, such as iLabs, WebLab-Deusto, NetLab, Remote Labs, LabShare, and UNILABS, have also been developed by individual universities (Grout, 2017).

Currently, learning through experiments is a standalone learning strategy for STEM subjects; not all schools may have the capacity to give their learners this experience, and school students may not be able to perform experiments in person at laboratories for various reasons, such as the school closures caused by the COVID-19 pandemic (United Nations Educational, Scientific and Cultural Organization [UNESCO], 2021). However, technology-enhanced learning processes via remote laboratories can help to overcome these barriers (Tho & Yeung, 2016). Online experiments (OEs) (the design will be elucidated later) do not replace physical laboratory work but rather become an alternative way to conduct experiments, enabling students to experience real-time practical learning at their convenience (time and place) via the Internet. Students are able to follow all the steps of experimental learning via OEs, including initiating a real experiment, making peer observations, taking technical readings, plotting graphs automatically, analysing the data, and drawing conclusions. These process engagements and their responses before and after their participation in OEs will provide reliable data to answer the basic educational research question: How do students experience and evaluate the use of OEs in STEM/science learning?

Literature Review

Educational Research on Remote Laboratories

How popular are the educational use of remote laboratories and the relevant research on their educational effectiveness and impacts? To answer this basic question, Lowe et al. (2013) examined around 400 peer-reviewed publications on the educational use of remote laboratories over the previous decade and found that a great majority targeted students in tertiary education, many focused on engineering education (especially control and electrical engineering), and a few were on physics (including topics on mechanics, optics, and electronics). Their literature review showed that 85% of reviewed articles focused on the technical aspects (design and implementation) of remote laboratories, while the remaining 15% dealt with the conceptualization and evaluation aspects. Using several well-known papers in this field, we looked at their citations on the Google Scholar site and retrieved 167 journals, conference proceedings, or doctoral thesis publications from the period 2013–2018 that contained substantial components on educational research or the evaluation of remote laboratories (rather than the pure design and development of experiments). Around 14% were related to the use of remote laboratories in schools (two articles on primary education and 22 on secondary education), and many simply applied their original undergraduate physics or electrical experiments for use in high schools with few or no pedagogical modifications.

Comparing the learning effectiveness of remote laboratories and traditional hands-on laboratories, Ma and Nickerson (2006) found that they were equivalent to the examination of four relevant research papers involving 12–40 participants each. Their review also suggested that a mixture of approaches (e.g. remote and hands-on laboratories) might be superior to any single approach. Corter et al. (2011) reported a large-scale and multi-year study (with 458 students) involving two experiments in mechanics and found that a remote laboratory was better for individual data collection than group or individual hands-on experiments. Recently, Brinson (2015) carried out a review of the empirical research on learning outcomes in non-traditional (virtual and remote) versus traditional hands-on laboratories (2005–2015). He concluded that, in most studies (89%), students in virtual or remote laboratories demonstrated equal or higher achievement than those in hands-on laboratories across all categories of learning outcomes. However, most of the studies (95%) were related to content knowledge only. Another large-scale study (with 471 students) on the educational impact of a remote laboratory called VISIR was recently carried out by Viegas et al. (2018) on some undergraduate engineering courses. They found that their remote laboratory was more useful for basic courses than advanced ones, and that students with more learning needs could benefit more from VISIR.

Research Gaps and Research Questions

From our recent (as summarized above) and past (Tho et al., 2017) literature reviews, we identified the following major gaps in educational research into remote laboratories:

(G1) Insufficient studies have been performed on school students as most have targeted university students;
(G2) Few publications have evaluated learning effectiveness as most have focused on the technical issues arising from the design and development of OEs;
(G3) Far fewer OEs and related evaluations have been devoted to biology and chemistry than to physics or engineering subjects;
(G4) Most evaluations have been restricted to the assessment of students' content knowledge or cognitive understanding, with few findings on particular students' skills and attitudes.

To fill some of the above research gaps, this chapter sets out to provide answers to the following research questions:

- What are students' prior experiences of, current views on, and receptivity towards OEs?
- What is their cognitive development as induced by OE activities?

The findings are part of a larger study which also looked into whether there are significant differences between male and female students and between students from Hong Kong and Mainland China in their views on and receptivity towards OEs, and the likely reasons for those differences, if any. Due to the word limitation, the complete set of findings will be presented in a journal paper.

Learning Materials and Research Methods

This research study adopted a mixed methods approach to gain broader perspectives (Creswell, 2003; Johnson & Onwuegbuzie, 2004). The study followed diverse approaches of nomothetic and ideographic tasks in a sequence (Morse, 1991). It used two basic designs in its two distinct stages; the first stage adopted the design-based research (DBR) framework (see e.g. Amiel & Reeves, 2008; Anderson & Shattuck,

2012; Kong et al., 2009). The second stage followed a quasi-experimental design, collecting responses through a survey as well as interviews (Campbell & Stanley, 1963) after students had conducted the (two) newly designed OEs.

Framework for Design and Learning Material Preparation

The design and set-up process for OEs followed the design-based research (DBR) protocol for science education, as suggested by Tho and Yeung (2016), for designing and developing the hardware, software, and courseware of the relevant OEs as well as studying the context, identification, design, and development for school STEM learning. This DBR framework embraces cycles of design, enactment, analysis, and redesign (Collins, 1992; Collins et al., 2004; Tho & Yeung, 2016), and we iterated several cycles of pilot trials for the collection of data to redesign and refine the hardware, software, and courseware of the OEs. The experimental hardware and courseware delivery server were located at the Education University of Hong Kong (EdUHK). The present approach has unmatched advantages over other research designs for science education to address the gap between research and praxis in science education (Juuti & Lavonen, 2006). It was based on a newly formulated, innovative, low-cost, and open-source approach to developing new OEs (see Fig. 8.1 for the core ideas). It led the first author to win a gold medal and a special prize in the *2018 International Invention Innovation Competition in Canada,* confirming the STEM nature of our DBR work. We labelled our laboratory 'online' rather than 'remote' to highlight the almost exclusive use of open-source internet (or web-based) technology in its development and delivery. In this new approach, teachers and students can easily access OEs with a lightweight web browser using any hardware (personal computer, notebook, tablet, or smartphone) and any operating system (Windows, Linux, Unix, Android, iOS, etc.) without the need to install any additional software or apps. Learners can observe and/or control the experiments, collecting authentic data (over a very short or long time) for further analysis by plotting graphs and/or conducting calculations to obtain important results. The study started with the identification of relevant OE topics and activities for secondary school students as found in the national science curricula and trends of education reform (Curriculum Development Council [CDC] & Hong Kong Examinations and Assessment Authority [HKEAA], 2015, 2018; Lu & Liu, 2012; Ministry of Education [MOE], 2018a, 2018b; Wei, 2020). The crystallization and effect of light on the CO_2 uptake of plants were selected as relevant OEs for secondary students in both regions who should have some preconceptions of these topics from their early grades. The design of these OEs was different from that of conventional experiments conducted in schools as repeating the same experiments would not build a fresh experience of OEs. Controlling the experiment, making observations, noticing changes over time, and discussing experimental learning contributed to higher cognition, raising students' understanding of science to a new level (Havu-Nuutinen, 2005). The OE platform was installed with these steps and concepts for the development of the hardware, software, and courseware.

For the delivery of the OEs for students' learning, we constructed the experimental set-up and placed it in a laboratory on the EdUHK campus (the left-most step in Fig. 8.1). Students control these experiments with their own computing devices at home/school (the right-most step in Fig. 8.1) using any browser with internet access. Of course, only one learner is allowed to control one set of OEs at a particular time. The information is mediated online (as indicated by the arrows in Fig. 8.1), whereas actions are initiated by the Arduino platform and the corresponding sensor, forming an 'online experiment'. The Arduino platform consists of an Arduino Mega board, a LAN shield, some relays, and an electrical power supply with a general-purpose operating program for the following actions:

- taking measurements from sensors at specific time intervals
- controlling the relays for the power supply to the individual apparatus
- receiving and executing commands from the server
- returning and automatically uploading data to the server.

Different sensors can be incorporated into this platform for measuring specific physical properties, like temperature, humidity, pH level, light intensity, electrical current and voltage, concentration of gases, and so on. Besides, servo motors, step motors, pumps, and switches could be installed in this system for control purposes, such as powering on or off different coloured lights and electrical ovens, raising/lowering test tubes, and so on.

An open-source Apache web server was installed to function as a secure agent for the internal experimental set-up and Internet protocol (IP) camera and to relieve the workload and memory/storage requirement for the Arduino board. The EdUHK Intranet ensured maximal security and data transfer speed. The server also delivered all the learning activities, managed by several thousand lines of HTML and php codes, which were also specifically written by the project leader for each courseware. A learner with a PC or mobile device can use an ordinary web browser to access and control the Arduino system via the above-mentioned server for observing and analysing data from this OE anytime and anywhere.

The two OEs used in this study follow the same principle of design and similar ways of development as worked out by two master's degree students as part of their research projects under the supervision of the first author, who was also responsible for all the programming tasks. Besides, a research assistant helped the first author to carry out the initial testing of the OE idea and the final refinement of the experimental set-up. A brief description of each OE will be elaborated below.

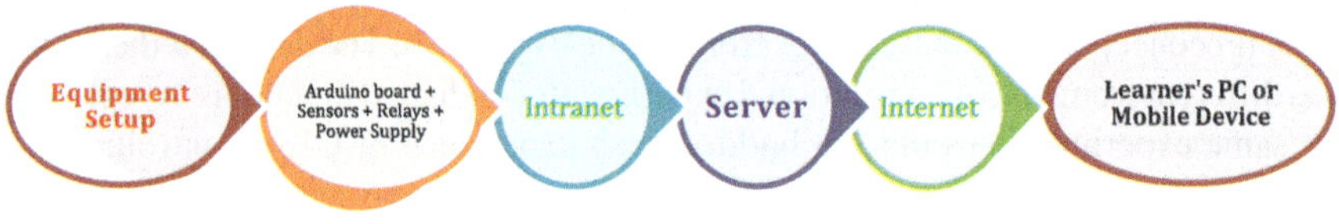

Fig. 8.1 A schematic diagram of the common framework of delivering an online experiment (Yeung et al., 2019)

Crystallization OE

The crystallization process from a supersaturated solution for sodium acetate is an exothermic process with the possibly of giving learners different experiences from most crystallization experiments that they know of (CDC & HKEAA, 2018; MOE, 2018b; Wei, 2020). The OE set-up followed the same architecture (Fig. 8.1). The set-up for the crystallization process OE is shown in Fig. 8.2.

In each experimental set-up, sodium acetate is put into the right-hand-side test tubes, while two types of control solution (saturated copper sulphate and brown sugar) are placed into the left-hand-side test tubes. Through remote control of the servo motors and relays, all the test tubes can be lowered into their respective beakers of water, which will be heated up by the heater underneath. When the test tubes are raised above the beakers, the two adjacent electric fans can be switched on to cool them further. A temperature sensor that takes a reading every 10 s is immersed into both (sodium acetate and control) solutions to record the change in their temperature. IP cameras capture photos of the ongoing experiment and make it available for the students to view the phenomenal results instantaneously and to help control/conduct the experiment (Tho & Yeung, 2015).

Effect of Light on the CO_2 Uptake of Plants

The participating students had a basic understanding of life processes in living entities before they entered secondary school (CDC & HKEAA, 2015; Lu & Liu, 2012; MOE,

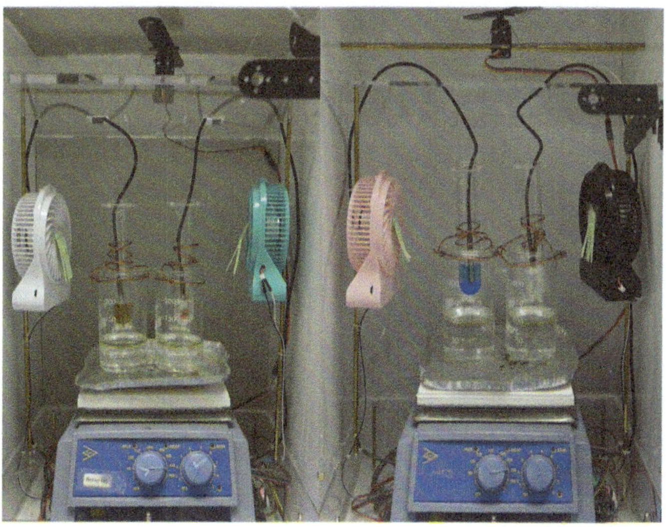

Fig. 8.2 Photo of two sets of OEs for investigating exothermic crystallization of sodium acetate from its supersaturated solution

2018a). This experiment aims to demonstrate the effect of different colours of light on the carbon dioxide (CO_2) absorption rates of plants. The set-up for this experiment is shown in Fig. 8.3, in which there are five LED lamps (red, green, blue, and white). Plants with green leaves and red leaves are placed inside individual jars. For direct comparison, two plants with the same leaf colour are housed in each set but one experimental jar is illuminated with a chosen colour of light (red, green, or blue) or no light while its counterpart for control receives white light. The results of the two set-ups can further be compared to investigate whether the colour of the leaves has any significant effect on the rate of CO_2 uptake.

To confirm that the OE set-up performed smoothly, a number of trials were undertaken by asking EdUHK undergraduate students to carry out experiments and give feedback based on their experiences. This helped to improve the OE settings. Multiple set-ups for each OE were added to facilitate more student participation in the experiment at the same time.

Research Instruments and Data Collection

A total of 172 students from four secondary schools (two from Mainland China and two from Hong Kong) participated in the research. The two public secondary schools from Mainland China are located in Zhejiang Province in the coastal region of Eastern China. They both completed the OE activities within 2 weeks in June 2019.

Fig. 8.3 Photo of two sets of OEs for investigating the effects of different coloured lights on the uptake of CO_2 by plants with green and red leaves

One of them is a key provincial high school in a first-tier city, and it often follows the roadmap of educational reform by piloting small-class teaching. The participants came from a class of 35 science students in Grade 10. The other one is an ordinary middle school in a second-tier city, and 37 students from a Grade 8 class participated in this study. Of the two participating secondary schools in Hong Kong, one is a band one co-educational school from which the participants consisted of 55 students from two Grade 8 classes (in August 2019) and 20 students from a Grade 10 class (in Nov 2019). The other one is a band two girls' school from which there were a total of 25 participants from different classes from Grade 7 to Grade 11 (in July 2019). Before conducting any OE activities, the aspects of integrity regarding voluntary participation as well as easy withdrawal of participation were communicated to the schools, facilitating the involvement of science teachers, students, and their parents. Then, students or their parents were required to complete and sign a consent form to confirm their voluntary participation in this study. Students participating in OEs undertook a sequence of tasks to conduct the OE activities and a related educational evaluation of their learning outcomes (see Fig. 8.4).

A short set of pre-quiz and post-quiz questions on the participants' conceptual understanding of the relevant science topic were specifically embedded into the online courseware for the participants to attempt before and after performing the core OE activities. The question set was adapted from an early study to contextualize undergraduate students' experiences for school students (Yeung et al., 2019). The original questionnaire was developed for evaluating students' technology-enhanced learning outcomes in science education, including the responses of the undergraduate students participating (Garfield, 1998) in the iteration cycle of OE development. The revised questionnaire instrument is composed of four sections. Section A was used to collect some basic information about the participants, such as their school or region, gender, and grade level. Sections B and C were key sections (two main constructs) of this study to measure participants' prior learning experiences/perceptions of and attitudes towards science or OE and their current learning experiences of OE, respectively. To represent students' responses for subsequent quantitative analysis, 5-point Likert scales (1—*strongly disagree*, 2—*disagree*, 3—*neutral*, 4—*agree*, and 5—*strongly agree*) were used for the items. Section D comprised two open-ended questions for the participants to give their views on the improvement of the OEs and an open remarks option. As part of the quantitative and qualitative research steps (McKenzie

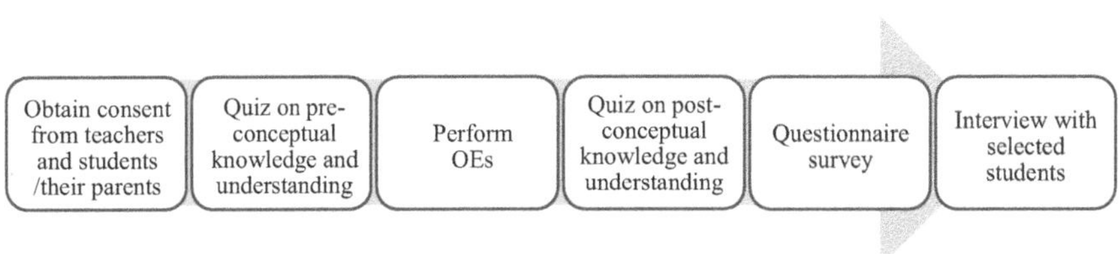

Fig. 8.4 Sequence of tasks for students to complete during the pilot implementation

et al., 1999), the authors and a few research project students in a Master of Education program formed a jury of experts in science education to review and ensure the face validity of the questionnaire instrument. The research tool had also undergone pilot studies in a teacher education university and a few secondary schools. We had employed the factor analysis method and Cronbach's alpha reliability to check the data as collected in the previous studies, and found that both the factor structure and internal consistency of the data were highly satisfactory.

All these data (Fig. 8.4) were collected through the OE learning platform. Randomly selected students (15 from Hong Kong and 13 from Mainland China) were invited to an interview after they had performed their OE. The interviews were conducted either face to face or through voice/video call over internet protocol applications (using either WeChat or WhatsApp) depending on the student's choice. Microsoft Excel was used to compute some basic statistical measures and to group the qualitative data. For advanced statistical analysis (such as principal component factor analysis, Cronbach's reliability, the test of normality, and the Mann–Whitney U test) and basic descriptive statistics, the SPSS (Version 25) software was employed.

Findings

Survey Responses: General Profile

The study measured two constructs through self-administered online questionnaires, namely students' prior learning experiences and related attitudes (B) before conducting the OE and their learning experience with the OE (C) after the experiment. Before moving to the main constructs, Table 8.1 provides a general profile (A) summary of the participating students.

The majority of the students were female (55.2%), and more students participated in the biology (the effect of light on the uptake of CO_2) experiment (55.7%) than in the chemistry (crystallization) experiment (only 44.3%). There were 210 instances of participation from the secondary levels (as requested, participation and confirmation were communicated via schools). The majority of the students were from Secondary

Table 8.1 General profile of the participating students

Demographic distribution and experiment type taken by the participants								
Region			Gender of participants			Type of experiment		
CHINA*	HK*	Total	Male	Female	Total	Crystallization	CO_2 uptake	Total
123	87	210	94	116	210	93	117	210
58.6%	41.4%	100%	44.8%	55.2%	100%	44.3%	55.7%	100%

* *Remark* Two schools each from Mainland China (CHINA) and Hong Kong (HK) participated in the study, and there were 72 and 100 students from Mainland China and Hong Kong, respectively, each of whom might carry out one or more OEs as indistinguishable participants. The figures in this table actually refer to the participation in different OEs rather than to individual students

2 (Grade 8) and Secondary 4 (Grade 10). A number of them had completed all the activities (including the questionnaire survey) of the two OEs and counted as two participants.

The Cronbach's alpha reliability for items in Sections B and C is 0.72 and 0.94, respectively. As shown in Table 8.2, the first construct aimed to measure prior learning experiences and was related to respondents' attitude towards or prior experience of five items (referred to as B1 to B5 later in this section). The study received 184 valid responses (Table 8.3) for the first construct. The mean value range, from 3.77 to 3.92, for B1, B2, and B4 indicated agreement with these item statements. For B3 and B5, the mean values were 3.10 and 2.92, respectively, so they indicated a neutral response to these item statements.

Table 8.2 Students' prior learning experience and attitude towards science and OEs

	N	*M*	*Md*	*Mo*	*SD*
B1: I like science more than other subjects	204	3.77	4	3	0.967
B2: I like to do scientific investigation activities	205	3.92	4	4	0.941
B3: I have rich experience of conducting scientific experiments before	203	3.10	3	3	0.972
B4: I have little or no prior experience of conducting OEs by myself	203	3.83	4	4	1.058
B5: I am familiar with the science topic related to this OE	186	2.92	3	3	0.873
Valid N (listwise)	184				

N—valid responses for each item, *M*—mean, *Md*—median, *Mo*—mode, *SD*—standard deviation from the mean

Table 8.3 Students' experience of science learning with OEs

	N	*M*	*Md*	*Mo*	*SD*
C1: I can conduct the OE activities smoothly, as I expected	203	3.61	4	4	0.970
C2: The objectives and the instructions of the tasks are clear	204	3.85	4	4	0.921
C3: The flexibility of conducting the OE anytime and anywhere is beneficial to me	200	3.91	4	4	0.906
C4: Without spending time on the experimental setup, I could focus my time and efforts on conducting more meaningful activities, such as the collection and analysis of the experimental data	203	3.74	4	4	0.977
C5: It is helpful that the system can plot the graphs for me automatically	203	4.19	4	5	0.887
C6: The OE can deepen my understanding of the science topic concerned	202	4.01	4	4	0.875
C7: I would like to have OEs offered in different courses	199	4.04	4	5	0.958
Valid N (listwise)	195				

N—valid responses for each item, *M*—mean, *Md*—median, *Mo*—mode, *SD*—standard deviation

The participating students in general have a high level of interest in science (B1; $M = 3.77$); more specifically, learners like engaging in investigation activities (B3; $M = 3.92$, $Mo = 4$) even though they have had limited exposure to experiments, including OEs (B4; $M = 3.83$, $Mo = 4$).

Students' experiences with OEs were the second and most important construct of the study. The construct had seven items (referred to as C1 to C7 below). Learners responded after conducting the OEs, and the responses are shown in Table 8.3.

The participating students in general considered the OEs to be highly supportive, fostering conceptual clarity and scientific observation skills (C6; $M = 4.01$). Interestingly, the feature of the OEs that the students most enjoyed was the automated graph summarizing the experiment (C5; $M = 4.19$, $Mo = 5$). This is consistent with the previous finding that manual graph plotting is actually a heavy burden and a great challenge for many students as they need to exert three to four times as much effort on their laboratory work to plot the graphs manually without using a computer, and they often make the wrong choice of coordinates and scale of coordinates (Barton, 2004).

The students clearly reported their interest in studying more lessons through OEs (C7; $M = 4.04$, $Mo = 5$). Factors like flexibility (C3; $M = 3.91$), instructional clarity (C2; $M = 3.85$), and the ready-to-operate experimental set-up (C4; $M = 3.74$), in decreasing order, still enriched their experiences positively, showing that this exposure helped them to develop their confidence (C1; $M = 3.61$) in OEs. A further discussion of these findings is presented in the discussion section.

Survey Responses: Comparative Statistics of OE Experiences

Besides looking for the general perspectives and learning experiences of students, we focused on the comparative perspectives among students of two different regions, students performing different OEs, and different genders' experiences of OEs. Normality tests (Ghasemi & Zahediasl, 2012) were conducted on the data collected and showed a non-normal distribution, largely due to the high degree of skewness found in the data. Therefore, a non-parametric Mann–Whitney U test was applied to compare the regions and genders (rather than the usual *t* test for bypassing the normal distribution assumption) in accordance with Ghasemi and Zahediasl's (2012) statement that 'assessing the normality assumption should be taken into account for using parametric statistical tests' (p. 489). The detailed results of the quantitative analysis will be presented in another journal paper which will also discuss the comparative perspectives for the two constructs: students' prior learning experiences and related attitudes and their learning experiences.

There are no significant differences, when students engaged in the two OEs, in their responses regarding their prior learning experience and attitude (construct B) or in their experiences (construct C) of OEs. The segregation by region and gender also does not show significant differences in prior and post experiences of conducting these two OEs. We carried out the relevant Mann–Whitney U test for equality of means, but the findings of null differences were not informative for presentation in this chapter.

Cognitive Performance in the Pre-quiz and Post-quiz

This study did not have a major focus on determining whether there had been any addition to content learning as it was conducted across a wide range of secondary level (Secondary 1–6) students. A short pre- and post-quiz on students' conceptual understanding of the science concepts related to the topic (four to five questions for each quiz) were available online for voluntary responses. Almost all the students attempted them, but there were just one to two questions for each OE that we could use to compare the overall knowledge gain of the participants. For the other questions, the content of the questions and their level of difficulty were so different that we could not make direct comparisons. The three questions for which we could compare the students' knowledge gain and the reduction in their lack of knowledge are presented in Table 8.4.

Question A was set for the OE on CO_2 uptake in plants, and in fact two separate questions were set in the pre-quiz, one for the green plants and another for the red plants. However, only one post-quiz question was posed to the respondents depending on the colour of the plants actually used in their OE activities. Questions B and C were used in both the pre-quiz and the post-quiz (with minor rephrasing of the sentences) of the OE on crystallization. The detailed findings on how the regional and gender

Table 8.4 Questions selected from the pre-quiz and post-quiz of the two OEs for direct comparison of students' cognitive performance

Q#	Question	Answer choices
A	When four different colours of lights (red, green, blue, and white) are separately shining on green (or red) plants, the largest decrease in the carbon dioxide level is due to:	a. Red light b. Green light c. Blue light d. White light e. I don't know
B	The temperature change during the crystallization of the supersaturated sodium acetate solution can be described as:	a. Liquid turned into solid 'condensation' due to the decrease in temperature b. Liquid turned into solid with the increase in temperature c. No way to predict the temperature change, which can increase or decrease d. I don't know
C	What is the reaction during the crystallization of the supersaturated sodium acetate solution?	a. Endothermic reaction b. Exothermic reaction c. No way to predict d. I don't know

factors affected the students' cognitive performance will be presented in a journal paper.

Qualitative Responses

Two instruments (open-ended questions in the online questionnaire survey and a semi-structured interview) were used to collect the qualitative responses of the students after the OEs. Open-ended questions were purposively designed to receive comments and feedback from the students with the aim of improving the experiment. Besides, two open-ended questions (Section D) were incorporated to elicit the students' view immediately following the set of post-OE experience items. While the interview data will be presented in a journal paper, the findings of the two open-ended questions are summarized below.

Open-Ended Questions

Many students gave a variety of feedback and comments on their experiences as well as suggestions for enhancing the learning effectiveness of the OEs. The responses of the students were thematically grouped (coded) under the following 10 codes, disaggregated by region and OE (Table 8.5).

Many of the students who responded to this open-ended section started their responses with compliments about their OE experiences. The responses indicated that improvements in the instructions (37%) with attractive and clear visuals (18%) on a user-friendly web interface (11%) would address most of the issues. They also suggested giving more and clearer information or instructions in the courseware. In addition to the identification of some technical problems, the students also complained of long waiting times (12%) as well as conflicts of time (1%) in their negative feedback.

Discussion

There are no controversies among teachers and students regarding laboratories as an effective venue and experiments as an appropriate tool for science teaching (Solomon, 1988). The use of technologies can provide ways to upgrade our existing educational models and bring innovation to pedagogy (Laurillard, 2008). OEs are a form of experimental learning strategy facilitated by innovative technology (Tho & Yeung, 2018; Tho et al., 2017). OEs can be an exciting learning opportunity at schools where the school curriculum time cannot accommodate all the learning domains of conventional experimental learning due to a wide variety of experiments covered in individual science subject curricula. From an educator's point of view, it is very

Table 8.5 Feedback and comments given by the participating students after conducting the OEs

SN	Feedback/comments for improvement	HK students		CHINA students		Sub-total	
		Crystallization	CO_2 uptake	Crystallization	CO_2 uptake	Count	%
1	Instructional clarity (good/easy/convenient/understandable, plus room for improvement)	7	24	17	42	90	37.3
2	Informative aspect (more/clearer information/instructions)	2	5	5	9	21	8.7
3	Adding visual effects (more photos/videos)	4	11	11	17	43	17.8
4	Technical aspect (bug fixing/login problems/internet connection)	3	1	4	7	15	6.2
5	Problem with a long waiting time	8	10	5	6	29	12.0
6	Design and functions (simplify the web page interface, adding a forum/platform to exchange experiences)	0	8	9	9	26	10.8
7	Time conflicts (more students to conduct OEs at the same time)	0	0	1	1	2	0.8
8	Request for a mobile application	1	1	1	0	3	1.2
9	Request for developing more OEs	1	0	7	2	10	4.1
10	Other	0	2	0	0	2	0.8
Sub-total		62	26	60	93	241	100

important to identify the most effective strategies that facilitate students' learning with OEs. Students prefer biology to chemistry OEs, which follows the sequence in which science is introduced in schools from biology to other (physics, chemistry, etc.) sub-disciplines of science (CDC & HKEAA, 2015, 2018). This substantiates that learning comfort, along with prior knowledge, is important to proceed further in teaching and learning (Garner, 2008). To stimulate students' interest in science, a feasible approach is to focus on the experimental designs and scientific explanations, including subject-specific interest, irrespective of students' prior exposure (Krapp & Prenzel, 2011; Tho et al., 2015). This could be one of the reasons for the mixed responses from the learners.

Based on the present findings, the OEs were a new experience for most school students as they learned both the science concepts and the design of new experiments in a new mode of learning—a remote laboratory. The findings also indicated that OEs can enhance students' self-learning confidence as they helped them to construct their own understandings of the science concepts underlying the OEs, rendering OEs an effective pedagogical method in STEM education for secondary students.

Looking at the comparative perspectives, Hong Kong students gained richer learning experiences from conducting scientific experiments. Many studies and comparisons of scientific achievements, such as PISA 2018 (Organisation for Economic Co-operation and Development [OECD], 2019) on 'scientific literacy', have elucidated that Hong Kong students have performed well 'above the average' for a long time. The attitude of Hong Kong learners remains consistent with this (Yeung & Cheng, 2018). There is insufficient information to make a temporal comparison of science attitude and interest for students from Mainland China (Yeung, 2015). The OE exposure for Chinese students was a new experience, and it showed a greater and more favourable experience than for the Hong Kong students. Experimental learning remains an efficient way to learn science, as discussed in relation to the theoretical perspectives, so OEs could help to maintain this mode of learning during the class suspension resulting from the Covid-19 pandemic.

The gender comparison in general showed no difference in the attitude towards and experience of OEs. The regional segregation showed that the male students in Mainland China had better experience with OEs, whereas the reverse was found in Hong Kong. Besides, the learning experiences of the genders among Hong Kong students remained the same, as supported by a previous study about applying an innovative approach to science learning in a thematic park (Tho et al., 2015). This learning experience did not differ between genders in Mainland China either.

Regarding the cognitive development induced by the OEs, the students' knowledge gain was substantial and the reduction in students' lack of concepts in the OE topics was amazing. Hong Kong students often had a much better academic performance in the pre-quizzes as they might have had more chances to conduct experiments. However, there was no statistically significant difference in the knowledge gain between Hong Kong and Mainland China students, while Hong Kong students had a slightly larger reduction in lack of concepts than that of Mainland China students. Regarding the gender difference, girls seem to have acquired much more knowledge than boys in the chemistry OE, while the reverse situation holds for the biology OE. A likely reason is that the chemistry OE involved some complicated procedures and instructions, and boys were often not patient enough to complete the experiment with careful observation of the changes in the temperature. However, further study will be required to confirm whether patience is a likely reason. On the other hand, there was some variation in the students' change in lack of concepts, but the gender difference was not significant.

Limitations

The study captured the scenarios of only four schools (two in each region) and thus has limited generalizability. A scaled-up and/or context-based study will produce more research findings to further define OEs' types, forms, and compatibility for different levels of secondary school students across different regions. Besides, the cognitive questions set in the pre-quiz and post-quiz did not match well and had different levels of difficulty. Consequently, we could not use all the questions from the quizzes of each OE to evaluate the students' knowledge gain and reduction in the lack of science concepts related to the OE topic concerned.

Conclusions and Implications

The application of technology in science education is advancing so rapidly that it has become an essential tool for improving the quality of science/STEM education. Various traditional scientific experimental tools are being replaced by digital automated tools, and one of their forms is OEs, which already had a fairly widespread level of acceptance in higher education (Yeung, 2020). The design and development of OEs obviously require an intensive integration of various knowledge and skills in various STEM disciplines and so the OEs themselves could effectively induce students to appreciate and understand the integrative nature of STEM education. OEs' processes may differ from those of virtual experiments and real experiments. This study profoundly showed that secondary students accepted the OEs and received good learning experiences from them; for instance, the instructional clarity, interactive web interface, and legible images are the most important factors to boost students' interest in OEs. Additionally, the main assets are no set-up hassles, auto data recording and archiving, and safety assurance, which were also well recognized by many students. The present study revealed that OEs are especially suitable for experiments that require extra precautions; take a long time for completion; are expensive to conduct in multiple settings; are being conducted in times of crisis, like the Covid-19 pandemic, when laboratory access is restricted; and for schools that are unable to offer laboratory practice for students' learning. The study substantiates experimental learning as an effective strategy in STEM learning (as the above discussion section showed). This study also considered the level of the students when designing the

OEs, such as the curriculum, medium of instruction, OE platform design, pedagogic culture, pupils' pre-conceptual knowledge, and so on. Students' cognitive development in the relevant science topic was clearly enhanced, with significant knowledge gain and a great reduction in the lack of concepts. Despite OEs taking learners through the engagement and reasoning in science and STEM learning, not all experiments may be compatible with OEs to facilitate the same level of learning. Educational communities are required to develop further and extend research on OEs for school education, which would undoubtedly make OEs an alternative learning strategy for STEM education at the secondary level. Furthermore, the study must be scaled up for generalizability and to outline the list of experiments (for coursework) that can be undertaken as OEs within the aim of the curriculum.

Acknowledgements The financial support from the Dean's Research Fund of FLASS (DRF/IDS-3) and the Internal Research Grant from the Committee on Research and Development, the Education University of Hong Kong, is gratefully acknowledged. The Research Grants Council of Hong Kong is thanked for the provision of a Hong Kong PhD Scholarship to Parbat. Besides, we would like to express our gratitude to the following persons for their different contributions: Ms XU Chang and Ms WANG Jingjing for developing the OEs and Dr SUN Daner, Mr XUE Song, Dr LEE Lit Hong, Ms CHENG Tsz Ki, Mr WU Qi, and Mr FENG Jianfeng for their help in facilitating the school implementation.

References

Abrahams, I., & Reiss, M. J. (2012). Practical work: Its effectiveness in primary and secondary schools in England. *Journal of Research in Science Teaching, 49*(8), 1035–1055.

Amiel, T., & Reeves, T. C. (2008). Design-based research and educational technology: Rethinking technology and the research agenda. *Educational Technology & Society, 11*(4), 29–40.

Anderson, T., & Shattuck, J. (2012). Design-based research: A decade of progress in education research? *Educational Researcher, 41*(1), 16–25.

Barton, R. (2004). *Teaching secondary science with ICT*. McGraw-Hill Education (UK).

Brinson, J. R. (2015). Learning outcome achievement in non-traditional (virtual and remote) versus traditional (hands-on) laboratories: A review of the empirical research. *Computers & Education, 87*, 218–237.

Campbell, D. T., & Stanley, J. C. (1963). *Experimental and quasi-experimental designs for research*. Houghton Mifflin Company.

Collins, A. (1992). Toward a design science of education. In E. Scanlon & T. O'Shea (Eds.), *New directions in educational technology* (pp. 15–22). Springer.

Collins, A., Joseph, D., & Bielaczyc, K. (2004). Design research: Theoretical and methodological issues. *Journal of the Learning Sciences, 13*(1), 15–42.

Corter, J. E., Esche, S. K., Chassapis, C., Ma, J., & Nickerson, J. V. (2011). Process and learning outcomes from remotely-operated, simulated, and hands-on student laboratories. *Computers & Education, 57*(3), 2054–2067.

Creswell, J. W. (2003). *Research design: Qualitative, quantitative, and mixed methods approaches* (2nd ed.). Sage Publications, Inc.

Crippen, K., Archambault, L., & Kern, C. (2012). The nature of laboratory learning experiences in secondary science online. *Research in Science Education, 43*(3), 1029–1050. https://doi.org/10.1007/s11165-012-9301-6.

Curriculum Development Council & Hong Kong Examinations and Assessment Authority. (2015). *Biology curriculum and assessment guide (secondary 4–6) 2007 (with updates in Nov 2015)*. The Education Bureau HKSARG.

Curriculum Development Council & Hong Kong Examinations and Assessment Authority. (2018). *Chemistry curriculum and assessment guide (secondary 4–6) 2007 (with updates in June 2018)*. The Education Bureau HKSARG.

Dintsios, N., Artemi, S., & Polatoglou, H. (2018). Acceptance of remote experiments in secondary students. *International Journal of Online Engineering, 14*(5), 4–18.

Di Paolo, T., Scanlon, E., Colwell, C., Cooper, M., Uren, V., & Jelfs, A. (2003). Redesigning practical work: Web-based remote experimentation. In R. Holliman & E. Scanlon (Eds.), *Mediating science learning through information and communications technology* (pp. 169–187). Routledge.

Dziabenko, O., Orduna, P., & Garcia-Zubia, J. (2013). Remote experiments in secondary school education. In *Proceedings of the 2013 IEEE Frontiers in Education Conference* (pp. 1760–1764). Institute of Electrical and Electronics Engineers. https://www.dropbox.com/s/jqbxg7073s3mz9p/FIE2013_Proceedings.pdf?dl=0.

Ertugrul, N. (2000). Towards virtual laboratories: A survey of LabVIEW-based teaching/learning tools and future trends. *International Journal of Engineering Education, 16*(3), 171–180.

Garfield, J. B. (1998). The statistical reasoning assessment: Development and validation of a research tool. In *Proceedings of the 5th International Conference on Teaching Statistics* (pp. 781–786). The International Association for Statistical Education (IASE). https://iase-web.org/documents/papers/icots5/Topic6u.pdf?1402524957.

Garner, B. K. (2008). When students seem stalled: The missing link for too many kids who don't "get it?" Cognitive Structures. *Educational Leadership, 65*(6), 32–38.

Ghasemi, A., & Zahediasl, S. (2012). Normality tests for statistical analysis: A guide for non-statisticians. *International Journal of Endocrinology and Metabolism, 10*(2), 486.

Grout, I. (2017). Remote laboratories as a means to widen participation in STEM education. *Education Sciences, 7*(4), 85. https://doi.org/10.3390/educsci7040085.

Harward, V. J., Del Alamo, J. A., Lerman, S. R., Bailey, P. H., Carpenter, J., DeLong, K., Felknor, C., Hardison, J., Harrison, B., Jabbour, I., Long, P. D., Mao, T., Naamani, L., Northridge, J., Schulz, M., Talavera, D., Varadharajan, C., Wang, S., Yehia, K., Zbib, R., & Zych, D. (2008). The iLab shared architecture: A web services infrastructure to build communities of internet accessible laboratories. *Proceedings of the IEEE, 96*(6), 931–950. IEEE. https://ieeexplore.ieee.org/document/4527087.

Havu-Nuutinen, S. (2005). Examining young children's conceptual change process in floating and sinking from a social constructivist perspective. *International Journal of Science Education, 27*(3), 259–279.

Hoxley, D., Catennazi, G., Anderson, C., Pawlowski, P., McDonald, A., O'Connell, C., Atsikidis, P., Pilgram, P., & Abbey, B. (2014). FARLabs: Enhancing student engagement via remote laboratories. In *Proceedings of the Australian Conference on Science and Mathematics Education (formerly UniServe Science Conference)*.

Johnson, R. B., & Onwuegbuzie, A. J. (2004). Mixed methods research: A research paradigm whose time has come. *Educational Researcher, 33*(7), 14–26.

Juuti, K., & Lavonen, J. (2006). Design-based research in science education: One step towards methodology. *Nordic Studies in Science Education, 2*(2), 54–68. https://doi.org/10.5617/nordina.424.

Kong, S. C., Yeung, Y. Y., & Wu, X. (2009). An experience of teaching for learning by observation: Remote-controlled experiments on electrical circuits. *Computers & Education, 52*(3), 702–717.

Krapp, A., & Prenzel, M. (2011). Research on interest in science: Theories, methods, and findings. *International Journal of Science Education, 33*(1), 27–50.

Laurillard, D. (2008). Technology enhanced learning as a tool for pedagogical innovation. *Journal of Philosophy of Education, 42*(3–4), 521–533.

Lowe, D., Newcombe, P., & Stumpers, B. (2013). Evaluation of the use of remote laboratories for secondary school science education. *Research in Science Education, 43*(3), 1197–1219.

Lu, Q., & Liu, E. (2012). Alignment between high school biology curriculum standard and the standardised tests of four provinces in China. *Journal of Biological Education, 46*(3), 149–164.
Ma, J., & Nickerson, J. V. (2006). Hands-on, simulated, and remote laboratories: A comparative literature review. *ACM Computing Surveys (CSUR), 38*(3), Article 7. https://doi.org/10.1145/1132960.1132961.
McKenzie, J. F., Wood, M. L., Kotecki, J. E., Clark, J. K., & Brey, R. A. (1999). Establishing content validity: Using qualitative and quantitative steps. *American Journal of Health Behavior, 23*(4), 311–318. https://doi.org/10.5993/AJHB.23.4.9.
Ministry of Education. (2018a). *Pǔtōng gāozhōng shēngwù xué kèchéng biāozhǔn (2017 niánbǎn) [National standards of general senior secondary school biology curriculum 2017]*. People's Education Press. http://www.jwc.ecnu.edu.cn/_upload/article/files/f7/28/dc6ae6dc46faa43b343da2b24d7a/80b73839-04a1-4548-8149-487ecec5ab26.pdf.
Ministry of Education. (2018b). *Pǔtōng gāozhōng huàxué kèchéng biāozhǔn (2017 niánbǎn)* [National standards of general senior secondary school chemistry curriculum 2017]. People's Education Press. http://www.jwc.ecnu.edu.cn/_upload/article/files/f7/28/dc6ae6dc46faa43b343da2b24d7a/70bc6036-a93c-40e8-a4e3-af165ea53534.pdf.
Morse, J. M. (1991). Approaches to qualitative-quantitative methodological triangulation. *Nursing Research, 40*(2), 120–123.
Organisation for Economic Co-operation and Development. (2019). *PISA 2018 results. What students know and can do* (Vol. I). OECD Publishing. https://doi.org/10.1787/5f07c754-en.
Scanlon, E., Colwell, C., Cooper, M., & Di Paolo, T. (2004). Remote experiments, re-versioning and re-thinking science learning. *Computers & Education, 43*(1–2), 153–163.
Solomon, J. (1988). Learning through experiment. *Studies in Science Education, 15*, 103–104.
Tawfik, M., Sancristobal, E. S., Martin, S., Gil, R., Diaz, G., Colmenar, A., Peire, J., Castro, M., Nilsson, K., Zackrisson, J., Hakansson, L., & Gustavsson, I. (2013). Virtual instrument systems in reality (VISIR) for remote wiring and measurement of electronic circuits on breadboard. *IEEE Transactions on Learning Technologies, 6*(1), 60–72.
Tho, S. W., Chan, K. W., & Yeung, Y. Y. (2015). Technology-enhanced physics programme for community-based science learning: Innovative design and programme evaluation in a theme park. *Journal of Science Education and Technology, 24*(5), 580–594.
Tho, S. W., & Yeung, Y. Y. (2015). Innovative IP camera applications for scientific investigation. *School Science Review, 96*(356), 58–62.
Tho, S. W., & Yeung, Y. Y. (2016). Technology-enhanced science learning through remote laboratory: System design and pilot implementation in tertiary education. *Australasian Journal of Educational Technology, 32*(3), 96–111.
Tho, S. W., & Yeung, Y. Y. (2018). An implementation of remote laboratory for secondary science education. *Journal of Computer Assisted Learning, 34*(5), 629–640.
Tho, S. W., Yeung, Y. Y., Wei, R., Chan, K. W., & So, W. W. M. (2017). A systematic review of remote laboratory work in science education with the support of visualizing its structure through the HistCite and CiteSpace software. *International Journal of Science and Mathematics Education, 15*(7), 1217–1236.
United Nations Educational, Scientific and Cultural Organization. (2021). *COVID-19 educational disruption and response*. http://www.iiep.unesco.org/en/covid-19-educational-disruption-and-response-13363.
Viegas, C., Pavani, A., Lima, N., Marques, A., Pozzo, I., Dobboletta, E., Atencia, V., Barreto, D., Calliari, F., Fidalgo, A., Lima, D., Temporão, G., & Alves, G. (2018). Impact of a remote lab on teaching practices and student learning. *Computers & Education, 126*, 201–216.
Wei, B. (2020). The change in the intended Senior High School Chemistry Curriculum in China: Focus on intellectual demands. *Chemistry Education Research and Practice, 21*(1), 14–23.
Yeung, Y. Y. (2015). Characteristics of Chinese learners as revealed from their affective domain and choices of science learning in China. In M. S. Khine (Ed.), *Science education in East Asia* (pp. 123–146). Springer International Publishing AG.

Yeung, Y. Y. (2020, August 20). A practical guide to digital teaching and learning: Establishing remote laboratories (special section). *Times Higher Education*. https://www.timeshighereducation.com/features/practical-guide-digital-teaching-and-learning.

Yeung, Y. Y., & Cheng, M. M. H. (2018). Hong Kong students' characteristics of science learning in relation to ROSE. In C. Buntting, M. M. H. Cheng, & J. Aliste (Eds.), *Studies in science education in the Asia-Pacific region* (pp. 61–74). Routledge.

Yeung, Y. Y., Wang, Y., Lee, L. H., Lee, H. C., Chan, K. C. K. & Cheang, F. C. C. (2019). Educational use of an innovative mobile logger and evaluation of students' learning effectiveness in STEM education. In J. Theo Bastiaens (Ed.), *Proceedings of EdMedia + Innovate Learning 2019* (pp. 261–267). Association for the Advancement of Computing in Education (AACE).

Yau Yuen Yeung is a multidisciplinary professor in two distinct academic disciplines, namely science and education. As a scientist, his research areas include the atomic spectroscopy of transition metal and rare earth elements, computational science and materials science. As a teacher educator, he focuses on science/STEM education, and technology-enhanced learning (including 3D visualization, virtual reality, remote-controlled experiments). He is also the Chief Editor of the journal, Asia-Pacific Forum on Science Learning and Teaching. He has won multiple awards for the invention of affordable educational technology devices to facilitate effective STEM education.

Parbat Dhungana is currently pursuing a Ph.D. in STEM education at the Department of Science and Environmental Studies, The Education University of Hong Kong. He is a recipient of the Hong Kong Ph.D. Scholarship provided by The Research Grants Council of Hong Kong. He received an M. Phil. in Education in 2011 and a Master's degree in Science in 2001 from universities in Nepal. He was a faculty member in Kathmandu University before his Ph.D. study. His interest and works focus on making science and STEM learning interesting and effective to young learners.

Siew Wei Tho is an associate professor at the Department of Physics, Faculty Science and Mathematics, Sultan Idris Education University, Malaysia. He received a Bachelor of Science Education degree (2008) and a Master of Physics Education degree (2010) from Sultan Idris Education University. He was awarded a Ph.D. degree in the field of science education from The Hong Kong Institute of Education in 2015 (retitled as The Education University of Hong Kong in 2016). His research focuses on educational technology, technology-enhanced science learning, STEM education, laboratory-based learning via the Internet, and open source-based data acquisition.

Chapter 9
Enhancing STEM Education in Malaysia through Scientist–Teacher–Student Partnerships (STSP)

Rohaida Mohd Saat and Hidayah Mohd Fadzil

Abstract Malaysia is envisioned to become a developed nation and emphasizing science, technology, engineering, and mathematics (STEM) workforce, to meet the challenges and demands of a STEM-driven economy. However, STEM education is facing a great challenge as students are no longer interested in STEM-related subjects. This is apparent in the decrease in the number of students enrolled in STEM-related fields. Our professional responsibility as educators is to ensure that we offer students the knowledge they need to be effective in the future. A change in the education setting is essential to address the emerging demands of the information age. To keep abreast of the changing times as well as to mitigate the issues in STEM-related fields, STEM education needs to be re-evaluated so that students will be more receptive attitude towards STEM-related subjects. An alternative is to demonstrate 'real' or 'authentic' science to students through Scientist–Teacher–Student Partnerships (STSP), where teachers, as well as students, acquire skills in 'real' scientific investigations. In this chapter, STSP refers to collaboration among upper secondary science students, teachers, and university scientists, and involves mutual learning via a partnership. This chapter will discuss the collaboration and contributing factors to the effective implementation of STSP in the Malaysian context. This innovative approach has demonstrated that the interest in STEM education could be enhanced by establishing more effective communication and understanding between scientists, students, and teachers, thus promoting better articulation of STSP as an innovative mechanism for education reform in STEM-related fields.

Keywords STEM education · Scientist–Teacher–Student Partnership (STSP) · Upper secondary level · Innovative approach

R. M. Saat · H. M. Fadzil (✉)
Faculty of Education, University of Malaya, 50603 Kuala Lumpur, Malaysia
e-mail: hidayahfadzil@um.edu.my

M. M. H. Cheng et al. (eds.), *Concepts and Practices of STEM Education in Asia*,
https://doi.org/10.1007/978-981-19-2596-2_9

Introduction

As Malaysia enters the Twelfth Malaysian Plan (2021–2025), more emphasis has been placed on the science, technology, engineering, and mathematics (STEM) workforce, under one of the three dimensions, specifically the economic empowerment dimension. This is to ensure more inclusive socioeconomic development and a more prosperous society, but one that needs to meet the demands of a STEM-driven economy. It is predicted that 50% of the current jobs with skill shortages are in the STEM fields, and the demand for professional, scientific, and technical services will also rise in the next 5 years (Department of Education, Government of Western Australia, n.d.).

However, STEM education in Malaysia is facing a great challenge as students are no longer interested in STEM-related subjects. This is apparent in the decreasing number of students enrolled in STEM-related disciplines at secondary schools (refer Table 9.1). Among the possible factors that have contributed to this issue are a decline in interest in science that has contributed to STEM talent depletion, ineffective teaching methodology, ad-hoc changes in policies, and a low level of awareness of the demand for specialized talent in STEM. Despite the government's commitment to various STEM initiatives, student attitudes toward STEM still vary. Thus, it is not only important to understand how to spark student interest in these fields, but also how to sustain such interest.

One of the ways to address the issue is to design effective teaching methodologies and strategies for use in the classroom. STEM educators need to initiate innovative and challenging teaching strategies. A number of teaching and learning strategies have been introduced such as inquiry-based learning, problem-based learning, project-based learning, and scenario-based learning. The main focus of these strategies is to make STEM instruction engaging and interesting as well as to make STEM lessons real and authentic.

Table 9.1 Percentage of students in the STEM streams in grade 10 (2013–2020)

Year	Science stream	Technical & vocational stream	Percentage 60:40
2013	31.80	10.79	42.59
2014	31.15	9.34	40.49
2015	26.84	14.31	41.15
2016	26.59	13.99	40.58
2017	25.18	13.80	38.98
2018	26.85	14.51	41.36
2019	25.30	14.80	40.10
2020	STEM stream		34.70

STEM Education in Malaysia

STEM education has received a great deal of attention all around the world (Fadzil et al., 2019; Roehrig et al., 2021). Malaysia, like other countries, places a high value on STEM education in order to meet the challenges and demands of a science and technology-based economy. Despite the growing popularity of the term STEM education, there is still some confusion about what it entails and what it means in terms of curriculum and student outcomes, as discussed by previous researchers (e.g. Bybee, 2013; Holmlund et al., 2018; Lamberg & Trzynadlowski, 2015). According to Vasquez et al. (2013), STEM education is an interdisciplinary approach to learning that eliminates the traditional barriers separating the four disciplines of science, technology, engineering, and mathematics, and integrates them into real-world, rigorous, and relevant learning experiences for students. STEM education can be viewed as a single or multi-disciplinary field, and in the latter case, there is no clear agreement on the nature of the content and pedagogical interplay among the STEM fields.

Integrated STEM education can be defined as an approach to teaching STEM content within an authentic context for the purpose of connecting these subjects to enhance student learning (Kelley & Knowles, 2016). It includes effort to combine some or all of the four disciplines of science, technology, engineering, and mathematics into a related set of activities that is based on connections between the subjects and real-world problems (Moore et al., 2014). In the context of Malaysia, integrated STEM education usually addresses at least two of the four STEM disciplines (e.g. science and mathematics; science and engineering). For an integrated STEM approach, it is important to determine how to help students both build knowledge in individual disciplines and learn to make connections among them (English, 2016; Marginson et al., 2013).

STEM in Malaysia is reflected through education policies and school curriculum choices in an effort to increase competitiveness in science and technology for students. It encompasses three aspects, namely (i) areas of learning, (ii) STEM streaming, and (iii) teaching and learning approaches. The first aspect is STEM as an area or field of learning in schools, and a field of study at tertiary level. Examples of STEM subject areas are Science, Chemistry, Mathematics, and Computer Science. Examples of STEM courses at the tertiary level are Mechanical Engineering, Medicine, Bio-Chemistry, and Computing and Information Systems.

The second aspect is STEM as subject packages or streaming where upper secondary students in the school system are given the opportunity to choose packages or streams according to their inclinations. Starting from 2020, students have the option of choosing STEM Core Subject Packages and Arts and Humanities Core Subject packages beginning in Grade 10. In addition to the core and compulsory subjects, students are able to choose subjects from either the STEM electives or the Arts and Humanities electives. Students are considered as being in the STEM stream when they choose to take at least one subject from the STEM elective subjects or any two subjects from the Applied Science and Technology subjects or Vocational subjects (refer to Fig. 9.1).

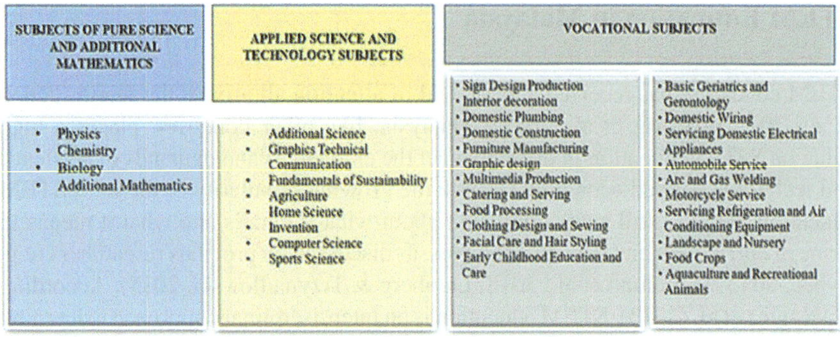

Fig. 9.1 List of upper secondary STEM elective subject groups

STEM in Malaysia can also be defined as a teaching and learning approach that involves the application of STEM knowledge, skills, and values to solve problems in everyday life, society, and the environment. This can be done by using teaching and learning strategies such as inquiry-based, problem-solving, and project-based learning. Students are encouraged to ask questions and investigate their surroundings through inquiry and problem solving in the real world. STEM in the primary and secondary school encompasses the three domains of (i) knowledge, (ii) skills, and (iii) scientific attitudes and values in all STEM-related subjects. STEM knowledge refers to the ideas, concepts, beliefs, and understandings in the STEM field that are formulated in the curricula of all STEM-related subjects. STEM skills, on the other hand, are the abilities to investigate, solve problems, design, and produce products. These abilities can be learned through events, tasks, or assignments that are part of the curriculum of any STEM subject. STEM ethics are positive morals and rules that STEM students must adhere to. STEM ethics are critical in developing students who are not only intelligent and professional, but also have strong personalities. In order to enhance the students' knowledge, skills, and value in STEM education in Malaysia, teachers are expected to have the capabilities to deepen students' understanding in STEM disciplines by contextualizing scientific concepts, to broaden student understanding of STEM disciplines by exposing students to socially and culturally appropriate STEM contexts, and to increase student engagement in STEM disciplines.

Initiatives to Enhance STEM Education in Malaysia

Among the efforts made by the Ministry of Education Malaysia (MOE) to increase the resources of skilled manpower and experts in the field of research and industry is the strengthening of STEM education. This has been done through various policies and initiatives. Accordingly, the Malaysian government established the

60:40 Science/Technical: Arts Education policy in 1967. The strategy refers to the Ministry's goal regarding the ratio of students with major science education to those in the arts stream. The policy was implemented in 1970. However, due to different factors such as a content-heavy curriculum, inconsistent teaching and learning quality, and minimal and obsolete infrastructure, this policy goal has never been achieved (Fadzil et al., 2019). Thus, the government is very committed to strengthening STEM education through the formulation of the Malaysian Education Development Plan (PPPM) 2013–2025. This development plan outlines the policies, strategies, and initiatives.

Three waves have been proposed in the Malaysian Education Development Plan (PPPM) 2013–2025 to improve the delivery of STEM through the educational system. The first wave focuses on improving the foundations of current programmes and promoting students from upper secondary and post-secondary schools to participate in the Science Stream. The second wave builds on these foundations by engaging in the support of a wider group of stakeholders, including those in the informal learning sector. To build a roadmap for more innovation, the third wave will see the review of such initiatives (MOE, 2013). Based on the development plan, actions have been taken that include raising the interest of students through new learning approaches and an improved curriculum, incorporating higher order thinking skills (HOTS), expanding the use of hands-on teaching tools, and increasing interest by making the science content more relevant to daily life. Besides that, intensive training for teachers to sharpen their skills and abilities in teaching science has also been highlighted in the policy. Passive teaching methods in the classroom need to be transformed into active learning that emphasizes creative ideas.

The MOE has also collaborated and worked together with other ministries and organizations to ensure that Malaysia has a sufficient number of trained STEM graduates to meet the workforce needs of the industries that power its economy. For example, there is a special committee that includes representatives of the Ministry of Higher Education and Ministry of Science, Technology, and Innovation (MOSTI) to implement a STEM Education Development Plan that is more holistic and inclusive by focusing on eight key areas. The key areas are policy, teaching and learning, facilities, career awareness programmes, strategic cooperation, data and research, commercialization, and innovation (MOE, 2016). The Academy of Sciences Malaysia (ASM) is a statutory body that works hand-in-hand with the ministries with the aim of pursuing excellence, and fosters development in the fields of science, engineering, and technology for the benefit of all. Among the initiatives are joint research and development (R&D) projects between researchers in Malaysia and collaborators from other countries, and expansion and enhancement of research capacity. The Ministry of Higher Education (MOHE) also provides special scholarships related to STEM fields such as the Malaysia International Scholarship and MyBrainSc Scholarships that cover levels of study from First Degree to PhD to meet the needs of the field of Pure Science. Despite various government efforts to promote STEM education in Malaysia, STEM education still faces issues such as being an unpopular field of study, which will affect the nation's agenda of addressing the challenges and demands of a science and technology-driven economy.

The Scientist–Teacher–Student Partnership (STSP) Approach to Enhancing STEM Education

Although the idea of STEM education has been widely implemented in Malaysia, only a few teachers seem to know how to operationalize STEM education. One of the most recent issues recognized is the problem of the lack of coherent effort and training for teachers in conceptualizing STEM integration in science classrooms. According to Bybee (2018), only a small number of STEM activities have been designed to help students make and demonstrate connections between ideas across disciplines. Thus, the scientist–teacher–student partnership (STSP) is proposed as one of the innovative teaching and learning approaches to move STEM educators forward by creating a learning experience of STEM integration for research and practice, through the development of STEM content by scientists and teachers.

Previous findings (e.g. Fadzil et al., 2019; Shein & Tsai, 2015; Ufnar & Shepherd, 2019) regarding Scientist–Teacher–Students Partnerships (STSP) showed that students appeared to learn more as a result of having hands-on experience as they had more opportunities for authentic learning with scientists. Collaboration or partnership between the scientific community and science educators has grown in popularity as a strategy for science education reform and is widely implemented in developed countries such as the United States and Australia (e.g. Maina et al., 2021; McClusky & Farland-Smith, 2021; Munson et al., 2013; Rushton, 2021). Scientist–Teacher–Student Partnership (STSP) in this context refers to a collaboration among upper secondary science teachers and university scientists which involves mutual learning via a partnership. This partnership might offer the flexibility to provide students with opportunities to explore science topics that encourage the development of students' interest and skills. Thus, extensive knowledge on how this partnership works is essential.

Previous studies on STSP (e.g. Fadzil et al., 2019; Hasanah & Tsutaoka, 2019; Saat et al., 2021) found that students' conceptions changed in a positive direction after the students had the opportunity to be involved in a scientists–teachers partnership programme. Students who were given the opportunity to learn from the scientists became more motivated to learn STEM-related subjects. As a result, the significance of this collaboration in our efforts to reform science education should not be underestimated. While this mutual learning appears to be ideal for science and STEM education, there are only a few mechanisms in place to support this cross-institutional collaboration. The majority of the studies have focused on the collaboration of scientists and teachers, whereas little work has been done to understand how such partnerships influence science learning and the mechanisms to ensure effective partnerships in science/STEM learning.

Further research in this area may provide information to STEM educators on how to best help students learn integrated STEM with understanding and make connections across STEM-related disciplines, as this partnership might offer the flexibility to provide students with opportunities to explore integrated STEM topics that encourage the development of students' interest and to focus on the development of scientific skills in science. The following section will elaborate the factors that need to be taken into consideration in ensuring that scientist–teacher–student partnerships are implemented within the Malaysian context.

Collaboration Factors: The Roles of Teachers, Scientists, and Students

Previous studies (e.g. McClusky & Farland-Smith, 2021; Munson et al., 2013; Rushton, 2021; Saat et al., 2022) have shown that the tripartite collaboration brings educational benefits to all groups. For teachers, this partnership provides insights into the scientific inquiry process, extending their content knowledge and pedagogical strategies, and renewing their teaching. The term "scientist-educator" has been proposed to represent science teachers who have experience in both professional cultures, that is, science and science education, with better understanding and conceptualization of the cutting-edge of STEM knowledge. Specifically, the teachers in the partnership played their roles as (i) mediators, (ii) instructional designers, and (iii) synchronizers of scientific concepts and terminologies with the scientists and students (Fadzil et al., 2019; Saat et al., 2022).

According to Fadzil et al. (2019), the teachers view their role as mediators to enhance the communication of knowledge between the scientist and students. This can happen while conducting the STEM activities and also when designing the teaching and learning materials with the scientists. With their experience of teaching students, teachers mediate the scientists' knowledge by linking new knowledge to the students' prior knowledge. Teachers also serve as instructional designers who design the instructional materials as well as the lessons. They work together with the scientists to plan appropriate activities for the students and also to synchronize the relevant scientific concepts and terminologies so that they will be comprehensible to the secondary school level students. For example, scientists may find it difficult to explain certain Chemistry concepts. It is the role of teachers to explain the concepts using familiar terminology as well as making the explanation simpler so that it is comprehensible for secondary school students.

In the partnership, scientists are seen as role models to the teachers, as well as to the students. They also act as trainers to teachers and students, as well as mentors to the teachers and students. They are regarded as experts in terms of scientific content and skills. The students see the scientists as their icons who motivate them to be more interested in pursuing their study and careers in STEM-related fields. From the perspective of the teachers, scientists serve as the trainer and coach, particularly for transferring scientific skills and knowledge. After graduating from their pre-service training, teachers lack the opportunity to update their knowledge and skills. In this partnership, teachers have the opportunity to upgrade their skills with the scientists at university including the latest scientific research skills such as extracting stem cells from specimens and the latest techniques in chromatography.

The students in this partnership, on the other hand, act as the end receivers of the skills and knowledge transferred by the teachers and scientists. Students in this partnership also provide feedback regarding the STEM activities. The partnership enriches their learning experiences by providing access to the scientific community and content knowledge gains, which in turn might highly improve their science performance. We argue that the students' role in the professional scientist–teacher–student partnership can be maximized by enhancing the appropriate time and commitment of all parties, updating the support system, and creating more resources that align with the science/STEM-related curriculum content. These contributing factors influence the scientist–teacher–student partnership and will be further explained in the following section.

Contributing Factors for Effective Scientist–Teacher–Student Partnerships

Saat et al. (2022) outlined the contributing factors for an effective partnership. The three factors are internal factors which consist of two sub-categories, that is, the suitability of time and commitment; external factors, that is, the support system and availability of resources; and the institutional factor, that is, the application of curriculum-related activities. The following (Table 9.2) explains the factors that affect the partnership, and the sub-categories.

The internal factors include suitability of time and commitment of the scientists, teachers, and students to this partnership. STSP activities with students can be conducted either during formal or informal classroom sessions. The teachers in the study by Saat et al. (2021) suggested that simple STEM activities should be conducted during formal classroom sessions in school. However, it is preferable to conduct more complex activities that take longer to achieve the teaching and

Table 9.2 Explanation of contributing factors in STSP (Saat et al., 2021)

Contributing factors	Sub-categories	Explanation
(i) Internal factors	Suitability of time	Suitability of time of the activities during formal or informal classroom sessions
	Commitment	Commitments of teacher, student, and scientist
(ii) External factors	Supporting system	Supporting system at school and university
	Availability of resources	Resources such as modules or guidebook to enhance the transfer of knowledge
(iii) Institutional factors	Curriculum related activities	The alignment of the activities to the science curriculum

learning objectives during informal classroom sessions. For example, they can be done after school or on weekends as co-curricular activities. It is easier for teachers to manage the students during informal sessions. A suitable time for students would also provide more space for them to investigate the complex tasks involved in STEM-related activities. According to Hasanah and Tsutaoka (2019) and Shein and Tsai (2015), in dealing with the STEM education approach, suitability of time is necessary for exploration such as developing a detailed plan and design, and finishing the prototypes of a project.

For the second sub-category, which is the commitment factor, we acknowledge that the partnership requires a significant and positive relationship since the teachers' role in implementing a new pedagogical approach via hands-on learning approaches is reasonably demanding. Thus, commitment from all the parties in terms of understanding their roles in this collaboration are important for ensuring an effective partnership. Moreover, most of the teachers and scientists within this partnership recognize the value of commitment of all members in the multifaceted interaction. This can be achieved during the scientist–teacher interaction such as during the scientist–teacher training session and when designing the STEM activities together. If all parties are committed, the ongoing exchange of ideas between professional scientists and teachers can be activated, resulting in the development of new professional knowledge as well as the transformation of values and beliefs.

The second contributing factor is the external factors. These include the support system at school and university, and the need for resources such as modules or a guidebook to enhance the transfer of knowledge during and after the partnership. A support system, such as the management, support staff, and laboratory assistants, is needed to ensure an effective partnership not only in school but also in the university. For example, the presence of laboratory assistants could encourage a comprehensive laboratory experience for the teachers during training sessions at university laboratories with scientists, and also for students in school laboratories. In addition, teachers need resources to assist them in conducting the activities in laboratories. A typical Malaysian teacher usually does not have much time to develop their own activities which are outside the textbooks or activity books. They need some form of guide to assist them in broadening their teaching scope, compounding ideas of STEM concepts, and helping mentor students while conducting activities or projects related to STEM.

The third factor is the institutional factor, that is, the alignment of the activities to the science curriculum. Compared to previous studies on STSP that were 'one-off' in nature (e.g. Maina et al., 2021; Shein & Tsai, 2015; Ufnar & Shepherd, 2019), this proposed model (Fadzil et al., 2019; Saat et al., 2021) has provided opportunities for the teachers to engage in the development of learning resources and to be used within the school setting. Such a model is more sustainable in nature. The resources included the STEM activities that cater for the needs of the students as well as the teachers. Thus, the strength of this partnership is that it connected teachers and scientists with the current science curriculum and extension to the content in the curriculum. This is important especially to the teachers as the teachers in the study admitted that they do not prefer to conduct STEM-related activities on topics that are not directly related

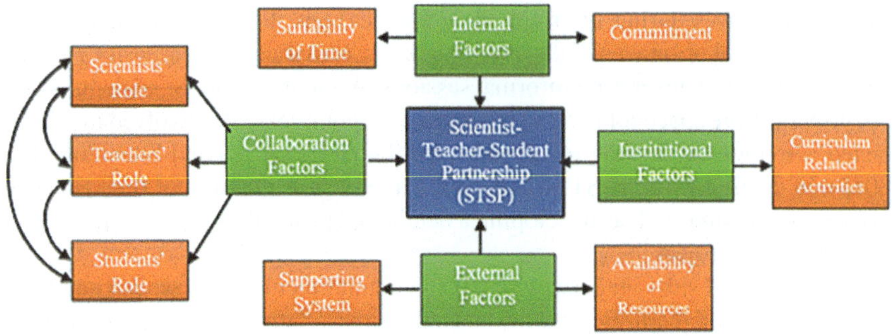

Fig. 9.2 Factors affecting Scientist–Teacher–Student partnerships (Saat et al., 2022)

to the school's science syllabus which therefore will not be asked in the national examination. Figure 9.2 depicts the collaborating factors and the contributing factors in the partnership (Saat et al., 2022).

Moving Forward: Integrated STEM

Current practices in STEM education focus on improving STEM subjects as isolated disciplines with not much integration (Breiner et al., 2012; Fadzil et al., 2019; Hoachlander & Yanofsky, 2011; Wang et al., 2011). Furthermore, most studies of STEM learning consider each discipline singly and do not measure students' ability to make connections across disciplines (Bybee, 2018). Implementing an integrated STEM approach in an educational system that has a very established segregated and discipline-based structure such as in Malaysia requires profound restructuring of the curriculum and lessons (Nadelson & Seifert, 2017).

Therefore, our current study on scientist–teacher–student partnerships aims to create integrated STEM instructional materials that involve crosscutting STEM concepts from the Biology, Chemistry, and Physics fields. According to Bybee (2018) and Kelley and Knowles (2016), most of the crosscutting connections in the practices in STEM education remain implicit or can be missing all together. Thus, scientists and teachers in our study work together to decide on the suitable Chemistry, Physics, and Biology concepts to be integrated and implemented as integrated STEM. It is important for researchers to come up with well-integrated STEM instruction that may provide opportunities for students to learn more relevant and stimulating content by encouraging the use of higher order thinking skills which may enhance their problem solving skills, as suggested by Stohlmann et al. (2012). Building a strategic approach to integrating STEM concepts requires strong conceptual and foundational understanding of how students learn and apply the integrated STEM content.

Conclusion

It is well recognized that STEM education is facing a great challenge as students are no longer interested in STEM-related subjects. This will impact the demands of a science and technology-driven economy. This chapter introduces readers to the scientist–teacher–student partnership as an innovative approach to reforming STEM education, particularly for addressing the low enrolment of students in the secondary school STEM stream. It aims at making STEM instruction real and authentic. Through the synergy of the scientists, teachers, and students in this partnership, each party is aware of its roles. With suitable support from the management and a proper environment, the STSP can be implemented effectively. Students will become motivated and interested in pursuing STEM as their career.

STSP is one of the approaches that can be adopted as well as adapted. Other innovative approaches and strategies can be developed and implemented. Furthermore, this effort must be sustained in order to meet the needs of the STEM discipline, as it is a dynamic and fast-changing field with uncertainties in order to meet future demand.

References

Breiner, J. M., Harkness, S. S., Johnson, C. C., & Koehler, C. M. (2012). What is STEM? A discussion about conceptions of STEM in education and partnerships. *School Science and Mathematics, 112*(1), 3–11. https://doi.org/10.1111/j.1949-8594.2011.00109.x.

Bybee, R. W. (2013). *The case for STEM education: Challenges and opportunities*. National Science Teachers Association.

Bybee, R. W. (2018). *STEM education now more than ever*. Victoria, Australia: Hawker Brownlow Education.

Department of Education, Government of Western Australia. (n.d.). *What is STEM?* https://www.education.wa.edu.au/what-is-stem.

English, L. D. (2016). STEM education K-12: Perspectives on integration. *International Journal of STEM Education, 3*, Article 3. https://doi.org/10.1186/s40594-016-0036.1.

Fadzil, H. M., Saat, R. M., Awang, K., & Adli, D. S. H. (2019). Students' perception of learning STEM-related subjects through scientist–teacher–student partnership (STSP). *Journal of Baltic Science Education, 18*(4), 537–548.

Hasanah, U., & Tsutaoka, T. (2019). An outline of worldwide barriers in science, technology, engineering and mathematics (STEM) education. *Jurnal Pendidikan IPA Indonesia, 8*(2), 193–200. https://doi.org/10.15294/jpii.v8i2.18350.

Hoachlander, G., & Yanofsky, D. (2011). Making STEM real: By infusing core academics with rigorous real-world work, linked learning pathways prepare students for both college and career. *Educational Leadership, 68*(6), 60–65.

Holmlund, T. D., Lesseig, K., & Slavit, D. (2018). Making sense of "STEM education" in K-12 contexts. *International Journal of STEM Education, 5*, Article 32. https://doi.org/10.1186/s40594-018-0127.2.

Kelley, T. R., & Knowles, J. G. (2016). A conceptual framework for integrated STEM education. *International Journal of STEM Education, 3*, Article 11. https://doi.org/10.1186/s40594-016-0046-z.

Lamberg, T., & Trzynadlowski, N. (2015). How STEM academy teachers conceptualize and implement STEM education. *Journal of Research in STEM Education, 1*(1), 45–58. https://doi.org/10.51355/jstem.2015.8.

Maina, F., Smit, J., & Serwadda, A. (2021). Professional development for rural stem teachers on data science and cybersecurity: A university and school districts' partnership. *Australian and International Journal of Rural Education, 31*(1), 30–41. https://journal.spera.asn.au/index.php/AIJRE/article/view/291.

Marginson, S., Tytler, R., Freeman, B., & Roberts, K. (2013). *STEM: Country comparisons*. Melbourne: Australian Council of Learned Academies.

McClusky, S. L., & Farland-Smith, D. (2021). The role of the knowledgeable other in student–scientist–teacher partnerships: Examining triads in second grade classrooms. In D. Farland-Smith (Ed.), *Enhancing learning opportunities through student, scientist, and teacher partnerships* (pp. 23–42). IGI Global. https://doi.org/10.4018/978-1-7998-4966-7.ch003.

Ministry of Education Malaysia. (2013). *Malaysia Education Blueprint 2013–2025*. https://www.moe.gov.my/menumedia/media-cetak/penerbitan/dasar/1207-malaysia-education-blueprint-2013-2025/file.

Ministry of Education Malaysia. (2016). *Implementation Guide for Science, Technology, Engineering, and Mathematics (STEM) in Teaching and Learning*. http://smksyedsira.edu.my/wp-content/uploads/2016/12/7.-Panduan-Pelaksanaan-STEM-Dalam-PP.pdf (in Bahasa Malaysia).

Moore, T. J., Stohlmann, M. S., Wang, H.-H., Tank, K. M., Glancy, A. W., & Roehrig, G. H. (2014). Implementation and integration of engineering in K–12 STEM education. In Ş. Purzer, J. Strobel, & M. E. Cardella (Eds.), *Engineering in Pre-College Settings: Synthesizing Research, Policy, and Practices* (pp. 35–60). Purdue University Press. https://doi.org/10.2307/j.ctt6wq7bh.7.

Munson, B. H., Martz, M. A., & Shimek, S. (2013). Scientists' and teachers' perspectives about collaboration. *Journal of College Science Teaching, 43*(2), 30–35. https://www.jstor.org/stable/43631069.

Nadelson, L. S., & Seifert, A. L. (2017). Integrated STEM defined: Contexts, challenges, and the future. *The Journal of Educational Research, 110*(3), 221–223. https://doi.org/10.1080/00220671.2017.1289775.

Roehrig, G. H., Dare, E. A., Ring-Whalen, E., & Wieselmann, J. R. (2021). Understanding coherence and integration in integrated STEM curriculum. *International Journal of STEM Education, 8*(1), 1–21.

Rushton, E. (2021). *Science Education and Teacher Professional Development: Combining Learning with Research*. Palgrave Macmillan.

Saat, R. M., Fadzil, H. M., Adli, D. S. H., & Awang, K. (2021). STEM teachers' professional development through Scientist–Teacher–Students Partnership (STSP). *Jurnal Pendidikan IPA Indonesia, 10*(3), 357–367.

Saat, R. M., Piaw, C. Y., & Fadzil, H. M. (2022). Creating a grounded model of performance quality of Scientist–Teacher–Student Partnership (STSP) for STEM education. *International Journal of Science and Mathematics Education* (Published online: 03 January, 2022). https://doi.org/10.1007/s10763-021-10236-4.

Shein, P. P., & Tsai, C. Y. (2015). Impact of a Scientist–Teacher Collaborative Model on students, teachers, and scientists. *International Journal of Science Education, 37*(13), 2147–2169. https://doi.org/10.1080/09500693.2015.1068465.

Stohlmann, M., Moore, T. J., & Roehrig, G. H. (2012). Considerations for teaching integrated STEM education. *Journal of Pre-College Engineering Education Research, 2*(1), Article 4. https://doi.org/10.5703/1288284314653.

Ufnar, J. A., & Shepherd, V. L. (2019). The scientist in the classroom partnership program: An innovative teacher professional development model. *Professional Development in Education, 45*(4), 642–658.

Vasquez, J. A., Sneider, C., & Comer, M. (2013). *STEM lesson essentials, grades 3–8: Integrating science, technology, engineering, and mathematics*. Heinemann.

Wang, H., Moore, T. J., Roehrig, G. H., & Park, M. S. (2011). STEM integration: Teacher perceptions and practice. *Journal of Pre-College Engineering Education Research, 1*(2), Article 2. https://doi.org/10.5703/1288284314636.

Rohaida Mohd Saat is an honorary professor at the Department of Mathematics and Science Education, Faculty of Education, University of Malaya. She was the former dean at the faculty. Her research interests and specialisations are in science education, teacher education, and qualitative research methodology. Her recent study focuses on the scientist-teacher-student partnership (STSP) in science teaching and learning. Her most recent publication is "STEM Education Research in Malaysia" a chapter written with Dr. Hidayah Mohd Fadzil, in a book titled "STEM Education from Asia Trends and Perspectives", and this book is part of the Routledge Critical Studies in Asian Education series.

Hidayah Mohd Fadzil is a senior lecturer at the Department of Mathematics and Science Education, Faculty of Education, University of Malaya. Engaging students in scientific discussions and maintaining strong bonds with them are her priorities. Her research interests include STEM education, environmental education, and qualitative research methodology. Her most recent publication is "Creating a grounded model of performance quality of scientist-teacher-student partnership (STSP) for STEM education" in the International Journal of Science and Mathematics Education with Prof. Dr. Rohaida Mohd Saat.

Chapter 10
Mentor–Mentee Outreach Programme: Promoting University and School Partnerships to Revitalize STEM Education in Rural Secondary Schools in Malaysia

Nyet Moi Siew

Abstract STEM education has drawn increasing attention internationally in recent years. In Malaysia, efforts to encourage students to take up STEM subjects have increased, but upper secondary school enrolments in almost every STEM subject area have continued to fall over the last decade. The situation is even more challenging in Sabah, an East Malaysian state where 72% of schools are located in rural areas with basic utilities and limited infrastructure. Therefore, a STEM Mentor–Mentee outreach programme through a university-school partnership was developed to increase STEM education attainment in the participating rural secondary schools. The programme targeted Form Four students (aged 16 years) to help them understand STEM by relating it explicitly to their local environment. STEM activities were guided by the engineering design process, which takes mentees from identifying a problem and designing a solution, to developing, creating, testing, and evaluating a prototype to solve daily life problems in their environment—while harnessing and developing their twenty-first century skills. Mentors, including in-service and pre-service teachers, provided guidance, support, and assistance to mentees. Data were captured through mentees' responses to a close-ended questionnaire, mentors' field notes, focus group observation and interviews, and open-ended questions. A total of 732 students, 342 in-service and 99 pre-service teachers were involved in the programme from 2015 to 2019. Findings suggest that the programme was able to develop creativity, problem solving, critical thinking, and teamwork skills among rural secondary school students; and to support students in achieving gains in STEM knowledge or skills, positive attitudes, and practices.

Keywords Twenty-first century skills · Mentor–mentee · Outreach programme · Rural schools · STEM education · University-school partnerships

N. M. Siew (✉)
Faculty of Psychology and Education, Universiti Malaysia Sabah, Jalan UMS, 88400 Kota Kinabalu, Sabah, Malaysia
e-mail: sopiah@ums.edu.my

M. M. H. Cheng et al. (eds.), *Concepts and Practices of STEM Education in Asia*,
https://doi.org/10.1007/978-981-19-2596-2_10

Challenges in Revitalizing STEM Education in Rural Schools

STEM (Science, Technology, Engineering, Mathematics) education has been given priority in many countries in order to produce a young generation that is able to stand out in the competitive job market. The integration of the disciplines of knowledge related to science (physics, chemistry, and biology) and mathematics with technology and engineering is known as STEM education (Bybee, 2013). In the Malaysian Education Development Plan (MEDP) 2013–2025, STEM is mentioned explicitly as a specific initiative to be implemented by the Kementerian Pendidikan Malaysia [Ministry of Education Malaysia, MOE], 2016a). The initiative is to strengthen STEM education so as to produce high quality and sufficient human capital in STEM who have higher order thinking skills, and who are innovative, prudent, independent, technologically literate, able to create, as well as being able to solve problems and make decisions (Kementerian Pendidikan Malaysia, 2016a).

The demand for a STEM-driven workforce in Malaysia has increased as the economy has evolved from a production-based to a knowledge-based economy. It has been estimated that Malaysia would need 500,000 scientists and engineers by 2020 to cope with the challenges of the Fourth Industrial Revolution (Academy of Sciences Malaysia, 2015). However, at that point, it had only 70,000 registered engineers. Undeniably the supply of the STEM-related workforce is highly dependent on new entrants into STEM-related programmes at upper secondary as well as at tertiary level. However, reporting has shown that only 22.5% of students enrolled in the science stream and in technical and vocational secondary school classes in 2017, which is far from the desired ratio of 60:40 Science/Technical: Arts Policy set in 1970 (Academy of Sciences Malaysia, 2018).

The challenge of achieving the 60:40 Science/Technical: Arts Policy is even tougher for the vast rural areas of Malaysia, where there is limited infrastructure, lack of good schools, and a small population (Ling et al., 2015). For example, Sabah, an East Malaysian state with a relatively high proportion of students in rural schools, is facing a challenging situation with respect to its efforts to keep pace with STEM education. Many rural primary and secondary schools are located in wide and isolated areas with unique topography (Edgeprop, 2011). Some schools, for example, are located in areas with limited road access and, as is often the case, water transport such as boats is used. According to the Sabah Development Corridor Blueprint (The Institute for Development Studies (Sabah), 2007), 72% of Sabah's schools were located in rural areas. In terms of infrastructure and basic utilities, most rural primary and secondary schools in Sabah do not have a 24-h electrical connection or clean water, and access to good teaching and learning resources, internet, and science laboratories is very limited. It is apparent that these limited opportunities and facilities impact negatively on STEM education attainment, with a gap in attainment between rural and urban schools in Sabah and in Malaysia as a whole.

In its report about Malaysian rural schools, the World Bank (2010) noted that:

> 'Potentially as a result of less favorable conditions in rural schools, students from rural and remote schools perform significantly worse on tests than their peers in urban areas. Disparities within states between rural and urban areas are most prevalent in poorer states like Sabah' (p. 92).

Specifically, the World Bank (2010) reports a disparity between urban and rural secondary schools' achievement in mathematics among students in Grade 9 (aged 15 years). It is clear that many rural school students have lagged behind their peers from urban schools in academic performance due to inadequate infrastructure, utilities, and learning resources. As a consequence, despite many new initiatives aimed at transforming rural schools, this will be difficult to achieve in the near future. Similarly challenging is the revitalization of rural secondary schools in STEM education.

In conclusion, the existing limited infrastructure, utilities, and learning resources in Sabah rural schools have made it difficult to provide new STEM education opportunities in the rural schools. Thus, innovative strategies need to be sought to fill this gap by creating after-school programmes that provide engaging STEM learning experiences for rural school students. In other countries, such as Colombia and the United States, after-school outreach programmes have been designed to help and encourage disadvantaged students of rural schools to increase their STEM literacy and enthusiasm. These after-school STEM outreach programmes aim to improve the quality of science education (Laursen et al., 2012), motivate school students to choose STEM subjects in the future, and generate more graduates who have the capacity to pursue science-based careers (Moskal & Skokan, 2011; Office of the Chief Scientist, 2013).

Brookshire (2014) highlighted that guidance from the right mentor in mentoring programmes can expand students' ideas about the possible careers in STEM fields and can trigger a passion for STEM. Cutucache et al. (2016) proposed a 'layered' approach to STEM mentoring programmes, in which senior faculty members supervise or mentor undergraduate mentors, who in turn work with high school or adolescent mentees. Cutucache et al. found that these mentor–mentee programmes have the potential to both spark STEM interest and efficacy in younger students, while also strengthening the undergraduate experience of STEM majors. These observations raised crucial questions:

- How would rural secondary school students' knowledge or skills about, attitudes towards, and practices in STEM evolve as a result of their participation in a STEM mentor–mentee outreach programme?
- How would a mentor–mentee outreach programme help students in rural secondary schools develop their twenty-first century skills related to STEM?
- What challenges would mentees potentially face in implementing a STEM outreach programme?

Tackling questions such as this, particularly in rural settings, often requires a more integrated approach to STEM education.

According to Essex (2001), school-college partnerships hold significant promise for renewal and improvement in education. Essex pointed out that successful partnerships allow both the school and the university to work together in an environment in which synergy leads to better decision making, thus having a positive effect on student learning. Therefore, school-university partnerships may have mutual benefits for all involved in the STEM mentor–mentee outreach programme, and would enable the study of what rural school students would learn and what challenges would be faced during the outreach programme. Information about effective practices could additionally be used to revitalize STEM education in the future.

To support inclusive and equitable STEM education, a STEM mentor–mentee outreach programme through a university-school partnership approach was developed to support disadvantaged and marginalized students in the rural areas of Sabah. Specifically, this outreach programme aimed to improve the reach of STEM education in schools geographically distant from the cities. It was based on the concepts of contextualization and collaboration. This enabled rural students to understand STEM by relating it explicitly to their local environment and to increase their twenty-first century skills. It was intended that exposing students to STEM and giving students opportunities to explore STEM-related concepts would help them to develop twenty-first century skills such as critical thinking, problem solving, creativity, collaboration and communication skills, and values and ethical applications as the desired outcomes of the Ministry of Education Malaysia (Kementerian Pendidikan Malaysia, 2016b). In an effort to form a Malaysian national identity, values and ethical applications are included as an added value to the twenty-first century skills framework. Among these values and ethical applications are: being thankful to God, a spirit of teamwork, honesty, being diligent and persevering, and responsibility.

As long as an appropriate approach is put in place, the quality of rural secondary schools can be improved, and the gap in STEM education attainment between rural schools and their urban counterparts can be reduced. As such, the objectives of the STEM mentor–mentee outreach programme were:

(1) to improve the knowledge and skills about, attitudes towards, and practices in STEM among students in rural secondary schools,
(2) to develop twenty-first century skills related to STEM among students in rural secondary schools, and
(3) to find out the challenges potentially faced by students when implementing a STEM mentor–mentee outreach programme.

The Design and Development of the STEM Mentor–Mentee Outreach Programme

The STEM mentor–mentee outreach programme was designed and developed collaboratively by a science lecturer as programme coordinator and school science teachers, and was supported by in-service and pre-service teachers. The pre-service teachers were second and third-year undergraduates aged between 22 and 23, training in physics and mathematics education, who had no teaching experience in schools. However, they were taking courses on pedagogy, including the teaching and learning of STEM. The in-service teachers were qualified teachers with degrees in science and social science education who were undertaking a Master's course at the time. Science teachers from the participating secondary schools were consulted to identify the STEM activities which could accommodate the specific contexts of the students' daily lives.

Mentors-Mentees Prior to the programme, the faculty recruited mentors among in-service and pre-service teachers, each of whom signed up voluntarily in exchange for experience and increased competence. Mentors attended a one-day training course in peer mentoring arranged by the programme coordinator. Mentors were trained to conduct in-person facilitation and assessment with their mentees. The mentors were empowered to carry out their role as motivators, facilitators, and evaluators during the programme. After completion of their mentor role, mentors were awarded a Certificate of Contribution by the faculty dean.

The Form Four Science Stream students (aged 16 years) from 16 participating secondary rural schools were chosen as the mentees of the programme. Mentees received guidance, support, and assistance from mentors in finding solutions to problems, using materials, and designing and building prototypes.

Ill-Defined Problems and STEM Activities King and Kitchener (1994) claimed that an effective technique for developing problem-solving and critical-thinking skills is to expose students to "ill-defined" problems in their field. Greenwald (2000) characterized an ill-defined problem as one that is unclear and raises questions about what is known, what needs to be known, and how the answer can be found. Because the problem is unclear, there are many ways to solve it, and the solutions are influenced by one's vantage point and experience (p. 28).

Thus, the problems were designed to be 'ill-defined' and were introduced to students within the context of their daily life. In other words, students were engaged in connecting their everyday experiences to solving ill-defined problems. Thus, the problems became better defined and more contextualized as they were worked on. Students were also asked to consider the constraints of the materials and time; to think about what they already knew; and to design, plan, construct, test, and evaluate a physical prototype of their design.

Different STEM activities were introduced in each school, with the local context used to enhance learning and understanding of the STEM concepts. Examples of ill-defined problems embedded in STEM activities were:

(1) Ali saw a bird perched on a tree branch. A question arose in Ali's mind, 'How can it perch for such a long period of time?

 Your task: Create a balancing toy that can stand stably on your fingers like a bird. Each student must produce at least one balancing toy that meets the requirements.

(2) After a year of construction, the Mesilou River bridge that collapsed due to the earthquake has been completed. This news is quite exciting for the villagers who want to cross the Mesilou River. But some are wondering: How much weight can the bridge support? Is it really safe? You and your friends have

been given the opportunity to show the villagers that the Mesilou bridge is actually strong enough to support the weight of vehicles crossing it.

Your task: You and your friend are asked to prove it by designing and building a bridge that crosses a river which is one meter wide. The bridge is strong enough to hold at least 3 cans of coca cola.

Engineering Design Process

Hynes et al. (2011) noted that an engineering design process (EDP) that focuses on solutions and construction of prototypes requires students to use creative and critical thinking as well as problem solving skills. Hence, the engineering design process would offer an effective route as an instructional framework for fostering twenty-first century skills among rural secondary school students in the STEM mentor–mentee outreach programme. The EDP, adapted from the Massachusetts Department of Education (2006) (Fig. 10.1), was employed to guide the implementation of the STEM mentor–mentee outreach programme.

Science teachers in previous programmes had noted several potential challenges while implementing a STEM project-based learning approach in their rural school classrooms in Sabah (Siew et al., 2015). These included inadequate materials, limited facilities, and limited allocation of classroom time. Accordingly, the EDP employed in this programme removed the 'redesign' step proposed by the Massachusetts Department of Education (2006). This modification was made to ensure that students could produce workable prototypes that made best use of the materials and time provided in the programme.

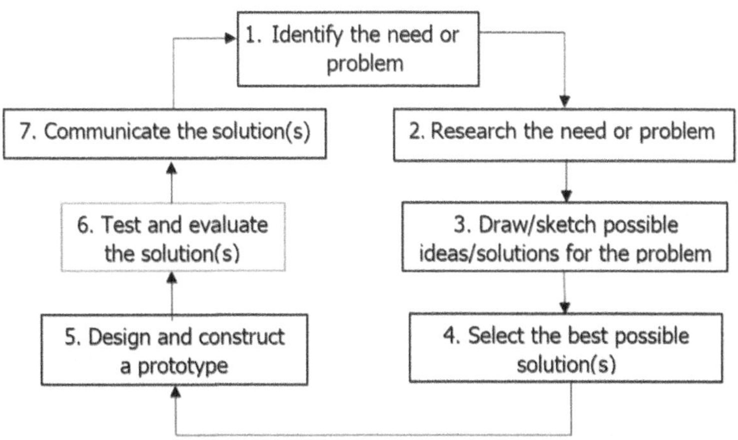

Fig. 10.1 The seven steps of the engineering design process (adapted from Massachusetts Department of Education, 2006, p. 84)

The EDP provides a flexible process that takes mentees from identifying a problem and designing a solution, to developing, creating, testing, and evaluating a prototype to solve daily life problems in their environment using inexpensive materials. Students learn important scientific concepts and their application in engineering and technology, as well as their relationship and application in daily life or real-world contexts. Students could look for connections by engaging with activities or material in 'real-world' contexts to establish relevance.

The EDP allows students to realize that there are many ways to find solutions, as they engage in brainstorming to identify problems and propose solutions. The process of finding the optimal solution based on the constraints requires participants to engage in critical thinking, creativity, imagination, and collaboration, thus developing their communication and problem-solving skills.

Higher Order Thinking (HOT) questions Students answered Higher Order Thinking (HOT) questions that were not strictly required by the curriculum. In a way, answering HOT questions inspired students to acquire new competencies. Anderson et al.'s (2001) taxonomy was used as a guide to develop a blueprint for the HOT questions, which belonged to the 'Analysis and Evaluation' category of Anderson et al.'s cognitive domain. Some samples of HOT questions are:

- In your opinion, if buildings were constructed identical to this prototype, would they be safe to inhabit? If yes/no, please explain why' (Evaluation).
- How can your prototype be modified in order to improve its results in the future? (Analysis).
- Explain why there is a difference in the two bottle submarines' speeds? (Analysis).

The HOT questions were specifically designed to evaluate students' critical thinking skills in connecting STEM activities with their daily life.

The University-School Partnership

The Transformational Learning Community (TLC) model proposed by Bernay et al. (2020) suggests that teaching practices in schools and universities should be extended into the real-world for children and student teachers to develop skills, knowledge, and dispositions. The TLC model utilizes perspectives and aspirations from the wider community as the drivers of future-oriented education programmes. Bernay et al. argues that TLCs should focus on new solutions, new roles, and new ways of working together to reinvent schools for transformational learning by involving children, student teachers, teachers, university lecturers, and the wider community. Recognizing the advantages of the TLC model as one form of school–university partnership to cater to the needs and challenges faced by students in rural areas, the TLC model was adapted in designing the mentor–mentee outreach programme. A school-university partnership would help to address the needs of rural schools by

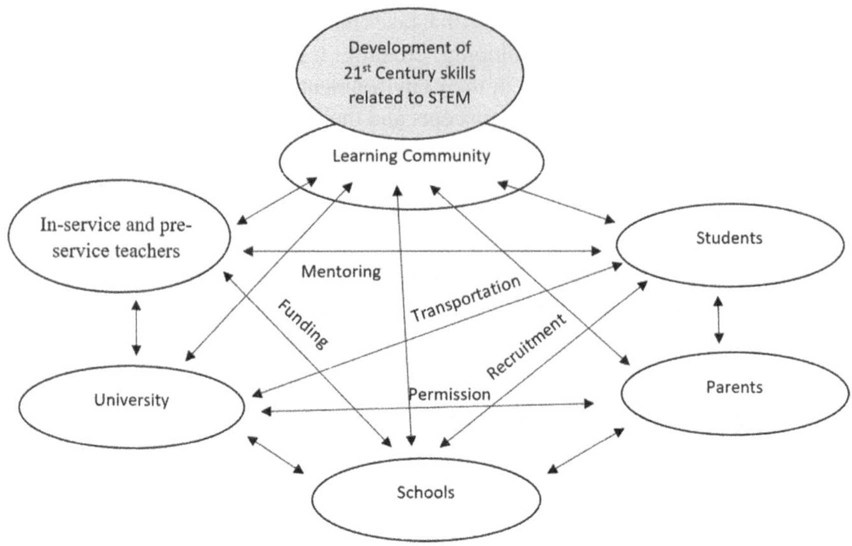

Fig. 10.2 The transformational learning community model adopted by the STEM mentor–mentee outreach programme (adapted from Bernay et al., 2020, p. 141)

providing adequate support, resources, and funding in the development, delivery, and implementation of the programme. Figure 10.2 illustrates the range of relationships and responsibilities in the adapted TLC model.

In order to work in an authentic partnership with rural schools in Sabah, a community of practice was created with staff from the University Malaysia Sabah (UMS). This enabled the programme to draw on the expertise and experience of university science lecturers and in-service and pre-service teachers who were studying at the University, along with teachers, principals, and parents. In other words, UMS science lecturers, UMS in-service and pre-service teachers, school heads of science departments, school principals, and parents worked as a team to implement the programme. The parents granted permission to allow their children to participate on Saturdays. School principals and heads of science departments collaborated to provide the school hall and public address system (PA system) to be used for the activities. The university provided a free bus service to transport the mentors and materials to the schools, which were located 16–216 kms from Kota Kinabalu, Sabah.

The resources used for running the programme were funded by the schools and university. The University's Centre for External Education allocated RM50 to each of its Master's students to run the programme outside the classroom, which was used to support the costs needed to buy materials and equipment. The in-service teachers also sought sponsorship from the local community, such as local businesses, to support the cost for food and drink for mentees and mentors. The in-service teachers also borrowed science apparatus from their school laboratory, such as crocodile-clips, hot glue guns, and glue gun sticks to support the activities. Recycled materials such as plastic bottles, empty boxes, newspapers and egg trays were collected by mentors and school teachers, and were used as the main resources for the activities. Some tools and materials, such as scissors, rulers, and cutters, could be re-used in subsequent programmes. Administrative costs were extremely low as free messaging platforms such as Telegram and WhatsApp were used to facilitate communication among participants.

Monitoring and Evaluation Tools The monitoring and evaluation were done through multiple qualitative and quantitative means: a scoring rubric for activities, participants' responses to open-ended questions, mentors' field notes, open-ended questionnaires, focus group observation, and interviews.

Scoring Rubrics Scoring rubrics were developed for mentors to evaluate the prototypes produced by the group during their presentation and testing. Aspects assessed were product functionality, sketches, group collaboration, and understanding and application of scientific concepts. Scoring rubrics were constructed based on analytical scoring. The quality of student responses and products was assessed from "Poor" (lowest level) to "Very Good" (highest level). Appendix 1 shows an example of a scoring rubric used to assess one of the STEM activities.

Field Notes Mentors wrote their field notes based on observations made during the STEM activities, and the semi-structured focus group interviews with mentees. Focus group observations were collected using an observation form adapted from scoring guides developed by Wang et al. (2015). The quality of the students' responses was ranked from '0' (lowest level) to '3' (highest level). The interview questions were open-ended and the students were explicitly encouraged to draw from their learning experiences of working on the STEM activities. Each focus group interview was conducted in groups consisting of four to five mentees after the completion of each STEM activity.

Close-Ended Questionnaires and Open-Ended Questions The pre- and post-programme questionnaires were adapted from Illinois Valley Community College (2011) and the Knowledge, Attitude and Practices survey (Kaliyaperumal, 2004). The pre-programme questionnaire consisted of 20 items while the post-programme questionnaire consisted of 12 items that could be grouped into three different dimensions:

- Knowledge or skills about STEM—Example: 'A scientist, technologist, engineer, or mathematician needs to be creative' (pre-test); 'The STEM activities today improve my ability to generate new ideas' (post-test).
- Attitudes towards STEM—Example: 'My parent(s) have suggested that I consider a career in science, technology, engineering, or mathematics' (pre-test); 'After today's activity, I would consider a career in science, technology, engineering, or mathematics' (post-test).
- Practices in STEM—Example: 'I was able to complete hands-on tasks with a team' (pre-test); 'I am able to complete hands-on tasks with a team' (post-test).

All responses were entered on a 6-point Likert scale: '*Strongly Agree*', '*Agree*', '*Slightly Agree*', '*Slightly Disagree*', '*Disagree*', and '*Strongly Disagree*'. The pre- and post-programme questionnaires measured shifts in mentees' knowledge or skills about, attitudes towards, and practices in STEM over the duration of the programme. Open-ended questions such as 'Something new I have learned today was…' were added to the questionnaire. The open-ended questions offered the respondents an opportunity to clarify their meaning, and to contribute additional information not captured by the close-ended questionnaire.

Implementation of the STEM Mentor–Mentee Programme

The mentor–mentee programme was implemented in 16 rural secondary schools in the districts of Tenom, Tambunan, Ranau, Tuaran, Kota Marudu, Kudat, Penampang, Putatan, Telipok, Sipitang, and Kiulu throughout 2015–2019. The one-day programme was carried out over a period of 11 h in each school, between the hours of 7.00 am and 6.00 pm. The mentors and mentees took an hour meal, prayer, and rest break in the middle of the day. Throughout the 5-year programme, a total of 732 students, 342 in-service, and 99 pre-service teachers were involved in the STEM mentor–mentee outreach programmes. Table 10.1 shows the details of each STEM mentor–mentee programme.

In these programmes, mentees worked in teams of four or five to solve an ill-defined problem by designing and building workable solutions in the form of prototypes which could be tested in relation to the criteria set in the problem. Mentees took different approaches, made mistakes, accepted input and learned from group members, and then tried again. The focus was on communicating and developing solutions based on the experience and environment of the mentees, enabling them to integrate their learning experiences inside and outside school. Aspects of science and mathematical concepts and communication skills were also emphasized and evaluated during the presentation of prototypes by group members to their peers and facilitators. A total of three STEM activities were conducted in each school, each taking about 3 h and 20 min. Appendix 2 shows the mentees participating in the activities.

Mentoring of each student group involved one or two adult mentors forming a supportive and caring relationship with four to five mentees. The mentors played the role of facilitator and made a commitment to interact with the mentees over the period of the one-day programme. The interaction was guided by activity worksheets, which allowed time for two-way discussion of the STEM activities between the mentees and mentors.

Table 10.1 The STEM mentor–mentee programme (2015–2019)

School	Name of rural secondary school (Distance from Kota Kinabalu in kilometer)	Implementation date	No. of students	No. of teachers	
				Pre- service	In- service
A	SMK Badin,Tuaran (42.1)	09 May 2015	62	3	19
B	SMK Kemabong, Tenom (216)	23 May 2015	37	9	21
C	SMK Mat Salleh, Ranau (129)	10 Oct 2015	48	14	20
D	SMK Abdul Rahim II, Kudat (178.7)	23 Oct 2015	35	5	27
E	SMK Tambunan, Tambunan (73.3)	23 April 2016	50	17	33
F	SMK Kota Marudu, Kota Marudu (120)	26 April 2016	48	9	26
G	SM ST Peter Telipok, Telipok (23.0)	22 Oct 2016	50	15	18
H	SMK Agama Tun Said, Kota Belud (64.4)	22 April 2017	50	11	14
I	SMK Putatan, Putatan (16)	28 Oct 2017	48	–	36
J	SMK Limbanak, Penampang (17)	02 Nov 2017	50	–	28
K	SMK Mat Salleh, Ranau (129)	21 April 2018	36	–	27
L	SMK Tun Fuad Stephens, Kiulu (55.1)	07 Oct. 2018	37	–	20
M	SMK Sindumin, Sipitang (161.4)	20 Oct. 2018	36	–	13
N	SMK Matunggong, Kudat (136)	13 April 2019	45	10	9
O	SMK Kanibongan, Pitas	02 Nov 2019	50	–	20
P	SMK Kampung Contoh, Petagas	09 Nov 2019	50	–	11
		Total	**732**	**99**	**342**

Data Analysis

The analysis of quantitative data used descriptive statistics in the form of frequency, percentage, and means with the aid of the Statistical Package for Social Sciences. The difference in the percentage between the pre- and post-programme questionnaires was computed as a measure of change in participants' knowledge or skills about, attitudes towards, and practices in STEM.

Mentors' field notes and mentees' responses to open-ended questions were analysed using thematic analysis. Thematic analysis is a pattern recognition technique that requires searching through the data for emerging themes.

Outcomes of the Programme

The main outcomes of the mentor–mentee outreach programme in relation to knowledge and skills about, attitudes towards, and practices in STEM, twenty-first century skills, and challenges encountered are discussed below.

Attainment of Knowledge or Skills about Attitudes Towards, and Practices in STEM

The questionnaire data revealed gains in knowledge or skills about, attitudes towards, and practices in STEM after mentees participated in the STEM outreach programme. The mentees perceived that they became highly knowledgeable about STEM through the programme (mean difference = 0.48). The programme approach, which focused on integration of STEM through an engineering design process, was also found to promote a positive change in mentees' attitudes towards STEM (mean difference = 0.20). In addition, the mentees' participation in the programme contributed to an increased capability of carrying out practices in STEM (mean difference = 0.49). With the support of the mentors, mentees were able to understand the underlying concepts and possible applications of STEM, thus enhancing their knowledge or skills about, attitudes towards, and practices in STEM.

Application of Scientific Knowledge in Solving Daily Life Problems

Almost every mentee (98%) noted that they benefitted from the STEM activities as they were exposed to real-life situations where scientific knowledge was applied for solving daily life problems. More importantly, STEM activities succeeded in providing a platform for them to apply scientific knowledge in solving problems. The scientific concepts the mentees identified related to water and air pressure, equilibrium of force, base area, balanced force, surface tension, stability, water density, and the buoyancy force in a submarine. Mentors from one group confirmed that interviews with mentees revealed that mentees found the need to apply the concept of impulse in order to create an innovation that helped to absorb the impact of an egg being dropped from a high place.

Application of Scientific Knowledge in Answering Higher Order Thinking (HOT) Questions

A significant number of mentees (88%) reported that a deep understanding of scientific knowledge helped them answer questions that required higher order thinking, and to be creative in reapplying knowledge learnt in the design and production of prototypes. Mentors observed that HOT questions provided mentees with an opportunity to think critically about the answers and make connections with scientific concepts they had learnt in class.

Connecting STEM Activities with Daily Lives and Scientific Concepts Learned

A large percentage (93%) of the mentors noted in their field notes that participants made connections between the STEM activities with their daily life. For example, mentors observed that mentees could relate how ships or boats function and why they could float on the surface of water by making comparisons with their own boat models. Another example related to answering the HOT questions: mentees could relate the floating needle and paper clip activity with the water strider bug, a floating log, water lilies, floating ants, and others. Mentors supported these claims. For example, one noted that:

> 'Scientific knowledge is not only for answering exam papers but is also useful in helping students create connections and explain situations faced in their daily lives. In this case, it is observed that students applied scientific concepts they learned during physics lessons in solving problems given to them. Students not only applied the science principles and laws they learnt but also used them in practical forms.' (Mentor, School A)

HOT Questions Sparked Critical Thinking

A large percentage (93%) of the mentees expressed through the open questions on the post-programme questionnaire that they were challenged to think critically when answering the HOT questions in the STEM programme. The mentees felt that the HOT questions were difficult but they tried their best to answer them and to link them with their prior knowledge.

According to the mentors, mentees were capable of giving rational answers to the HOT questions. For example, one of the group members gave an excellent answer and showed that he/she understood the concept and was able to give a suggestion to improve the existing prototype if given the opportunity to design it with extra materials. The sharing of answers added knowledge collectively to the group, in addition to developing students' critical thinking skills. According to the mentors, the STEM activities tested and challenged mentees to think "outside the box" using higher order thinking skills.

Designing and Building Something New and Practical

A large percentage (96%) of the mentees expressed in the open questions that the STEM activities gave them an opportunity to create many new, interesting and practical science-related products using everyday materials. They stated that making the balloon-powered car made from plastic bottles was a new experience for them. They were fascinated with finding new ways to make a powered car that could be moved by air, using readily available materials such as glue, bottles, pencils, and so on. Another activity was making a boat. The mentees said they realized that play dough can float when shaped into a boat. Others noted that finding the gravity centre through making the balancing toys was a new activity. A few mentees commented that they discovered new ways of floating needles and paper clips.

When participants were asked why they were excited by participating in the STEM activities, the hands-on approach was highlighted. For example, a typical response was: 'Because we got the chance to design and build a new model which we only see in textbooks'. In addition, students showed interest in the STEM activities because they could become 'designers' of their own boat in the future (Mentor, School F). Mentors observed that the mentees could design egg protection tools and that every group member worked together the whole time by contributing ideas and carrying out the projects as they had planned. Thus, according to mentors, the STEM activities seemed to provide a very good start to stimulating mentees' interest in learning STEM.

Thinking Creatively through the Combination of Ideas

A considerably large percentage (78%) of the mentors noted in their field notes that mentees exhibited creativity above what they had anticipated. They reported that the group work seemed to enhance mentees' ability to produce different kinds of products using limited materials as a result of the combination of ideas contributed by group members. This could be seen during the construction of a bridge using newspaper. Many new ideas and views were brought up during brainstorming within the group. With a combination of ideas from group members, students were able to build a paper bridge that could carry the load of 3 cans of drinks. Thus, mentors noted that teachers need to acknowledge students' potential and use the right tools and mentorship to enhance their hidden potential.

Ill-Defined Problems Inspired Creativity and Thinking

A significant number of mentors (93%) reported that mentees faced complexities posed by the ill-defined problems in the programme. Developing effective responses to the challenging tasks inspired creativity and thinking. For example, the mentees needed to be creative when designing and building a bottle car that was powered by a balloon. Participants had to figure out ways to move the 'car' using only air within a balloon, and to think of a method of reducing the car's weight and decreasing its tyre resistance.

Sketching, Designing, and Constructing Models Fostered Creative Thinking and Problem-Solving Skills

A considerably large percentage (88%) of the mentees reported that the STEM activities encouraged creative thinking and also problem-solving skills. This was because each activity needed them to sketch and design models, utilizing the creativity of each group. Mentees noted that they had to think of a way to design models that worked and at the same time possessed creative elements. For example, the 'balancing toys' activity successfully inspired creativity and imagination within mentees as almost every one of them was able to build a balancing toy, but with different designs. Unexpectedly, mentees in one group were able to create nine balancing toys with different designs. Furthermore, this activity also encouraged mentees to use their thinking skills. By looking at the other solutions, mentees gained ideas on how to create toy designs. This was supported by observations made by mentors who noted that: *Besides creating one 'balancing toy', students can think of ways to merge a few 'balancing toys' in a stable condition.*

Another example in which creativity was evident was in the boat making activity. From the provided materials, mentees designed two different kinds of boats, one from play dough and another from straws. These new designs increased the number of marbles carried by the boat, as long as the boat was stable enough to carry them.

Working Cooperatively

A significant number of mentors (93%) observed that mentees not only came up with some thoughtful ideas but also showed a spirit of teamwork during the STEM activities. Mentors described that mutual understanding and cooperation boosted the confidence of each mentee to do his/her best work in order to construct a workable prototype. Mentees made attempts to cooperate as a group to solve the problems, and brainstorming within the group helped mentees to use their critical and creative thinking. This was proven when a group of five was able to create nine balancing toys with different designs. In another group, group members divided up some tasks, such as rolling up a newspaper and stitching together some newspapers, in order to produce a paper bridge, which in their opinion was a very difficult and challenging activity.

Solving Problems with a Determined Effort

A large percentage (92%) of the mentors reported that the STEM activities challenged the mentees to think of many ideas and make many attempts without giving up. Mentees tested their prototypes through many attempts, and modified and improved their original ideas through the process of trial and error. For example, mentees made modifications to the boat several times so that the boat they built could hold up to 26 marbles. They proved their determination to solve the problem despite having to repeatedly test the boat's ability to accommodate large quantities of marbles.

Challenges

A number of challenges were faced by mentees participating in the STEM outreach programme.

Time Constraints Time constraints were a major concern during the programme, causing some mentees to not complete their prototypes according to plan. Mentees agreed that it was tough to design and create three prototypes or complete three activities in one programme. However, mentees also said that their problem-solving skills were highly stimulated because they had to solve problems in the stipulated time, including to create a working model out of the materials prepared. For this reason, the number of activities was reduced from three to two in subsequent outreach programmes.

Mentees Had Limited Understanding of Relevant Scientific Concepts Mentors noted in their observations and interviews that the mentees experienced difficulty in applying scientific and technical knowledge in the implementation of the STEM activities. For example, the mentees were weak in mastering physics concepts, and the mentors needed to put in extra effort to relate physics principles to designing and building activities. This led the mentees to be less creative in creating something unique for the STEM activity. For example, in one activity, mentees were unable to link the scientific concepts such as buoyancy, force, and density to their design. Mentees faced difficulty in stating and explaining the concept of buoyancy connected with large ships made out of steel. It was also noted that students chose inappropriate materials and assembly techniques for the parts of the boat prototypes.

Discussion, Conclusion, and Future Directions

The STEM mentor–mentee outreach programme represented a new idea to address the specific challenge of supporting STEM education in rural schools in Sabah, adding value to rural school students, teachers, and the University. The programmes used a university-school partnership model to reach out to rural schools to improve the quality of STEM education and make it more relevant to students' local contexts. The programmes were executed successfully in a cost-effective way. The personnel and resources were creatively mobilized through the collaboration of different partners from the university and participating schools. Mentors included in-service and pre-service teachers, both studying at the University. Schools contributed to the programme success through the provision of facilities to carry out the activities. The University provided transport for the mentors to get to the schools. All partners benefited from contributing to and being able to use their knowledge, skills, funds, and resources effectively in this university-school partnership.

The programme supported the development of participating students' twenty-first century skills related to STEM, and supported increases in students' knowledge or skills about, positive attitudes towards, and practices in STEM. Evidence from school students suggests that the programme enabled them to apply STEM knowledge in solving daily life problems, designing and producing everyday products, and answering HOT questions. Students were also able to connect the STEM activities with their daily lives and with scientific concepts learned in the classroom, and to create new and practical products using everyday materials. These findings indicate that the STEM mentor–mentee outreach programme can help students to relate and apply STEM knowledge to their real-world problems and contexts.

The STEM mentor–mentee outreach programme not only allowed students to gain and integrate STEM knowledge, but also provided an avenue to boost their creativity, critical thinking, problem solving skills, and team work. Students' creative and critical thinking was sparked through solving HOT questions and responding to ill-defined problems posed in the STEM activities. In addition, activities such as sketching, designing, and constructing prototypes helped students to foster their creative thinking and problem-solving skills. New ideas were generated through the combination of ideas of group members as well as through trial-and-error. Students as a group responded effectively even with limited materials and time in terms of organizing their thoughts to choose the best possible solution for their prototype using related scientific concepts.

While the students described many positive learning experiences gained in this programme, they also pointed out several challenges that influenced their success in the STEM activities. The two most commonly mentioned challenges were the limited amount of time, and their need for additional understanding of scientific concepts and technical knowledge in order to construct prototypes. This was especially evident when the STEM activities required the use of a wide range of cognitive abilities, involving higher order thinking skills such as applying, analysing, evaluating, and creating from Anderson et al.'s (2001) Taxonomy.

Feedback from students and mentors indicated that reducing the time pressure by negotiating or extending the execution time would help some students complete the activities. Feedback also highlighted that students who were equipped with sufficient knowledge of scientific concepts were able to answer the HOT questions. Adequate classroom opportunities are therefore needed to strengthen students' scientific and technical knowledge.

The STEM mentor–mentee outreach programmes opened the eyes of many rural secondary school students about their potential to pursue careers in STEM such as engineers, scientists, and technologists. STEM activities have the potential to develop a variety of STEM-related twenty-first century skills, including problem solving, critical thinking problem solving, critical thinking, creativity, communication skills, and teamwork, which can be used across a range of STEM jobs. The findings confirm the usefulness of STEM mentor–mentee outreach programmes facilitated by university-school partnerships in offering a meaningful way to develop twenty-first century skills among rural high school students.

The STEM mentor–mentee outreach programme described in this chapter is a cyclic process for promoting the interests and dispositions of students in selecting STEM-related fields for their future careers. As part of this process, the STEM activities which were evaluated for their suitability and applicability for rural secondary schools have been improved and compiled as a learning module for use in other rural schools. Information about the implementation from this programme can also be used as a reference to grow the STEM movement in other rural schools. Attributes such as being able to think and solve problems critically and creatively in the STEM domains, as addressed in the learning module, could help prepare youth for STEM-related careers. Efforts have also been made to improve the content of the STEM activities so that they are relevant to the local conditions of rural school students by introducing 'localized' topics and activities. For example, in later iterations of the STEM activities, students were engaged to make predictions about and produce new prototypes that could be available in the future. Some school science teachers have also begun using the learning module to conduct STEM activities in extra-curricular Science Clubs. In Malaysia, students are able to choose from three categories of after-school activity: clubs and associations, uniform bodies (such as Scouts. St. John's Ambulance, and Fire Brigade), and sports. Thus, the idea of the STEM mentor–mentee outreach programme and its activities can be adapted and fine-tuned by any educators to be delivered to new groups of students based on their local context.

To ensure the sustainability of the STEM outreach programmes and the university-school partnership, efforts have been made to expand the programmes to primary rural schools in the interior of Sabah since 2019. Entrepreneurial thinking skills have been introduced in the programme so that primary school children could apply these skills in addressing 'localized' problems and issues and turning problems into marketable products. Therefore, the STEM outreach program can be a holistic programme that aims to increase the number of students participating in STEM, especially among rural students. This programme can therefore be considered as a catalyst supporting various initiatives undertaken by the Malaysian government to achieve developed country status by 2050.

Appendix 1

Sample of Scoring Rubric for Product Presentation

Criteria	Poor	Acceptable	Good	Very good	Marks obtained
	2.5	5.0	7.5	10.0	
Product functionality (1) Ability of the water sprinkler model to produce a powerful water jet	The water sprinkler does not emit water directly	Only a few water channels can emit water / holes are not drilled properly	The water sprinkler squirts water in a small area/ all holes can emit water	The water sprinkler is very stable, squirts water to a large area/ all holes squirt water strongly	
Product functionality (2) Ability of the water sprinkler model to wet the soil over a long distance	Failed to moisten the soil	Soil gets wet in the original place only (0.2 m - 0.4 m)	Water wets the soil covering a moderate distance. (0.5 m- 0.7 m)	Water wets the soil covering a long distance. (0.8 m- 1.0 m)	
Group collaboration	There was no discussion and only one member did the job	Discussions between members only occurred once and only 2 members carried out the task	Discussions between members were conducted regularly, but only 3 members carried out the task	Discussions were conducted all the time. Members worked well together and respected each other's ideas	

(continued)

(continued)

Criteria	Poor	Acceptable	Good	Very good	Marks obtained
	2.5	5.0	7.5	10.0	
Understanding and explanation of scientific concepts	Students are not confident in explaining concepts, and speak in a way that is poorly understood (stuttering voice)	Students are less confident in explaining concepts, and speak in a way that is poorly understood (voice stutters but still try to explain)	Students are very confident in explaining concepts, in an easily understood way (clear voice) but cannot answer all questions posed by the panel	Students are very confident in explaining the concepts, in an easily understood way (loud voice) and can answer questions posed by the panel	
Sketch	Incomplete sketch, no label	Semi-complete sketch, few labels	Complete sketch and some labels	Detailed sketches, label statements and materials used are detailed	

Appendix 2

Mentees working together to complete the "The Great Green Octa" model

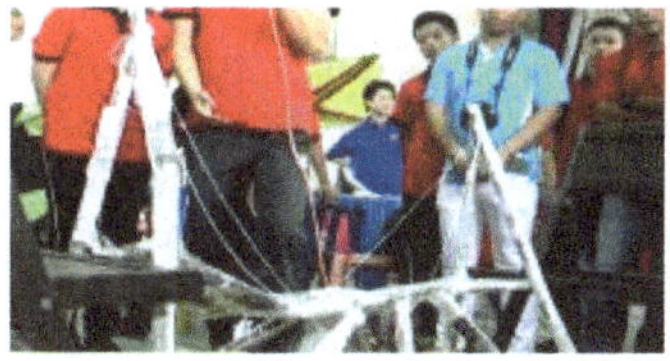

Mentees presenting and testing their paper bridge in front of the judges

Mentees presenting their sketch of a future town in the year 3030

All groups ready to make a presentation

Mentees testing their wind mill with a milliammeter while observed by mentors

Mentees testing their future boats in a pond

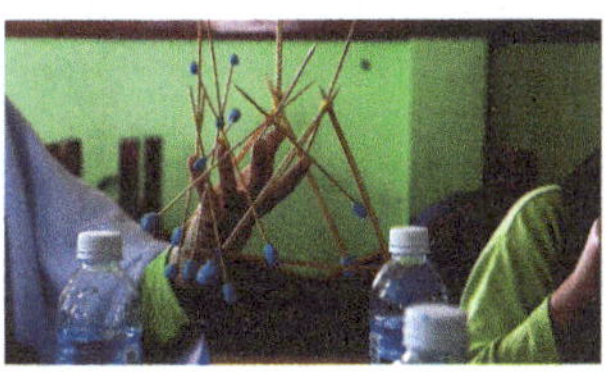

Mentees proud to show their balancing toys

The judge listening to the presenters introducing their green building

STEM activities in action

References

Academy of Sciences Malaysia. (2015). *Science outlook: Action towards vision (Executive Summary)*. Academy of Sciences Malaysia. https://issuu.com/asmpub/docs/science_outlook_2015_exec_summary.

Academy of Sciences Malaysia. (2018). *Science outlook 2017: Research and policy recommendations document*. Academy of Sciences Malaysia. https://issuu.com/asmpub/docs/so2017.

Anderson, L. W., Krathwohl, D. R., & Bloom, B. S. (2001). *A taxonomy for learning, teaching, and assessing: A revision of Bloom's Taxonomy of educational objectives*. Longman.

Bernay, R., Stringer, P., Milne, J., & Jhagroo, J. (2020). Three models of effective school–university partnerships. *New Zealand Journal of Educational Studies.*, *55*, 133–148. https://doi.org/10.1007/s40841-020-00171-3.

Brookshire, B. (2014). *A teacher's guide to mentoring in STEM: How to find and be a good mentor in STEM*, November 26. Science News for Students. https://www.sciencenewsforstudents.org/article/teachers-guide-mentoring-stem.

Bybee, R. (2013). *The case for STEM education: Challenges and opportunities*. NSTA Press.

Cutucache, C. E., Luhr, J. L., Nelson, K. L., Grandgennett, N. F., & Tapprich, W. E. (2016). NE STEM 4U: An out-of-school time academic program to improve achievement of socio-economically disadvantaged youth in STEM areas. *International Journal of STEM Education, 3*, 6. https://doi.org/10.1186/s40594-016-0037-0.

Edgeprop. (2011, March 2). *Bringing 'transformation' to Sarawak rural schools*. https://www.edgeprop.my/content/bringing-transformation-sarawak-rural-schools.

Essex, N. L. (2001). Effective school-college partnerships, a key to educational renewal and instructional improvement. *Education (chula Vista), 121*(4), 732–736.

Greenwald, N. L. (2000). Learning from problems. *The Science Teacher, 67*(4), 28–32.

Hynes, M., Portsmore, M., Dare, E., Milto, E., Rogers, C., Hammer, D. & Carberry, A. (2011). *Infusing engineering design into high school STEM courses*. National Center for Engineering and Technology Education. https://files.eric.ed.gov/fulltext/ED537364.pdf.

Illinois Valley Community College. (2011). *STEM activities for middle school students: Special focus on girls* .https://www.coursehero.com/file/37173728/STEM-Activities-Handbookpdf/.

Institute for Development Studies (Sabah). (2007). *Sabah development corridor blueprint, 2008–2025: Harnessing unity in diversity for wealth creation and social well-being*. Institute for Development Studies (Sabah). https://www.sedia.com.my/SDC_Blueprint.html.

Kaliyaperumal, K. (2004). Guideline for conducting a knowledge, attitude and practice (KAP) study. *AECS Illumination, 4*(1), 1–9.

Kementerian Pendidikan Malaysia [Ministry of Education Malaysia]. (2016a). *Panduan pelaksanaan Sains, Teknologi, Kejuruteraan dan Matematik (STEM) dalam pengajaran dan pembelajaran* [A guide to the implementation of Science, Technology, Engineering and Mathematics (STEM) in teaching and learning]. http://smksyedsira.edu.my/wp-content/uploads/2016a/12/7.-Panduan-Pelaksanaan-STEM-Dalam-PP.pdf.

Kementerian Pendidikan Malaysia [Ministry of Education Malaysia]. (2016b). *Pembelajaran abad ke-21* [21st century learning]. https://anyflip.com/qqfu/zmlg/basic.

King, P. M., & Kitchener, K. S. (1994). *Developing reflective judgment: Understanding and promoting intellectual growth and critical thinking in adolescents and adults*. Jossey-Bass.

Laursen, S. L., Thiry, H., & Liston, C. S. (2012). The impact of a university-based school science outreach program on graduate student participants' career paths and professional socialization. *Journal of Higher Education Outreach and Engagement, 16*(2), 47–78.

Ling, S. E., Mahdib, R., Mohamadin, M. I., & Manaf, B. A. (2015). Second chance science education for school leavers. *Procedia - Social and Behavioral Sciences, 167*, 288–292.

Massachusetts Department of Education. (2006). *Massachusetts science and technology/engineering curriculum framework*. https://files.eric.ed.gov/fulltext/ED508413.pdf.

Moskal, B., & Skokan, C. (2011). Supporting the K-12 classroom through university outreach. *Journal of Higher Education Outreach and Engagement, 15*(1), 53–75.
Office of the Chief Scientist. (2013). *Science, Technology, Engineering and Mathematics in the national interest: A strategic approach.* Australian Government. https://www.chiefscientist.gov.au/sites/default/files/STEMstrategy290713FINALweb.pdf.
Siew, N. M., Amir, N., & Chong, C. L. (2015). The perceptions of pre-service and in-service teachers regarding a project-based STEM approach to teaching science. *Springerplus, 4*, 8.
Wang, C. C., Ho, H. C., & Cheng, Y. Y. (2015). Building a learning progression for scientific imagination: A measurement approach. *Thinking Skills and Creativity, 17*, 1–14.
World Bank. (2010). *Malaysia economic monitor: Inclusive growth.* http://documents1.worldbank.org/curated/en/956321468087840820/pdf/639430WP0Malay00Box0361533B0PUBLIC0.pdf.

Nyet Moi Siew is an associate professor at the Faculty of Psychology and Education, Universiti Malaysia Sabah, Malaysia. Her primary research interests are in the area of scientific and mathematical creativity and thinking, and STEM Education. She has written three book chapters published in IGI Global, Springer, and Taylor and Francis Group, and over 45 research papers published in international peer reviewed journals. She currently served as program coordinator for outreach community programmes about STEM for rural secondary school students in Sabah (2015-2021) and she received an Honourable Commendation for the Wenhui Award for Educational Innovation in 2019.

Chapter 11
STEAM Education in Korea: Enhancing Students' Abilities to Solve Real-World Problems

Jiyeong Mun and Sung-Won Kim

Abstract The Korea Foundation for the Advancement of Science and Creativity (KOFAC) funds a valuable programme called STEAM R&E for research and education as one of its programmes offered to high school students; the aim of the programmes is to enhance creative problem-solving ability and peer cooperation skills through STEM/STEAM education. STEAM R&E consists of student-centred research tasks that aim to improve students' real-world problem-solving abilities and research capabilities. The programme follows the scientific research process: students make a proposal, apply for a grant, receive consultation from experts, and make final presentations orally, with poster displays, or both. Through the programme, students experience for themselves constructing research questions, designing and conducting experiments, and writing reports, and even failures are allowed. Participants reported enjoying the presentations and the opportunities to communicate with each other. In this chapter, we report on the quantitative and qualitative achievements of the participating students through, respectively, the gains in cognitive, affective, and practice domains during the programme and essays the students wrote on their experiences in the STEAM R&E programme. Drawing on the results of the study, we make suggestions to improve students' learning in STEM/STEAM programmes through student-led research.

Keywords STEAM R&E · Student research task · Creative problem-solving ability · Peer cooperative competency

Introduction

Background and History of STEAM Education in Korea

In the IEA's Trends in International Mathematics and Science Study, Korean students are always among the highest achievers in science and mathematics, but they also rank

J. Mun · S.-W. Kim (✉)
Ewha Womans University, 52, Ewhayeodae-gil, Seodaemun-gu, Seoul 03760, Korea
e-mail: sungwon@ewha.ac.kr

M. M. H. Cheng et al. (eds.), *Concepts and Practices of STEM Education in Asia*,
https://doi.org/10.1007/978-981-19-2596-2_11

among the lowest in interest and self-confidence (IEA, 2011). It is not common for top-ranked students in Korea to voluntarily show interest in science and mathematics.

In 2011, Korea's Ministry of Education, Science and Technology (MOEST) put forth the Second Scientific and Technical Personnel Development Assistance Plan (2011–2015), the first phase of which strengthens scientific and technical personnel training by incorporating arts into STEM education, creating STEAM (MOEST, 2011). Systematic approaches will be needed to secure the supply of trained personnel who will be able to lead industries of the future, and STEAM education attracted great interest when it was introduced.

In the second phase of the plan, in 2017, the Ministry of Education (MOE, 2017) announced a mid- to long-term plan for convergence personnel education (2018–2022). The plan's vision was the 'Cultivation of creative and convergent personnel through internalization of convergence personnel education', which had three goals: (1) future-oriented education that fosters students' and teachers' convergence capabilities, (2) classroom innovations, and (3) STEAM education with real-world applicability. The plan included four strategies for achieving these goals: expanding STEAM education to elementary, middle, and high school students to 'nurture' their interest in STEAM careers through specific strategy; 'making teachers happy' through STEAM, creating 'STEAM everyone enjoys', and creating 'shared STEAM platforms'. The focus of the MOE's plan was on establishing a research-based STEAM support system and strengthening domestic and overseas STEAM partnerships.

In 2020, the MOE issued the 2nd Comprehensive Plan for Convergence Education (2020–2024), which can be seen as the third phase of the 2017 long-range plan (MOE, 2020). The aim of the most recent plan is to cultivate global creative and convergence personnel, and to transition to a future education system. The main goals are (1) for convergence education between subjects centred on science, mathematics, and information education to be settled in the field; (2) to design learning spaces that enable multidimensional projects and combine cutting-edge technologies and tools; and (3) to establish an ecosystem for science, mathematics, and information education based on the linkages and cooperation between individual-school-society. Moreover, the concrete strategies of the plan are as follows: first, not only achieving basic subject literacy but also solving real-life problems using that subject knowledge; there should also be subject-oriented project learning that integrates and reinforces multiple subjects. The plan also promotes curriculum-linked project classes in which students choose their own learning topics and design their learning processes. Second, the phase-three plan supports the development of a teacher convergence learning community in which teachers of various subjects can cooperate and research, develop, and share convergence education content. Third, the plan aims to establish spaces and environments where convergence education is realized. By developing digital textbooks, advanced learning tools (AR/VR, IoT) and online learning environments, the government aims to create a future convergence learning space without textbooks and desks. Finally, the government will work to expand the learning space by linking various resources and content outside of schools such as libraries and research institutes, and to build a convergence education network ecosystem by establishing cooperative relationships with schools and local communities.

Concept of STEAM Education in Korea

The Korean Ministry of Education defined STEAM as 'education for increasing students' interest and understanding in scientific technology and for growing STEAM literacy based on scientific technology and the ability to solve problems in the real world' (KOFAC, 2021). The United States (Sanders, 2009) and the United Kingdom (Bell, 2016; Harrison, 2011) are implementing STEM education for a similar purpose. The stated goal of STEM is to build a STEM-literate society by developing a general workforce with 21st-century competencies and an advanced research and development workforce focused on innovation (Bybee, 2010, 2013). In Korea, the movement to enhance creativity by adding the elements of humanities and art to STEM is gaining attention (Kim, 2015; KOFAC, 2021). It is recognized nowadays that arts can be the backbone of real-world problem solving through design, expression, and student inquiry and expression across disciplines (Quigley & Herro, 2019).

STEAM classes cannot be guided solely by students' interest levels. Instruction must focus on helping students understand the principles of scientific technology and reach the achievement standards of science and mathematics. When students realize the meaning and purpose of their learning, they become better equipped to solve real-world problems through having designed, explored, and tested solutions for themselves. STEAM has introduced technological and engineering elements to connect the real world to classroom teaching and to increase students' science and mathematics understanding (Jeong et al., 2020; Kim et al., 2012).

Organizations Supporting STEAM and Education Programmes in Korea

The government is the main organization for enhancing STEAM education. The MOE creates new policies and supports municipal offices of education, and KOFAC manages STEAM initiatives financially and administratively. Municipal education offices are responsible for encouraging teachers and schools to implement STEAM programmes as regular or after-school classes.

KOFAC supports many STEAM education programmes both inside and outside the school. There are three categories of inside-school programmes: school programmes, teacher programmes, and student programmes, and the school programmes are STEAM Leader School and Infinite Imagination Maker Rooms in School. The former programme is the model for broad application of STEAM education according to the school curriculum, and the latter is to build spaces for students to realize their creative ideas and cultivate self-directed learning skills. The student programme is STEAM R&E for high school students, as discussed below.

STEAM R&E

Among the STEAM education programmes, STEAM R&E is the most suitable for enhancing students' ability to solve problems in the real world. The programme has supported student R&E activities since 2015 and awarded students who were outstanding in integrating their knowledge and experiences of different subjects to solve real-life problems on their own (KOFAC, 2021). Students select their own research topics and conduct real investigations that follow the procedures real-world scientists follow: write a proposal, apply for a grant, begin the research, communicate through progress reports incorporating reflections and expert feedback, prepare the final report, etc.

Every April, KOFAC usually announces that STEAM R&E is open to applications from high school students, and selects 130 teams (120 teams in 2021 due to the pandemic) from among the applicants. Each team consists of three to seven students (usually in grade 10 or 11) and a teacher or teachers. KOFAC supports all teams during the progress of the research as well as with financial and administrative matters.

Start-up workshops for the participating teachers improve their research ethics and leadership skills and help them develop student leadership skills. In addition, professionalism is enhanced through special lectures on career guidance and research competence. Students are guided through workshops, camps, and lectures on research methods and research ethics. During the research process, each team receives blended research consultation via either online or face-to-face mode from an assigned expert who checks their progress and makes suggestions for improving presentations. At the end of each programme cycle, a professional evaluation team judges the students' final presentations.

In this chapter, we report on the students' experiences throughout the project. We collected survey data and also analysed the students' responses in their individual essays. Finally, we examined the students' experiences against whether STEAM R&E had achieved the programme goals.

Research Method

Developing the Performance Indicators

We developed the STEAM R&E performance indicators to attempt to measure the students' performance. We developed the indicators according to the following process: First, we conducted a perception survey to identify indicators that teachers reported following the 2019 programme wave, and we were able to categorize what the teachers considered important into three areas: affective, cognitive, and practical competence. Second, to identify outcome indicators, we conducted a Delphi survey with eight professors, researchers, and teachers with R&E and STEAM expertise to investigate the suitability, determining the factor weights using analytic hierarchy processing. Based on the weights, we composed an initial set of performance indicators including names and definitions, and revised these based on comments from the experts.

As we noted above, we could categorize the indicators as affective, cognitive, or practical competence. A total of eight items measured affective competence, divided into two subdomains: science and technology-related self-efficacy and science and technology-related job interests. The cognitive competence subdomains were creative problem-solving ability–divergent thinking, creative problem-solving ability–critical thinking, and inquiry subject exploration ability; we measured cognitive competence with 12 questions. The practical competence subdomain consisted of 10 questions, and the relevant subdomains were inquiry performance ability, collaboration ability, and communication ability.

Participants

We administered the developed STEAM R&E performance indicators to students who participated in STEAM R&E 2020 using an online survey. A total of 174 students completed the pretest survey, and 148 responded to the posttest survey; for the data analysis, we selected the 100 students who had completed both surveys, and we excluded all non-responses. As demographic information, students were asked about their school type (57 general high school, 41 science core HS/science HS/gifted HS, and 2 schools listed as "etc."), gender (71 males and 29 females), and grade (34 10th graders and 66 11th graders).

Data Collection and Analysis

Quantitative Data

The students rated all STEAM R&E survey items on 6-point Likert scales (1 = *strongly disagree*, 2 = *disagree*, 3 = *somewhat disagree*, 4 = *somewhat agree*, 5 = *agree*, 6 = *strongly agree*) as 6-point scales have shown greater reliability and discrimination than 5-point Likert scales (Chomeya, 2010). For the reliability of the subdomain indicators across the pre- and posttests, Cronbach's alphas ranged from 0.850 to 0.903 (see Table 11.1). We conducted paired-sample *t* tests to identify significant differences between the pre- and posttest survey responses, and we ran

Table 11.1 Cronbach's alphas for pre- and posttest subdomain scores

	Subdomain†							
	a	b	c	d	e	f	g	h
Pretest	0.870	0.880	0.818	0.841	0.869	0.844	0.895	0.841
Posttest	0.922	0.926	0.900	0.859	0.874	0.899	0.845	0.902
Total	**0.896**	**0.903**	**0.859**	**0.850**	**0.872**	**0.872**	**0.870**	**0.872**

†a: science and technology-related self-efficacy; b: science and technology-related job interests; c: creative problem-solving ability-divergent thinking; d: creative problem-solving ability-critical thinking; e: inquiry subject exploration ability; f: inquiry performance ability; g: collaboration ability; h: communication ability.

independent *t* tests to check for differences by school type.[1] We used the Shapiro–Wilk test and Levene's test to examine the differences between the students' pre- and posttest scores. We analysed all the pre- and posttest survey data using SPSS 21.0 and set significance at 5%.

Qualitative Data

We collected and analysed qualitative data in addition to the survey data to capture what students experienced and learned through participating in the STEAM R&E project. We announced that we were collecting essays on the STEAM R&E experience and invited students and teachers to submit at the end of the project; we received a total of 66 individual essays (43 students' and 23 teachers') from 2018 to 2020. We asked respondents to describe their experiences such as difficulties, feelings, and new knowledge related to the STEAM R&E activities. Using qualitative analysis methods, the analysis was to reveal the meaning of participating in the STEAM R&E for the students. We focused on how the students felt and experienced the project-based inquiry activities, looking for and analysing the evidence (Yin, 2009), and finally explored what the STEAM R&E activities meant for the students. The keywords in the sentences were found and categorized into several concepts.

Results

Quantitative Research by Performance Indicators

Table 11.2 presents the paired-sample *t*-test results for the students' STEAM R&E performance indicators; all overall mean scores were higher at the end of the STEAM R&E activities.

[1] For the school type analysis, we only compared general high schools and the science core, science, and gifted schools. Only two students attended schools we classified as etc., and we excluded their data owing to the small sample size.

Table 11.2 Paired-sample T-test results for STEAM R&E performance indicators

Area[†]	Subdomain	Questionnaire	Mean		t
A (8)	Science and technology-related self-efficacy (5)	1. I believe in getting good grades in science and technology 2. I am confident that I will do well in science and technology exams 3. I believe that I will be able to acquire knowledge and skills in science and technology 4. I am confident that I can understand science and technology 5. I am confident that I will do well in science and technology lab classes	Pre	4.82	2.391*
			Post	5.04	
	Science and technology-related job interest (3)	6. I am interested in a career in science and technology 7. I am interested in a career in science and technology 8. I will enjoy working in science and technology related professions	Pre	5.27	1.729
			Post	5.44	
B (12)	Creative problem-solving ability-divergent thinking (4)	9. I can solve a problem in a new way that is different from what is already known 10. I am able to express new and unique ideas 11. I am able to generate diverse and abundant ideas to solve problems 12. I can make connections between seemingly unrelated things	Pre	4.31	4.894***
			Post	4.78	

(continued)

The students' total mean score increased significantly from 4.89 on the pretest to 5.11 on the posttest ($t = 3.914$, $p < 0.01$). There were also statistically significant differences in the subdomains between before and after the project: science and technology related self-efficacy, $p = 0.019$; creative problem-solving ability-divergent thinking, $p = 0.000$; creative problem-solving ability-critical thinking, $p = 0.022$; inquiry subject exploration ability, $p = 0.017$; and inquiry performance ability, $p = 0.000$. The findings show that students' competence in all three areas – affective, cognitive, and practical – improved through the STEAM R&E project. The cognitive competence subdomains showed the most significant differences.

By individual items, 13 of the total of 30 items showed statistically significant increases at $p < 0.01$: 1, 9, 10, 11, 12, 16, 20, 21, 23, and 24, which includes all

Table 11.2 (continued)

Area[†]	Subdomain	Questionnaire	Mean		t
	Creative problem-solving ability-critical thinking (4)	13. I am able to distinguish between reality and imagination 14. I am able to refine my ideas or conclusions carefully 15. I can judge whether what others are saying is right or wrong in performing inquiry 16. Based on the evidence exchanged with colleagues, I can draw conclusions for solving problems on my own	Pre Post	4.87 5.07	2.325[*]
	Inquiry subject exploration ability (4)	17. I am able to explore research topics in collaborating 18. I am able to discuss ideas about inquiry with my colleagues 19. I am able to explore inquiry ideas through literature review 20. I am able to construct research questions	Pre Post	4.89 5.09	2.426[*]
C (10)	Inquiry performance ability (4)	21. I am able to design experiments aligned with an inquiry theme 22. I am able to construct proper inquiry hypotheses 23. I am able to clarify the research range based on the literature 24. I am able to evaluate published papers in terms of the design or analysis methods	Pre Post	4.64 4.97	3.798[***]
	Collaboration ability (3)	25. When I collaborate with others, I try to consider the position or the situation of others first 26. When I collaborate with others, I try to respect others' opinion as well as my own 27. For the team, I try to allow and encourage others' mistakes or new attempts	Pre Post	5.30 5.29	0.162

(continued)

Table 11.2 (continued)

Area†	Subdomain	Questionnaire	Mean		t
	Communication ability (3)	28. I am able to communicate with team members aligned to the purpose of the research 29. I am able to lead the conversation to develop the ideas 30. When I collaborate with others, I try to resolve conflicts in a reasonable way (e.g., discussion or consensus)	Pre	5.30	1.035
			Post	5.39	

$^{*}p < 0.05$, $^{**}p < 0.01$, $^{***}p < 0.001$
† A: Affective; B: Cognitive; C: Practical

four items in the creative problem-solving ability-divergent thinking subdomain. For example, for item 10, 'I am able to express new and unique ideas', the mean increased from 4.35 preintervention to 4.81 after the project ($p = 0.000$). Item 10 relates to coming up with creative ideas for problem solving, and we infer that scores increased because STEAM R&E gave students opportunities to brainstorm creative solutions to real-world problems; indeed, the project emphasizes that students should actively identify the problem to solve on their own. Other items that showed high mean score differences were item 20, 'I am able to construct research questions' (*Mpre-test* = 4.67, *Mpost-test* = 4.99, $p = 0.005$) under inquiry subject exploration ability, and item 23, 'I am able to clarify the research range based on the literature' (*Mpre-test* = 4.29, *Mpost-test* = 4.74, $p = 0.001$) under inquiry performance ability. Item 20 is about the ability to explore and construct a research theme on one's own, and item 23 is about the ability to clearly set a research scope as part of the experiment design. Both are necessary competences for conducting an inquiry, and therefore, we interpret these findings as indicating positive impacts of the STEAM R&E project on students' inquiry competences such as constructing research questions or designing experiments. Furthermore, STEAM R&E includes expert feedback for each research team, and we believe that this continuous feedback contributed to the increases in the students' inquiry competences.

Meanwhile, there were no significant increases in the means for the subdomains of science and technology related job interest, collaboration ability, and communication ability, although we note that their preintervention scores were already high at over 5 points. Table 11.3 shows the mean independent *t*-test scores for differences between the pre- and postintervention scores by school type; there were no significant pre-/post- score differences for the two school type groups.

As seen above, there was a significant positive change in only one subdomain: science and technology-related self-efficacy ($t = 2.600$, $p = 0.011$). The students from the science HS, science core HS, and gifted HS ($M = 5.27$, $SD = 0.688$) showed significant improvement in self-efficacy, in contrast with the general high school students ($M = 4.87$, $SD = 0.788$) following the STEAM R&E project. Among

Table 11.3 Independent T-test pre-/post- scores between general HS and science core, science, and gifted HS

Area	Subdomain†	Type of school	N	Mean	Standard deviation	t
Affective	a	Science HS, science core HS, gifted HS	41	5.27	0.688	2.600*
		General HS	57	4.87	0.788	
	b	Science HS, science core HS, gifted HS	41	5.46	0.756	0.164
		General HS	57	5.44	0.730	
Cognitive	c	Science HS, Science core HS, gifted HS	41	4.95	0.783	1.694
		General HS	57	4.64	0.948	
	d	Science HS, science core HS, gifted HS	41	5.15	0.760	0.971
		General HS	57	5.00	0.718	
	e	Science HS, science core HS, gifted HS	41	5.21	0.679	1.567
		General HS	57	4.97	0.814	
Practical	f	Science HS, science core HS, gifted HS	41	5.11	0.696	1.675
		General HS	57	4.85	0.857	
	g	Science HS, science core HS, gifted HS	41	5.34	0.709	0.729
		General HS	57	5.23	0.729	
	h	Science HS, science core HS, gifted HS	41	5.41	0.717	0.187
		General HS	57	5.38	0.668	

* $p < 0.05$; † a-h: same subdomains as in Table 11.1

the 30 items, there were significant differences in mean scores for five items: 2, 4, 5, 11, and 20, with the largest difference for item 5: 'I am confident that I will do well in science and technology lab classes' ($M_{general,\ n=57} = 4.89$, $M_{etc.,\ n=41} = 5.44$, $t = 3.269$, $p = 0.002$); this item is about confidence in science and technology experiments in classes. The item with the second highest mean difference was item 20, 'I am able to construct research questions' ($M_{general,\ n=57} = 4.79$, $M_{etc.,\ n=41} = 5.22$, $t = 2.358$, $p = 0.020$). We consider that the experience of conducting research through the STEAM R&E project might have had greater influence on the science and technology research confidence among the students from the specialized schools than among the general high school students. Table 11.4 shows the mean score comparisons in the science and technology-related self-efficacy subdomain of the students from general versus specialized schools before and after STEAM R&E.

The table shows that the mean postintervention scores for the students from the science, science core, and gifted schools were higher than those for the general high

Table 11.4 Comparison of the mean scores

Subdomain	School type	N	Mean		t
Science and technology-related self-efficacy	Science HS, science core HS, gifted HS	41	Pre	4.91	3.055**
			Post	5.27	
	General HS	57	Pre	4.76	0.374
			post	4.87	

** $p < 0.01$

school students' scores, and the general high school mean was not significant. Briefly, the findings indicate that the STEAM R&E project had a particularly educational effect on the science and technology self-efficacy of students from specialized high schools.

The STEAM R&E Experience Essays

We identified three main themes from the collected essays: (1) experience conducting research, (2) increased interest in STEM fields, and (3) learning from achievements and failures.

Experience Conducting Research as a Future Scientist

Improved Research Skills

Participants wrote that the greatest advantage of participating in the STEAM R&E was being able to experience designing and conducting research on their own. The students did not have access to similar opportunities for research design outside of the programme. In addition, the students felt that the programme had cultivated their skills by making public presentations before experts and receiving their feedback:

> 'While writing the research results as a report, **we learned how to organize our research report by organizing our contents into motives, processes, results, and conclusions**. As we prepared for our interim presentation, we learned how to effectively communicate our work within a short period of time. **STEAM R&E has made me voluntarily experience what I have not done before and what I have not experienced before**.' (S22)

> 'Even if we want to conduct biological research, there are many restrictions because we are students. There was no funding for research, and there were no opportunities to conduct research and publish its results. **Even if there was a hypothesis that I wanted to test, it was always regrettable that I could only complete the literature review process such as preceding theories and thesis search**. However, **I was really happy to have the opportunity to conduct the biological research that I wanted**.' (S2)

Authentic Experience of Scientific Research

STEAM R&E gave students opportunities to explore new areas of interest, and many reported that being able to investigate content they were not learning about in class had not only aroused their interest in science but also instilled in them new levels of research confidence. For example, S14 studied making filters related to biomimetics, and he was impressed with his opportunity to conduct real-world inquiry related to the theoretical knowledge he had been learning about biomimetics in class. Following is another student's reflections on new STEAM experiences:

> 'I have always wanted to improve my programming skills by learning and applying programming methods. [Through the research projects] **I designed new trapping machines and unmanned surveillance systems**. I coded and programmed by myself during the process. **The experiences I've had on my own are the most impressive.**
>
> STEAM R&E allows students to study the subjects they are curious about outside of lessons. **It is meaningful to explore a truly unique, diverse, and extraordinary subject.** … **Also, since there is no burden of evaluation, I do not have the compulsion to be perfect**, so it is one of the joys of school life.' (S18)

Improving Collaboration and Communication Skills

Many students mentioned that the STEAM R&E process was difficult but that they were proud of having improved their cooperation and collaboration skills. The students on some teams were already familiar with each other, but other teams comprised students who fit the interests of the research subject but were not close with each other. Some of these students reported difficulties collaborating with their colleagues during the project:

> 'Even if I just got angry, it would not improve at all. From that moment, **I desperately realized that harmony with team members is really important, above all, and that it is important to communicate well so that there is no friction no matter what happens.**' (S34)

> 'We met several times a week for research, and even during vacations, we came to the science lab at our school to conduct research. And we became closer with team members who were not very close. **Rather, the disagreement of opinion during the research served as a facilitator to devise and try a number of different methods.**' (S28)

One of the goals of the STEAM R&E project is to cultivate students' collaboration and communication skills, and we established that the participating students came to realize the importance and necessity of coordinating and collaborating with other students to complete assignments even when it is difficult.

Growing Interest in a Career Path in STEM Fields

Career Interests as Future Scientists

Students reported that participating in the STEAM R&E project had strengthened their existing career interest in science and engineering fields, but others reported gaining new interests from participating in the project. For instance, S1 reported feeling pleasure from conducting an experiment he had designed himself, and the process helped him think more clearly about his career path. Other student comments are given below:

> 'Even before I participated in STEAM R&E, I had thought that my learning ability and problem-solving ability as a future engineer was not bad, **but when I actually conducted the research, I felt that I was like a frog in a well. Through this activity, the career of a researcher in science and engineering was further clarified**, and I had a desire to conduct professional research in the field.' (S8)
>
> 'While participating in this project, **I learned about my attitude as a biologist in the future.** While planning and conducting research on my own, **I realized what meaningful research was in the process of drawing out research results.**' (S2)

Students who participated in STEAM R&E were able to experience aspects of being engineers or other scientists by designing research studies and conducting the research, and the activity stimulated them to think about or rethink careers in science and engineering.

Enhanced Interest in Science

Students grew interested or more interested in careers in science and engineering through participating in the self-directed STEAM R&E research activities. Many students chose topics they had been interested in or for which they had prior experience conducting research, but there were also students who gradually became interested in previously unfamiliar research topics:

> 'Before participating in the STEAM activity, **I thought that biotechnology was a very far-off science from us, which is conducted only in the laboratory.** However, an advisor professor gave us various examples of biotechnology being used in our real life. And he explained how close biotechnology is to real life. **His explanation cleared my prejudices against biotechnology, and this was a great opportunity to take my interest in biotechnology one step further.**' (S26)
>
> 'STEAM R&E has influenced not only my general interest in science, but also my interest in various subjects such as information science. ... **I was interested in not only life science, which I was usually interested in, but also physics and chemistry through the STEAM R&E project**, so I studied harder.' (S41)

Experiences of Achievement and Lessons from Failure

Experience of Achievement

STEAM R&E is a rather long-term project in terms of students' activity. Conducting collaborative research and generating results gave the project participants a sense of accomplishment. We confirmed from the students' essays that this sense of achievement not only aroused their interest in STEAM careers but also served as drivers of their own personal development:

> 'It feels like stuffing things one by one into an empty box with 'I' written on it. From planning to recording results, there were many failures, **but I feel like I have become the 'me' who has grown even more than before, by continuing to accumulate what I have gained from it.**' (S9)
>
> 'Considering so many points, **it was the first time I had crossed my limits.** Before STEAM R&E, I had made only small models or models used for decoration purposes. However, **when I made a large outer frame that was not even a small part of a robot in such a**

> **large project, it was burdensome at first, and I thought deeply about how to make it well.** But it meant a lot to me.' (S38)

The students' sense of accomplishment from completing a systematic research project to solve a real-world issue they themselves had come up with helped reduce their fear of research and increased their confidence.

Lessons Learned from Failure Experiences

Students' STEAM R&E activities are not always successful. The research topic can change, problems can arise, and the expected results might not be what the study findings reveal. For instance, one team had difficulties completing the research because they could not communicate with each other. Overall, the students' failures were difficult to get through, but overcoming the difficulties inspired in them a spirit of challenge and confidence:

> 'We have had successes and achievements from doing these activities, **but we have failed and suffered so much. Whenever that happened, our team tried to improve it.** I can bet it was a very valuable experience for us. ... **I thought about jumping over the wall and going back when it hit a wall.**' (S23)

> **'I have been doing various research activities so far, but it seems that trials and errors were rare like this.** I changed the target theme twice, practiced experimentation, and was clumsy in many ways, **but unexpected situations soon happened, and I was busy. ... But it was both difficult and fun.** It was more fun than any other exploration I had done before.' (S16)

The students experienced multiple difficulties related to the research process itself. Some students lacked research experience and were confused by the experiment process, some were attempting to tackle subjects that had not been specified for the subject, and a number of participants reported poor communication among team members. One student, S42, reported that he had learned the importance of the ability to face any situation and the importance of research design as his team revised their research plan through trial and error.

Conclusion

In this chapter, we have examined Korea's STEAM policy over the past 10 years. The first step was the conceptualization of STEAM education in Korea. In the second phase, relevant organizations and science educators tried to implement STEAM education as a national curriculum and expand it nationwide. The third phase began in 2020 with a comprehensive plan of science, mathematics, and information science. The keywords in the plan are artificial intelligence, high technology, and space outside the school.

In every phase, the goal of STEAM/STEM education is always to enhance problem-solving ability in the real world. A good example of pursuing this goal is the STEAM R&E programme supported by KOFAC. The programme is a project-based research activity for high school students that mimics real-world science research processes, and it increases students' research competencies based on analysis of their responses on a quantitative survey. Specifically, we collected data from participants in the 2020 wave of STEAM R&E using a performance indicator survey that we developed through a Delphi survey with experts incorporating previous teacher participants' opinions. We also collected and analysed semi-structured essays from students and teachers.

In 2020, because of the unprecedented circumstances surrounding the COVID-19 pandemic and related measures, students began research in July and conducted intensive research during a short period of about 4 months. In the analysis of the statistical data, specifically the differences between the students' scores on a number of STEAM subdomains before and after they participated in the project, STEAM R&E stimulated students' science and technology-related self-efficacy, reflecting affective competence; creative problem-solving and inquiry subject exploration ability, reflecting cognitive competence; and inquiry performance ability, which reflected practical competence. STEAM encompasses these three areas of competence, and the means in all subdomains of these areas increased significantly based on post-intervention indicators. In particular, the item with the highest mean score increase was about the ability to construct research questions and determine the scope of the research. We therefore concluded that STEAM R&E as a project gave students opportunities to direct their own research themes and processes in ways that they could not learn in class.

Second, the quantitative results showed no significant increases in science and technology-related career interest, communication ability, or collaboration ability. For career interest, it was possible that students who wanted to participate in this project already had at least some interest in science and technology, so the project itself would not have made drastic changes to their interest. However, analysis of the essay responses showed that participating in the project had further developed students' interest in the field of science and technology. Meanwhile, in terms of collaboration, STEAM R&E is a long-term team project that requires communication and cooperation among students, and some of the essays highlighted difficulties in working with unfamiliar team members including conflicts and their resolutions; students reported, however, that the process had increased their communication skills, providing "an opportunity for [them] to mature mentally as well as academically." Participating in the project appeared to have provided students with opportunities to collaborate and communicate with team members, although in the quantitative results, the findings were not statistically significant. It is possible that quantitatively, the changes were not significant because the mean pre-intervention scores for both interest in STEM careers and communication and collaboration ability were already high at 5.30.

Third, there were significant differences by school type between the mean indicator scores for the general high school students and those for students from specialized science and gifted schools in the STEM-related self-efficacy subdomain. In Korea, the science core high school curriculum can comprise more than 45% science-related subjects in general, whereas the percentage is only around 30% in general high school and, conversely, around 60% for the science high schools. Given these percentages, it is possible that students from the specialty high schools already have some science-related confidence that the programme only enhanced rather than creating, and hence the lack of significant changes after the project. Regardless of school type, we conclude that STEAM R&E activities improved students' inquiry competence.

Fourth, the students' essays reflected that the students had made meaningful gains in the science-related competences through participation in the STEAM R&E project. Students reported that their research abilities and their career interest in STEM fields had increased and that they had learned from both their achievements and their failures. Students said they benefited from the direct research experience they had acquired including applying classroom concepts in practice, which increased their grasp of these concepts. Separately, according to the essays, participating in the project in teams improved the students' cooperative spirit and communication skills. Although there were no significant increases in the performance indicators for STEM career interest or collaboration ability, we still found that participation in STEAM R&E could affect students' STEM-related career interests and their ability to collaborate with colleagues.

Despite the pandemic conditions, the intensive four-month project period showed increases in six of nine subdomains of affective, cognitive, and practical competence, and we conclude that STEAM R&E as a long-term project cultivated students' affective, cognitive, and practical inquiry capabilities. Even for students with little R&E experience, participating in STEAM R&E gave them opportunities to improve their inquiry skills.

Further investigations on the construct validity of the questionnaire items could be pursued. Moreover, in the future, statistical confirmation of the relationships among the domains and subdomains could guide the development of effective STEAM education programmes that develop students' confidence and multifaceted competence. Future studies may look into the role of teachers in the programme as we do note that the project could not succeed without the professionalism of teachers to guide the students through complicated research tasks and conflict resolution with other students.

References

Bell, D. (2016). The reality of STEM education, design and technology teachers' perceptions: A phenomenographic study. *International Journal of Technology and Design Education, 26*(1), 61–79.

Bybee, R. W. (2010). What is STEM education? *Science, 329*, 996.

Bybee, R. W. (2013). *The case for STEM education: Challenges and opportunities*. National Science Teacher Association.

Chomeya, R. (2010). Quality of psychology test between Likert 5 and 6 points. *Journal of Social Sciences, 6*(3), 399–403.

Harrison, M. (2011). Supporting the T and the E in STEM: 2004–2010. *Design and Technology Education: An International Journal, 16*(1), 17–25.

IEA. (2011). *TIMSS 2011: Trends in International Mathematics and Science Study 2011*. https://www.iea.nl/studies/iea/timss/2011#section-68.
Jeong, S., Kim, H., & Tippins, D. J. (2020). From conceptualization to implementation: STEAM education in Korea. In A. J. Stewart, M. P. Mueller, & D. J. Tippins (Eds.), *Converting STEM into STEAM programmes: Methods and examples from and for education* (pp. 241–257). Springer Nature Switzerland.
Kim, S.-W. (2015). STEAM Gyoyuk Gicho Irongwa Hyeonjang Jeogyong [Theory and Practice of STEAM Education]. *Ewha Education Monograph Series, 2015–01*, 52.
Kim, S.-W., Chung, Y. L., Woo, A. J., & Lee, H. (2012). Development of a theoretical model for STEAM education. *Journal of the Korean Association for Science Education, 32*(2), 388–401. https://doi.org/10.14697/jkase.2012.32.2.388 (in Korean).
Korea Foundation for the Advancement of Science and Creativity. (2021). *About STEAM*. https://steam.kofac.re.kr/?page_id=11269.
Ministry of Education. (2017). *Yunghap Injae Gyoyuk (STEAM) Jungjanggi Gyehoik (2018–2022)* [Convergence Talent Education (STEAM) Mid- to Long-Term Plan (2018–2022)].
Ministry of Education. (2020). *Yunghap Gyoyuk Jonghap Gyehoik An* (2020–2024) [Comprehensive Plan for Convergence Education (2020–2024)]. https://www.moe.go.kr/boardCnts/viewRenew.do?boardID=294&lev=0&statusYN=W&s=moe&m=020404&opType=N&boardSeq=80718.
Ministry of Education, Science and Technology. (2011). *Changuijeok Gwahak Gisul Injae Daegukeul Wihan 'Je 2cha Gwahakgisulinjae Yukseong Jiwon Gibon Gyehoik (2011–2015)'* [The 2nd Basic Plan for Nurturing Highly High Technological Talents for the Great World of Creative Science and Technology Talents (2011–2015)].
Quigley, C. F., & Herro, D. (2019). *An educator's guide to STEAM: Engaging students using real-world problems*. Teachers College Press.
Sanders, M. (2009). STEM, STEM Education, STEM mania. *The Technology Teacher, 68*(4), 20–26.
Yin, R. K. (2009). *Case study research: Design and methods* (4th ed.). Sage Publications.

Jiyeong Mun is a lecturer and a postdoctoral researcher in the Department of Science Education at Ewha Womans University, Seoul, Korea. She obtained her Ph.D. degree from the Department of Science Education at Ewha Womans University. She has been conducting research on science learning to improve students' inquiry skills, especially developing STEM/STEAM curriculum and programs, as well as an analysis of the process by which students construct research questions.

Sung-Won Kim is currently an emeritus professor at Ewha Womans University. He received his Ph.D. in Physics from KAIST, Korea in 1983. From 1985 to 2020, he served as a professor in the Department of Science Education at Ewha Womans University. His research interests include the imagination in science and STEAM education as well as theoretical astrophysics. He served as President of KOSSS (The Korean Society for School Science) and executive member of EASE (East-Asian Association for Science Education).

Chapter 12
Arts-Integrated STEM in Korean Schools

Hye-Eun Chu, Sonya Martin, Ei Seul Kim, and Hyeong Moon Lee

Abstract This chapter addresses the theme of innovative teaching approaches, reporting on three funded studies investigating the effects of a STEAM teaching approach where the arts are integrated into STEM teaching/learning activities. 'Arts' is defined as any event or product that reflects a community's sociocultural practices and values, for example, paintings, architecture, literature, leisure activities, and festivals. The STEAM approach has been adopted into Korea's national science curriculum. The chapter describes the STEAM teaching/learning process, including classroom activities: encouraging students to ask inquiry questions and hypothesize answers; engaging students in activities aimed at verifying their hypotheses; and inviting peer collaboration in testing, applying, and evaluating their hypotheses. Throughout the teaching/learning process, arts-related sociocultural events (e.g. a light festival) and/or products (e.g. 3D quasi holograms) are used to demonstrate to students how science concepts (e.g. light propagation) create or explain a sociocultural experience. Samples of students' work provide evidence of their developing understanding of science concepts. The chapter also reports on the effect of the STEAM approach on students' perceptions of and attitudes towards science and studying science, and presents some evidence of scientific creativity in STEAM lessons. The chapter concludes with a discussion of the potential benefits and challenges of the STEAM approach.

Keywords STEAM · STEM · Perceptions to science learning · Perceptions of science

H.-E. Chu (✉)
Macquarie University, Balaclava Rd, Macquarie Park, Sydney, NSW 2109, Australia
e-mail: hye-eun.chu@mq.edu.au

S. Martin · E. S. Kim · H. M. Lee
Seoul National University, 1, Gwanak-ro, Gwanak-gu, Korea

M. M. H. Cheng et al. (eds.), *Concepts and Practices of STEM Education in Asia*,
https://doi.org/10.1007/978-981-19-2596-2_12

Introduction

This chapter reports on the STEAM approach implemented in Korea that incorporates the arts into teaching/learning activities. The STEAM approach has been defined in the literature (Paik et al., 2012; Park et al., 2012; Yakman & Lee, 2012), with the letter 'A' standing for the integration of the arts into STEM lessons. The STEAM approach presented in this chapter integrated science, technology, engineering, and mathematics (STEM) with the arts or culture to afford students the experience of seeing science concepts operating in real-world contexts. In the sections that follow, we provide examples from three STEAM projects implemented in Korea to provide some evidence of the potential of STEAM education to develop students' understanding of science concepts in the context of sociocultural environments familiar to the students. We also present evidence that these STEAM learning experiences can support positive attitudes towards and perceptions of science and science learning, besides promoting scientific creativity in the science classroom. The Korean evidence was collected in an Australia-Korea Foundation funded project involving intercultural interactions between students in STEAM classes in Korean and Australian schools. A total of eight schools in Korea and Australia participated in this project from 2016 to 2020. In this chapter, however, only data from the Korean classrooms will be reported.

What is STEAM?

What distinguishes STEAM from STEM is the intentional use of the arts in the teaching/learning process. The term 'arts' refers to non-STEM school subjects such as art, history, and literature (Yakman & Lee, 2012), as well as sociocultural products and events familiar to students (e.g. photography, traditional foods, and musical instruments). In STEAM lessons, the arts activity or product is integral to the process of guiding students' understanding of STEM.

The theoretical framework underlying many STEAM initiatives is that of social constructivist theory (Chu et al., 2019; Chu et al., 2019; Holbrook et al., 2020). The social constructivist model of learning holds that learners construct knowledge and concepts through interaction with others and with their environment (Duit & Treagust, 1998; Ernest, 1998). Applied to the STEAM classroom, this means that science concepts and scientific ways of thinking, speaking, and reasoning about the world are learned through interacting with teachers and peers, and interacting with the environment, such as objects, sociocultural events, natural phenomena, and experimental results.

Through these interactions, learners verbalize their questions and nascent understanding and, in the process, form mental approximations of the science concept, and finally arrive at a scientifically acceptable concept. In the context of STEAM lessons, interaction with the environment includes asking questions and talking about the arts or any other sociocultural activity that occurs in the student's community. Integrating sociocultural elements with the teaching and learning of science allows students to see the role of science concepts in a sociocultural activity or product, for example, Korean science students investigating the working of the Korean flute (daegeum). Observing how science concepts help to explain phenomena in their sociocultural life can create in students' minds the perception that science is relevant to real life (Chu et al., 2019; Chu et al., 2019).

STEAM in Korean Schools

The initial inclusion of STEAM in Korean schools was in response to PISA and TIMSS reports about students' negative attitudes towards learning science. Although Korean students consistently ranked in the top achievement category (with an average science score of 561 in the 1999 TIMSS assessment), since 1999 few Korean students have displayed positive attitudes towards science (below 20%) (Mullis et al., 2020), and this has not changed substantially as of 2019. In these assessments, many Korean students disagree with statements such as, 'I enjoy learning science' and 'I learn many interesting things in science'.

In response to the substantial difference between students' high achievement in science and the lack of positive attitudes towards science, the Korean Foundation for the Advancement of Science and Creativity (KOFAC) and the Ministry of Education, Science, and Technology (MOEST) proposed modifications to the way science was being taught. Their proposals included the integration of non-science subjects into the teaching of science and mathematics, an approach known as 'STEAM' in the United States (MOEST, 2010; MOEST & KOFAC, 2012). Through STEAM lessons, KOFAC aimed to encourage students' curiosity about phenomena, engagement in inquiry, and participation in creative experiences to build an awareness of the connection between science and the students' real-world sociocultural lives (KOFAC, n.d.).

The STEAM approach is currently practised in many schools in Korea. A survey in 2015 revealed that of about 6,551 schools which responded to the survey, 54% of elementary schools, 47% of middle schools, and 32% of high schools offered STEAM lessons (Park et al., 2016). The STEAM programs generally take the form of integrating science lessons with students' out-of-school experiences, discussing socioscientific issues, and student-initiated projects in which students pose questions or define problems, and seek solutions. It is likely that more schools will adopt a STEAM approach to science teaching as the Korean Ministry of Education (MOE, 2020) has announced that from 2024 schools can choose to replace some parts of the national curriculum with school-designed integrated lessons, including STEAM lessons.

Three STEAM Research Projects

This section reports on the trial implementation of three STEAM programs in Korean schools with about 100 students in grades 5, 8, and 10 between 2016 and 2020. Project 1, which was related to the teaching of seasonal change to Year 5 students, illustrates the four-phase design based on a social constructivist view, and offers examples of teaching materials and activities. Project 2, which was implemented with the topic of light reflection for Year 8 students, shows how STEAM lessons differ from regular science classes and highlights the impact on student learning. Both projects describe STEAM lesson design and impact on students' concept development and attitudes towards and perceptions of science learning. Project 3, which focused on Year 10 lessons on the topic of thermal energy and energy efficiency, describes the impact on students' scientific creativity and critical thinking.

Project 1: A Four-Phased Design Adopting the STEAM Approach

Project 1 relates to the teaching of seasonal change with Year 5 (9–10 year-olds) and took place in 2016–2017. There were 23–25 students in each class from two schools. Arts-related elements were specifically incorporated throughout the STEAM lessons using a four-phased design: engagement, collaborative exploration, elaboration, and communication, and peer evaluation of revised explanations. Students paid attention to the design (colour, choice of symbols, visual representations, and organization of information) of their explanatory models about what causes seasonal change. Creating visual depictions of explanatory models has been recognized as not simply a passive communication tool, but also a generative reasoning process that helps students actively shape their knowledge (Tytler et al., 2020).

Engagement: Students are presented with a situation or phenomenon that invites them to be curious and to pose questions (why?/how?/what consequences?). Korean Year 5 students discovered through Skype conversations with Australian students that Christmas celebrations occur in winter in Korea but in summer in Australia. The different seasons occurring at the same time of the year triggered questions such as, 'Why is it winter in Seoul but summer in Sydney?' The teacher steered students to questions likely to lead to scientific inquiry, explaining that scientists observe the world around them, get curious about phenomena, and ask questions to start a process of investigation and discovery (Driver et al., 1996; Windschitl et al., 2008). The use of a sociocultural event (in this case, Christmas in summer/winter) was intended to show students that science can explain phenomena observed in our daily life.

Collaborative exploration: Group activities enhance construction of knowledge through social interaction. Students were asked to produce an explanatory model in the form of a drawing to explain why it was winter in one country but summer in the other at the same time of year (see Fig. 12.1). By getting the students to verbally

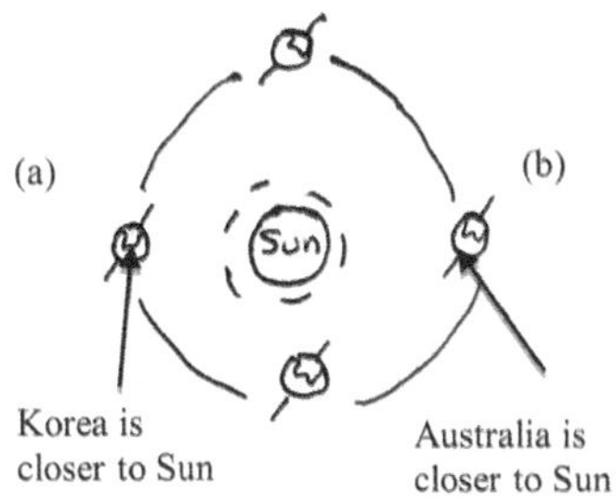

Written explanation:

The Earth's rotation axis is tilted, and this causes the different seasons in Korea and Australia at the same time of the year. In the diagram, the Earth revolves around the sun. When Korea is closer to the sun [see (a)], it is summer in Korea. When Australia is closer to the sun, it is summer in Australia [see (b)].

Fig. 12.1 An example of a group's initial model explaining seasonal differences in two countries at the same time of the year

present their hypothesized group answers or submit their written answers, the teacher can identify gaps in students' knowledge and/or alternative conceptions. This enables the teacher to prepare a learning activity to guide students towards building more accurate understandings of the targeted science concept.

Figure 12.1 shows that some students initially thought the main reason for seasonal change was the differential distance between different places of the Earth and the sun, caused by the tilt of the Earth's axis. To invite students to question this view, the teacher provided students with an experiment using two heat lamps, each directed at a tray of soil. One lamp was angled at 75 degrees and the other at 40 degrees, with both lamps positioned at the same distance from the soil. Using a thermometer, students measured the temperature of the soil in each tray, leading them to observe that the soil in the tray with the lamp positioned at 75 degrees above it was hotter than the soil in the other tray. This learning activity engaged students in the scientific practice of experimenting, together with observing, and collecting data. The aim was to help them to understand the importance of the angle of the sun on the temperature of the Earth.

Next, the teacher helped students interpret the data with reference to their previous models to help students develop a more complete scientific explanation. To do this, the teacher reminded students of a concept they learnt in Year 4 by measuring the shadow of a stick: that the angle at which the sun's rays strikes the earth changes from sunrise to sunset, and with the change the intensity of the sun's light (heat energy) will become stronger or weaker. Through guided discussion, the teacher showed students that when the heat lamp, representing the sun, is at a 75-degree angle to the tray of soil, the soil is hotter due to the higher intensity of heat reaching it. The teacher related this observation to the difference in the sun's altitude (the solar elevation angle relative to the horizon, for example, the elevation is 0 degrees at sunrise and sunset) in summer and winter, higher in summer and lower in winter. Students reviewed their hypothesized answers to the initial Engagement phase question and were able to revise their explanations (Chu et al., 2019; Chu et al., 2019).

Elaboration: From the students' revised answers, the teacher noted the shortcomings in the students' construction of the target science concept. The teacher planned another learning experience (e.g. a demonstration, an additional experiment, or a video clip) aimed at enabling students to make observations that could lead to their construction of a more scientifically accurate science concept. Students reviewed their revised answer, further refining it with the new understanding gained from the Elaboration activity. The teacher noted that the students seemed to have difficulty integrating the two phenomena that cause the seasons: the Earth's tilted axis, which affects the altitude of the sun, and the Earth's revolution around the sun.

The teacher demonstrated the effect of the Earth's tilted axis on the intensity of heat at two different locations (Korea and the south-eastern part of Australia) on a globe, using a lamp to represent the sun as the Earth moved around it (see ② in Fig. 12.2). Affixed to each of the two locations was a disc with a scale for measuring the length of the shadow of a pin one centimetre high planted in the middle of the disc (see ① in Fig. 12.2). The length of the shadow enabled students to observe the sun's altitude changing as the earth moved around the sun.

The teacher instructed the students to observe the sun's altitude at the two locations and drew their attention to the effect of the Earth's tilted axis on the angle at which the sun's rays reach the earth which determines the intensity of light received at each location. Next, to demonstrate the effect of the Earth's revolution around the sun, the teacher positioned the globe in two opposing locations with respect to the lamp (sun). Through questioning and guided discussion, students were led to see that the Earth's revolution and its tilted axis affect the number of hours of daylight and the altitude of the sun relative to the earth in each country, resulting in summer when there are more daylight hours and a greater intensity of light.

After the demonstration, students reviewed their group answer to the inquiry question a second time, integrating into their explanatory model the effect of the Earth's tilted axis and revolution on the sun's distance and altitude. They then checked whether their model was correct by performing tests using equipment similar to that

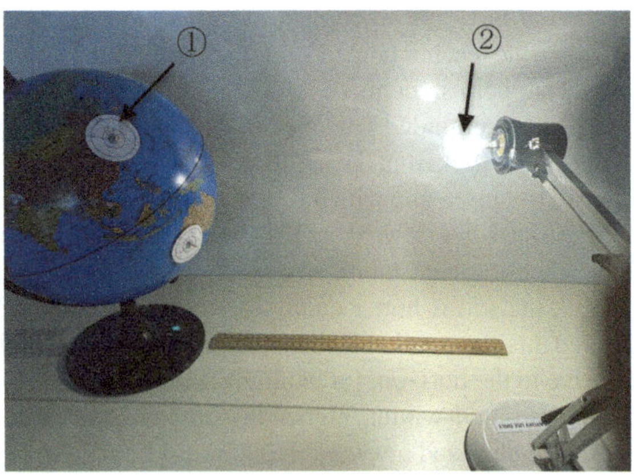

Fig. 12.2 Demonstrating the effect of the Earth's tilted axis on the sun's altitude and therefore the intensity of the heat from the sunlight

which the teacher had used. In doing so, the students moved gradually towards a closer approximation to the scientific concepts that account for seasonal change.

Communication and peer evaluation of revised explanations: Student groups next presented to each other and the teacher their final responses to the inquiry question. Students and the teacher asked questions to clarify their understanding. The teacher also pointed out limitations or alternative conceptions and corrected them. Students then selected the best presentation after appraising each presentation using rubrics provided by the teacher.

The rubrics related to the accuracy of the science concepts (e.g. Does the model include the role of the Earth's tilted axis?). Peer evaluation offered an additional opportunity to reflect on the students' current understanding of the science concepts (e.g. factors causing seasonal change) and to revise understandings with the aim of developing more accurate construction of those concepts.

The arts and culture element in the Project 1 STEAM lessons lies in the social interaction the students experienced in three video conferences over Skype with Australian students studying the same topic in Sydney. Despite some language difficulty, meeting with and talking to fellow students of a different culture heightened and sustained the Korean students' motivation in the science learning activities. They were thrilled to see the Australians dressed in summer clothes in December and to hear them speak about school and out-of-school activities. Excited to be engaging in the same science learning activities, they were driven to work enthusiastically so as to have learning artefacts (drawings, models) to show the Australian students via the project's online platform. At the conclusion of the project the Korean students described the intercultural sessions as "great" (Student, p. 201) and "helpful" (Student, p. 202).

Project 2: Promoting Students' Conceptual Understandings of Science, and Attitudes Towards and Perceptions of Learning Science through STEAM

The context for project 2 is a Year 8 class (12–13 year-olds) doing a unit on light in the 2016–2017 academic year. There were 30 students in the class. Here, we focus on the design of a quasi-3D hologram, with students using their mobile phones/iPads and a pyramid made of four trapezoid-shaped pieces of clear plastic placed over the phone (see Fig. 12.3).

The goal was to create an image (e.g. a butterfly or a cultural symbol) appearing to float above the phone or iPad screen in the middle of the pyramid. To achieve this effect, students needed to synthesize two concepts: light from the mobile phone travels through the pyramid and reflects into the eye, and the brain's interpretation of the pattern of reflected light as an image hovering above the phone. Whereas a traditional science lesson introducing these concepts is likely to deliver a teacher's explanation of the link between light reflection and seeing, the STEAM lesson engages

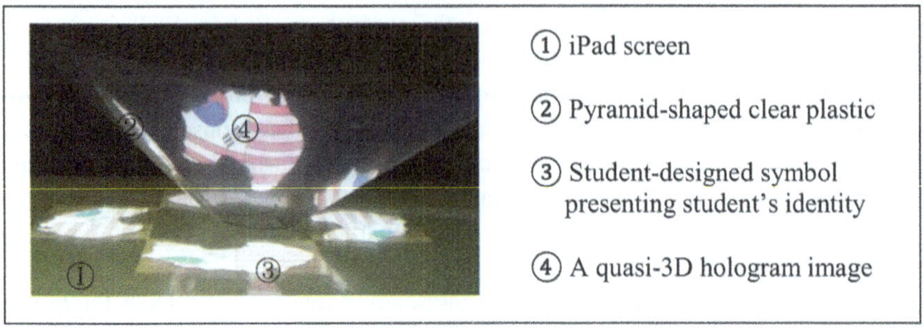

Fig. 12.3 Quasi-3D hologram using iPad and pyramid-shaped clear plastic

students in the experience of positioning the eye so that the created image can be clearly observed. The hologram creating activity took place in art classes with the science teacher present to provide guidance on the scientific aspects of how to make the image visible to the observer.

In this project, we collected students' learning artefacts, conducted a survey of 30 students, and interviewed four groups of three to four students before and after the project. The findings suggest that the STEAM approach appeared to facilitate students' movement from basic, and sometimes incorrect, understanding of science concepts, to more scientifically acceptable understanding. Interview responses revealed the development of students' favourable attitudes towards studying science and positive perceptions of science.

Songer and Linn (1991) posited that the development of scientific understanding is evidenced when science concepts are used to explain what might be viewed as unrelated events. In this case, the two seemingly unrelated events of seeing things, and light bouncing off the surface of objects was difficult to understand as over 70% of students failed to recognize that we see objects due to light being reflected off the objects with some of the light entering our eyes. After the STEAM lessons, students understood the concept.

The transcripts of the post-project interviews showed that students were appreciative of the teaching/learning methods of STEAM, such as arts integration, and the small-group collaboration that they reported helped them to 'think better'. Indicative comments, translated from Korean and made in the post-project interviews, included:

> 'It is amazing to learn science in this way, with knowledge from non-science subjects included.' (Student, p. 201)
>
> 'These science lessons taught me to think, to see other things besides science.' (Student, p. 205)
>
> 'We were not interested in the topic of light before but now I am, after observing its connection to light festivals.' (Student, p. 206)

Negative statements were markedly reduced in the post-project interviews, from five comments made pre-project to just one comment made post-project, with one student reporting that some members in her group had not participated fully.

The transcripts of the post-project interviews showed students' positive perceptions that when science is differently presented, it can be better understood:

> 'This program is very different from usual science programs. These subjects seem new to me. We realized that science is involved in many different aspects in our life and culture.' (Student, p. 201)
>
> 'We apply the same science concepts and theories to explain .... We could easily communicate our science explanations even though our languages are different.' (Student, p. 202)
>
> 'Science could be fun and not so difficult if we learn science related to cultural events or everyday life phenomena, like in this program.' (Student, p. 206)

The arts and cultural dimension integrated in the Project 2 lessons lay in the students' efforts to design the image for their hologram. For instance, the image in Fig. 12.3, consisting of an outline of the Australian continent holding elements of the Korean and United States flags, was created by Student p211 to express the multiple facets of her identity. She was Korean by birth but had lived in the US before her family moved to Australia. Expression of personal identity through art is one aspect of creativity in art recognized by art authorities such as the Museum of Modern Art in New York (https://www.moma.org/learn/moma_learning/themes/investigating-identity/). Other students in Project 3 designed or chose images expressing their thoughts and feelings about their world (e.g. traditional Korean costumes). It must be reiterated that the students' artistic expression of themselves in the making of the hologram involved them in learning the science concept of light and how we see.

Project 3: STEAM Learning and Creativity

Project 3 took place in 2020 in a Year 10 class in Korea, with 27 students aged 14–15 years. It focused on whether STEAM learning had any effect on scientific creativity. Scientific creativity is defined as a combination of divergent thinking, which is the ability to explore multiple possible solutions to generate creative ideas from an initial problem or reference point (de Vries & Lubart, 2017). Convergent thinking is the application of logical reasoning to deduce a solution from known information (de Vries & Lubart, 2017). In convergent thinking, the solution to problems must be preceded by the process of defining the problem space (de Vries & Lubart, 2017; Kocabas, 1993), which is a practice integral to the culture of science. To be creative in science, students must experience viewing things from different perspectives and then be able to generate new possibilities or alternatives (Franken, 1994).

In Project 3, scientific creativity was encouraged by engaging students in solving a problem that required them to apply their knowledge of science in the context of exploring the traditional architecture of houses in the two cultures of Korea and Australia. The Korean students' problem was to design a zero-energy school building for students in Sydney. The Korean students met a class of Australian students over Skype to exchange knowledge about the design of buildings in their respective countries. The conversation was directed at discovering what architectural features in each community's house were designed to accommodate the climate of the region. The Korean students learnt that many traditional Australian houses have high ceilings so that hot air will rise and rooms will be cooler in contexts where air conditioning is not available. In designing a zero-energy school, the science that the students had to consider was related to conduction of building materials and insulation/heat transfer via convection, radiation, and conduction.

The learning activities gave students scope to practise divergent and convergent thinking. One activity requiring divergent thinking had students examining how the structure of a traditional Korean house (*hanok*) allowed for energy efficiency. Using knowledge from the teacher's explanation of energy efficiency in house design, the students explored different characteristics of hanoks to work out how each feature facilitated heat conservation in winter and/or free flow of cooler air into the house in summer. Students discovered that heat channelling tunnels (*gorae*) were constructed so that smoke was prevented from escaping rapidly through the chimney, keeping the house warm long after the fire had died out.

Students were engaged in deductive logical reasoning (convergent thinking) by investigating the effect of roof overhangs on the amount of sunlight entering a house. Students cut windows in a box representing a house. The teacher gave the students the sun's altitude at noon in winter (39.5 degrees) and in summer (76.5 degrees). The students used a lamp, representing the sun, to simulate these altitudes and used a light meter app on a mobile phone to measure the amount of light entering the house through the windows. Then the students attached a cardboard overhang to the roof of the house and measured the amount of light entering the house again. Studying the data they collected, and using logical reasoning and deductive thinking, the students learnt that overhangs block some sunlight from streaming into a house in summer while still allowing low -angle winter sun to enter.

Students deployed divergent thinking when they explored strategies for reducing energy transfer between the interior of the building and its exterior, and maximizing the use of natural methods for cooling and heating. The different strategies students discussed included the use of double-glazed glass for windows, the choice of insulating material with reference to heat transfer, the direction of the windows with reference to whether the building is in the northern or southern hemisphere, and whether window location would maximize the entry of sunlight in winter and maximize airflow in summer. They also considered the L-shaped layout of the *hanok*. Students' discussion of these strategies was targeted to find the best solution: the design of an energy-efficient school building for students in a country situated in a different hemisphere from their own. Their discussion exemplified divergent thinking leading to creativity (Harvey, 2014).

The students' school building designs also showed thinking processes characteristic of scientific creativity. One group of students reasoned that an L-shaped building would allow windows to be placed on opposite walls, thereby maximizing air circulation in summer, reducing reliance on non-renewable energy for air conditioning (see

① The temperature of the soil in the yard rises due to solar heat.
② The air in the yard is heated and forms an upward airflow.
③ Cool wind flows from the backyard of the *hanok* due to differences in air pressure.
④ The backyard of the *hanok* is set facing a mountain or covered with trees and plants to provide cool air.

Fig. 12.4 L-shaped *hanok* and air movement in/out

Fig. 12.4). Thus, through evaluations and logical reasoning in convergent thinking (Brophy, 2001), the students were able to synthesize knowledge and ideas from different sources to generate a creative solution to an architectural problem.

Scientific creativity includes the ability to recognize the existence of a problem (Sternberg et al., 2020). In the case of the L-shaped design, the students realized that windows in summer would let sunlight in. The solar heat-reduction solution they arrived at was to plant trees near the west-facing windows, which, they reasoned would provide shade and reduce solar radiation onto the windows.

The thinking processes used in these STEAM lessons have the potential to promote students' scientific creativity. Three features of the STEAM method appear to play a role in encouraging scientific creativity. The first is presenting students with a situation and problem that interests them and which they perceive to be relevant to their lives. The second feature is the integration of arts- and culture-related elements in science teaching/learning activities. By bringing these areas together in the school curriculum, the STEAM approach opens students' minds to the generation of new possibilities (Franken, 1994). The third feature is group collaboration when discussing problems and working out solutions. Collaboration involves conversations in which students must articulate ideas to contribute to the group's endeavours. Communication and cooperation-oriented climates are significant factors in scientific inquiry leading to creativity (Hong & Song, 2020).

Discussion

Impacts of STEAM Approaches on Student Learning

STEAM teaching/learning has the potential to enhance students' attitudes towards science learning by helping them to experience the application of science in real-life contexts. Millar (1991) suggested that the abstract nature of science was a reason for students' perceptions of science as being difficult. STEAM learning activities engage students in constructing and applying their understanding of abstract concepts in real-life situations. For example, Project 3 challenged students to learn how to keep the interior of a building warm during winter and cool during summer by using natural systems rather than fossil fuel. When students focus on concrete issues (e.g. would eaves keep the inside of the building cooler in summer? How?), the science concepts involved do not seem as difficult as memorizing the definition of heat transfer and listening to a lecture on the different types of heat transfer. Having students consider the arts dimension of Korean traditional house design added to their appreciation of how an ancient method of achieving energy efficiency can be explained with modern-day science concepts. By demonstrating science concepts in real-life contexts and incorporating the arts into science learning, STEAM makes science interesting and relevant, countering the view that science is difficult.

In the 'learn to think' study (Hu et al., 2013), learning activities also incorporate daily life experiences. Hu et al. believe that requiring students to apply their learning to daily life accelerates the development of scientific creativity. In the STEAM approach, events or products from the students' sociocultural life provide not only the context for the application of science concepts but also form the context for identifying problems, hypothesizing initial explanations, and synthesizing disparate pieces of science knowledge in problem solving. As Project 3 demonstrates, the incorporation of a sociocultural product (e.g. a traditional Korean house) can have the effect of developing scientific creativity.

Another factor that appears to play a role in promoting scientific creativity is the classroom environment. Hu and colleagues (2013) argue that "a free, open, democratic…environment is a key factor for the development of students' scientific creativity" (p. 7). A collaborative, high-trust learning environment was created in their study by allowing students to make mistakes and take risks without fear of censure and by encouraging them to explore learning strategies by themselves. Similarly, STEAM approaches generally provide students with the space to verbalize their thoughts, prior knowledge, and hypothesized explanations or solutions without fear of correction from a teacher. The STEAM method also allows students the freedom to choose the form their explanation or solution will take (e.g. the colours, image, and symbols in a quasi-hologram).

Challenges of STEAM Implementation in Korean Schools

As with all innovations in pedagogy, there are challenges to address in the implementation of STEAM. The first is the gulf between the objectives and methods of STEAM and the traditional forms of assessing science learning in schools. Traditional pen-and-paper tests assess students' recall of facts, theories, or problem-solving solutions, and there is generally only one correct answer. Because the STEAM approach engages students in learning science in the context of real-life problems, assessment must evaluate students' application of science to sociocultural issues. There may not be a single correct answer. Rather, students are required to demonstrate understanding of the target science concepts with reference to the context in which those concepts are applied. In Project 3, a good answer would demonstrate students' understanding of the concepts of heat transfer and solar altitude, with reference to aspects of the design of a zero-energy school building. Until science examinations, especially national examinations, are aligned with the philosophy of the STEAM approach, teachers may hesitate to implement STEAM for fear of disadvantaging their students in classes preparing for a public examination.

Secondly, arts-science integration is challenging, as it requires science and non-science subject teachers to collaborate to plan what and how arts/culture-related content will be employed in science lessons, or vice versa. For example, in Project 2, the hologram creation activity occurred in art classes but required the science teacher to also be present to engage students in applying the principles of the science of light and vision when problems arose (e.g. the hologram image was not visible). The solution may be co-teaching, with science and non-science teachers planning and delivering integrated lessons in the same classroom. However, most Korean schools do not have a culture of team teaching, and the traditional structure of school timetables is an obstacle.

A third challenge for teachers is managing classroom discourse to provide as many opportunities as possible for students to think for themselves and to have some freedom in exploring alternative solutions for problems. The familiar traditional scenario of teacher-centred classrooms with teachers seeking correct answers must be replaced with teacher-student and student-to-student dialogue that encourages students to explore and question ideas, and to express their thoughts, however nascent. To maintain an encouraging learning environment in STEAM lessons, teachers' responses to students' contributions need to be affirming rather than deficit-focused. Teachers who are used to the teacher-centred classroom models may need professional development and support to shift away from prescriptive talk to a more affirming tone that opens up further dialogue and thinking.

In summary, teachers need help to guide classroom talk towards dialogic discourse between students and between students and the teacher, instead of the conventional IRE (initiate, respond, and evaluate) model typical of the traditional science classroom. It should also be noted that, in the quasi-3D hologram creation activity, there were times when students were so absorbed in making their hologram image visible to their classmates that there was no discussion of the science involved. The image was supposed to appear to be floating over the surface of their phone inside a transparent plastic pyramid with its sides at a 45-degree angle to the phone. The teacher had to constantly direct the discussion to the science concept of the angle of light reflection in relation to the eyes of the viewer. Dialogic discourse creates an environment in which teachers can exploit opportunities for students to probe each other's ideas, evaluate arguments relating to their ideas and evidence (Duschl & Osborne, 2002), and articulate dissenting views (Kelly et al., 2001). The importance of dialogic discourse in advancing knowledge construction in science has also been acknowledged by Ritchie (2001). Developing the skill of engendering dialogic discourse with and among students is a challenge STEAM teachers must address. Not surprisingly, there are ongoing challenges for Korean teachers and students to successfully bridge the gap between traditional science examinations that value the single correct answer and the focus in STEAM on demonstrating understanding of science concepts through their application in real-life sociocultural contexts. In response to this challenge, initiatives have begun in Korea to allow schools to opt for project-based assessment more aligned with the goals and methods of STEAM.

Conclusions

This chapter has reported the experiences of an arts-integrated approach (STEAM) in Korean schools. The trial implementation of three STEAM programs revealed positive impacts on students' conceptual understanding of the science concepts taught, attitudes towards and perceptions of science and the study of science, and indications of improved scientific creativity. The projects reported in this chapter showcase the potential advantages of moving away from the traditional teacher-fronted science classroom towards a collaborative student-centred teaching/learning approach. STEAM offers a path to promoting STEM learning in Asian science classrooms, which may translate to creative innovations in Asian industry and technology.

In 2022, the Korean education system began moving towards a theme-centred convergence model (MOE, 2021) so that school subjects like science and home economics can be thematically integrated so that students can learn about a topic like the science of cooking. The aim of the integration of STEM and non-STEM subjects in the Korean convergence model is to develop school graduates who are creative and able to respond flexibly to rapid social changes (MOE, 2021; Song et al., 2019) and to environmental and other problems of the twenty-first century. The outcomes of the STEAM projects presented here provide reason to believe that the convergence model may support positive attitudes towards learning, including science learning, and would encourage creativity in the next generation of citizens.

The authors acknowledge the support of the Australia-Korean Foundation (AKF), which provided funding for the design and implementation of Projects 1, 2, and 3 (AKF Grant 20,150,098 and AKF Grant 2,018,040).

References

Brophy, D. R. (2001). Comparing the attributes, activities, and performance of divergent, convergent, and combination thinkers. *Creativity Research Journal, 13*(3–4), 439–455. https://doi.org/10.1207/S15326934CRJ1334_20.

Chu, H.-E., Martin, S. N., & Park, J. (2019a). A theoretical framework for developing an intercultural STEAM program for Australian and Korean students to enhance science teaching and learning. *International Journal of Science and Mathematics Education, 17*, 1251–1266.

Chu, H.-E., Son, Y.-A., Koo, H.-K., Martin, S. N., & Treagust, D. F. (2019b). The potential of arts-integrated STEM approaches to promote students' science knowledge construction and a

positive perception of science learning. In Y.-S. Hsu & Y.-F. Yeh (Eds.), *Asia-Pacific STEM teaching practices* (pp. 17–38). SpringerNature.
de Vries, H. B., & Lubart, T. I. (2017). Scientific creativity: Divergent and convergent thinking and the impact of culture. *Journal of Creative Behavior, 53*(2), 145–155.
Driver, R., Leach, J., Millar, R., & Scott, P. (1996). *Young people's images of science*. Open University Press.
Duit, R., & Treagust, D. F. (1998). Learning in science—From behaviourism towards social constructivism and beyond. In B. Fraser & K. Tobin (Eds.), *International Handbook of Science Education* (pp. 3–26). Kluwer Academic Publishers.
Duschl, R. A., & Osborne, J. (2002). Supporting and promoting argumentation discourse in science education. *Studies in Science Education, 38*, 39–72.
Ernest, P. (1998). *Social constructivism as a philosophy of Mathematics*. SUNY Press.
Franken, R. (1994). *Human motivation* (3rd ed.). Brooks/Cole Publishing Co.
Harvey, S. (2014). Creative synthesis: Exploring the process of extraordinary group. *The Academy of Management Review, 39*(3), 324–343.
Holbrook, J., Rannikmäe M., & Soobard, R. (2020). STEAM education—A transdisciplinary teaching and learning approach. In B. Akpan & T. J. Kennedy (Eds.), *Science education in theory and practice: An introductory guide to learning theory* (pp. 465–477). Springer, Cham. https://doi.org/10.1007/978-3-030-43620-9_31.
Hong, O., & Song, J. (2020). A componential model of Science Classroom Creativity (SCC) for understanding collective creativity in the science classroom. *Thinking Skills and Creativity, 37*, 100698. https://doi.org/10.1016/j.tsc.2020.100698.
Hu, W., Wu, B., Jia, X., Yi, X., Duan, C., Meyer, W., & Kaufman, J. C. (2013). Increasing students' scientific creativity: The "learn to think" intervention program. *The Journal of Creative Behavior, 47*(1), 3–21.
Kelly, G., Crawford, T., & Green, J. (2001). Common task and uncommon knowledge: Dissenting voices in the discursive construction of physics across small laboratory groups. *Linguistics and Education, 12*(2), 135–174.
Kocabas, S. (1993). Elements of scientific creativity. *AAAI Technical Report SS-93-01*, 39–45. https://aaai.org/Papers/Symposia/Spring/1993/SS-93-01/SS93-01-006.pdf.
Korea Foundation for the Advancement of Science & Creativity. (n.d.). *Concept and definition of STEAM*. https://steam.kofac.re.kr/?page_id=11269.
Millar, R. (1991). Why is science hard to learn? *Journal of Computer Assisted Learning, 7*(2), 66–74. https://doi.org/10.1111/j.1365-2729.1991.tb00229.x.
Ministry of Education. (2020). *hagseub-ui paeleodaim-eul bakkueo ganeun yunghabgyoyug jonghabgyehoeg (an) 2020 nyeon ~ 2024 nyeon* [Plan for convergence education changing paradigm of learning (proposal) 2020–2024]. Curriculum and Policy Office, MOE. https://policy.nl.go.kr/search/searchDetail.do?rec_key=SH2_PLC20200251571.
Ministry of Education. (2021). *gugmingwa hamkkehaneun milae gyoyuggwajeong non-ui bongyeog chagsu* [A plan to promote the future curriculum with the people, April 20]. https://www.moe.go.kr/boardCnts/view.do?boardID=294&boardSeq=84176&lev=0&searchType=null&statusYN=W&page=1&s=moe&m=020402&opType=N.
Ministry of Education, Science and Technology. (2010). *2011 nyeon eobmubogo: chang-uiinjaewa seonjingwahaggisullo yeoneun milae daehanmingug* [2011 Work Report: Korea's future with creative talent and advanced science and technology]. MOE. https://www.moe.go.kr/sn3hcv/doc.html?fn=f22e94549941e423aaca35ab3726de3b&rs=/upload/synap/202202/.
Ministry of Education, Science and Technology, & Korea Foundation for the Advancement of Science and Creativity. (2012). *son-e jabhineun STEAM gyoyug - mueos-i aideul-eul jeulgeobge haneunga?* [Hands-on STEAM education: What makes children enjoyable?] https://steam.kofac.re.kr/steam-includes/download.php?mode=bbs&n=18.
Mullis, I. V. S., Martin, M. O., Foy, P., Kelly, D. L., & Fishbein, B. (2020). *TIMSS 2019 International Results in Mathematics and Science*. TIMSS & PIRLS International Study Centre, Boston College. https://timssandpirls.bc.edu/timss2019/international-results/.

Paik, S., Kim, J., Choi, S., Lee, Y., Choi, J., Yang, K., Jeong, K., Choi, J., Lee, S., Jeon, M., & Kim, K. (2012). *The development of curriculum for pre-service teachers of secondary school (Report no. 2012–27)*. Korea Foundation for the Advancement of Science and Creativity.

Park, H.-J., Baek, Y., Shim, J., Jeong, J.-S., Byun, S.-Y., Kim, N.-H., Min, B.-W., Song, M.-H., & Lee, H.-S. (2016). *2015 nyeon STEAM gyoyug-ui siltaejosa michhyogwaseong simcheung bunseog yeongu gyeolgwabogoseo* [Study on 2015 STEAM education status in Korea and its effectiveness, Research Report]. KOFAC. https://steam.kofac.re.kr/?p=6925.

Park, H.-J., Kim, Y., Noh, S., Lee, J., Jeong, J.-S., Choi, Y.-H., Han, H.-S., & Baek, Y. (2012). Components of 4C-STEAM education and a checklist for the instructional design. *Journal of Learner-Centered Curriculum and Instruction, 12*(4), 533–557.

Ritchie, S. M. (2001). Actions and discourses for transformative understanding in a middle school science class. *International Journal of Science Education, 23*(3), 283–299. https://doi.org/10.1080/095006901750066529.

Song, J., Kang, S.-J., Kwak, Y., Kim, D., Kim, S., Na, J., Do, J.-H., Min, B.-G., Park, S. C., Bae, S.-M., Son, Y.-A., Son, J. W., Oh, P. S., Lee, J.-K., Lee, H. J., Ihm, H., Jeong, D. H., Jung, J. H., Kim, J., & Joung, Y. J. (2019). Contents and features of 'Korean Science Education Standards (KSES)' for the next generation. *Journal of the Korean Association for Science Education, 39*(3), 465–478. https://doi.org/10.14697/jkase.2019.39.3.465 (in Korean).

Songer, N. B., & Linn, M. C. (1991). How do students' views of science influence knowledge integration? *Journal of Research in Science Teaching, 28*(9), 761–784. https://doi.org/10.1002/tea.3660280905.

Sternberg, R. J., Todhunter, R.J., Litvak, A., & Sternberg, K. (2020). The relation of scientific creativity and evaluation of scientific impact to scientific reasoning and general intelligence. *Journal of Intelligence, 8*(2), Article 17. https://doi.org/10.3390/jintelligence8020017.

Tytler, R., Prain, V., Aranda, G., Ferguson, J., & Gorur, R. (2020). Drawing to reason and learn in science. *Journal of Research in Science Teaching, 57*(2), 209–231.https://doi.org/10.1002/tea.21590.

Windschitl, M., Thompson, J., & Braaten, M. (2008). Beyond the scientific-method: Model-based inquiry as a new paradigm of preference for school science investigations. *Science Education, 92*(5), 941–967. https://doi.org/10.1002/sce.20259C.

Yakman, G., & Lee, H. (2012). Exploring the exemplary STEAM education in the US, as a practical education framework for Korea. *Journal of Korean Association for Science Education, 32*(6), 1072–1086. https://doi.org/10.14697/jkase.2012.32.6.1072.

Hye-Eun Chu is a senior lecturer at the School of Education, Macquarie University in Sydney, Australia. Before joining Macquarie University, she was an assistant professor in science education at Nanyang Technological University, Singapore; post-doctoral research fellow at Curtin University, Perth, Western Australia; and lecturer and researcher at Dankook University, Young-In, Korea. Her research interests include the investigation of students' conceptual development and attitudes towards learning science in arts-integrated STEM contexts.

Sonya N. Martin is a Professor in the Department of Earth Science Education at Seoul National University in Seoul, Republic of Korea. Sonya holds a bachelor's degree in Biology from Bryn Mawr College, and master's degrees in Elementary Education and in Chemistry Education from the University of Pennsylvania in the United States. She also holds a doctoral degree in Science Education from Curtin University in Australia. Her research focuses on identifying science teacher practices that promote learning for diverse students and on promoting professionalization of science teachers through classroom-based participatory research.

Ei Seul Kim is a Physics teacher at a science high school and a graduate student of the Department of Science Education at Seoul National University in Seoul, Republic of Korea. Her research focuses on exploring students' scientific creativity in STEAM classes. She is particularly interested in investigating students' competence development and reconstruction of scientific concepts. Her research interests also include doing comparative analysis of curricula from different countries with a focus on cultivating scientific competence.

Hyeong Moon Lee is a high school Physics teacher and a doctoral student in the Department of Science Education at Seoul National University in Seoul, Republic of Korea. He has studied students' understanding of scientific concepts using text network analysis. His research focuses on examining students' cognitive conflict situations in various contexts.

Part III
Preparing Teachers and Teacher Professional Development for STEM Education in Asia

Chapter 13
Towards Integrating STEM Education into Science Teacher Preparation Programmes in Indonesia: A Challenging Journey

Nurul F. Sulaeman, Pramudya D. A. Putra, and Yoshisuke Kumano

Abstract The science teacher preparation programme is vital as the early stage of developing science teacher competency. While the needs of STEM education are identified, it is challenging to adopt this movement to the programme. The complexity of the science teacher preparation program is intertwined with the context of each country. In this chapter, the context of Indonesia as one of the most populated countries in Asia with around 184 science teacher education programmes is introduced. The discussions start with the organization of science teacher education programmes, how STEM education is integrated, challenges and strategies that observable in Indonesia. The data was collected from university websites, programme curricula, observations of classes, interviews with key informants, and personal reflections. While the programmes require students to choose one specialty from the beginning, the challenge of integrating that subject with other STEM subjects was identified. We found several strategies to address this challenge, such as designing a new compulsory course in STEM education, integrating STEM into other pedagogical courses, elective courses, and extra-curricular activities. However, more significant efforts are needed to develop STEM PCK in Indonesia's science teacher preparation programmes. Integrating STEM education in pre-service science teacher programmes should be seen as aligned with the national goal of science education at the school system level and the needs of common goals of educational policy regarding STEM education among the programmes.

N. F. Sulaeman (✉)
Physics Education Programme, Mulawarman University, Jalan Muara Pahu Gunung Kelua Campus, Samarinda 75119, East Kalimantan, Indonesia
e-mail: nurul.fitriyah@fkip.unmul.ac.id

P. D. A. Putra
Science Education Programme, University of Jember, Kampus Tegalboto, Jalan. Kalimantan No. 37, 68121 Jember (East Jawa), Indonesia
e-mail: pramudya.fkip@unej.ac.id

Y. Kumano
STEAM Education Institute, Shizuoka University, 836 Oya, Suruga Ward, Shizuoka 422-8529, Japan
e-mail: kumano.yoshisuke@shizuoka.ac.jp

M. M. H. Cheng et al. (eds.), *Concepts and Practices of STEM Education in Asia*,
https://doi.org/10.1007/978-981-19-2596-2_13

Keywords Indonesia · Pre-service science teacher program · STEM education

Introduction

This chapter provides a description of the current science teacher preparation programmes in Indonesia, and the challenges of integrating STEM education into these programmes. While science education in Indonesia requires school students to engage in problem-solving and to explore integrated STEM issues, the science teacher preparation programmes face challenges in adapting to these goals. At present, no programme explicitly prepares pre-service science teachers to be STEM teachers. This chapter begins with a brief overview of Indonesia's context, goals of K-12 science education, the science teacher preparation programmes, and challenges of integrating STEM education into these programmes. Although all programmes follow the Indonesian national qualification framework implemented in 2012, some aspects of the programmes have developed differently in various universities. In this chapter, we discuss some of these variations. Our analysis focuses on general trends identified from a variety of sources. These include university websites, programme curricula, observations of classes, interviews with key informants, and personal reflections. The findings indicate that science teacher education in Indonesia generally follows the national standard. A pre-service science teacher should choose one specialty in integrated science, biology, physics, or chemistry. Pre-service teachers must attain a minimum of 144 credits during the 4-year curriculum, equal to around 216 in the European Credit Transfer and Accumulation System (ECTS), to build their pedagogical content knowledge (PCK) in science. The programmes prepare science teachers to teach at junior and senior high school levels. Various additional courses are offered, influenced by local needs and international trends. While the programmes require students to choose one specific subject (for example, biology education) from the beginning, the challenge of integrating that subject with other STEM subjects was identified. We found several strategies to address this challenge, such as designing a new compulsory course in STEM education and integrating STEM into other pedagogical courses, elective courses, and extra-curricular activities. However, more significant efforts are needed to develop STEM PCK in Indonesia's science teacher preparation programmes. Integrating STEM education in pre-service science teacher programmes should be seen as being aligned with the national goal of science education at the school system level and the needs of common goals of educational policy regarding STEM education among the programmes.

Background

Since 2012, the education system in Indonesia has been reformed to clarify the national qualification framework from elementary school up to post-graduate programmes (Directorate General of Higher Education, 2012). This framework clarifies the set of qualifications that should be achieved at each education level. Within the nine levels of qualification, Teacher Education Institutions (TEIs) are categorized as being at levels 6 and 7. The science teacher education programme involves a 4-year course-based programme (Level 6) plus a 1-year teaching practice-based programme (Level 7) followed by a national qualification framework. From the outset, pre-service teachers need to choose one subject as their major: physics, biology, chemistry, or integrated science. A teacher with a specialty in integrated science is qualified to be a science teacher at the junior high school level. A teacher with a specialty in specific science subjects (such as biology, physics, and chemistry) is qualified for senior high school level.

Discussions on the preparation of science teachers need to be understood in the context of the science learning goals in the schooling context of Indonesia. Globally, more attention is being paid to integrated STEM education, and the Indonesian science curriculum has adopted this approach as one of the goals of science education. However, this change has not been readily implemented by science teachers in schools (Nugroho et al., 2019; Permanasari et al., 2021). Although science teachers have stated that STEM education is in line with the science curriculum (Suwarma & Kumano, 2019), parents and teachers in Indonesia have reported a lack of effort to educate the youth about 21st-century skills (Nambiar et al., 2019). Understanding of integrated STEM approaches has also been found to be low among pre-service science teachers in Indonesia (Putra & Kumano, 2018). The discrepancy between the science curriculum and science teachers' implementation of STEM education could be understood by exploring the science teacher preparation programmes in more detail. This chapter introduces the pre-service science teacher preparation programmes in Indonesia, the relation between teacher preparation and the high school science curriculum, and analysis of STEM education in the teacher preparation programmes. Building from this discussion, this chapter addresses the challenges, complexities, and opportunities for enhancing pre-service science teacher programmes by including integrated STEM education.

The Indonesian Context: Education as a National Government Responsibility

Indonesia is an archipelagic nation located in Southeast Asia. Currently, 13,466 islands interconnected by straits and seas have been registered with valid coordinates. The five largest islands are Sumatra, Kalimantan, Sulawesi, Jawa, and Papua. A population of around 270 million is spread across the archipelago (Statistics Indonesia, 2021). When discussing Indonesia as a context, it is important to realize that it is a highly diverse country. Most Indonesians are Muslim, while others are Christian, Buddhist, or Hindu, among others. Moreover, there are more than 1,000 ethnic groups speaking nearly 500 different languages (Steinhauer, 1994). The founding fathers of

Indonesia recognized the need for a spirit of unity. The country's motto is *Bhinneka Tunggal Ika,* or Unity in Diversity, which is a reminder of the unity of all Indonesians. In every official meeting, including in schools, the language that should be used is the national language *Bahasa Indonesia.*

Indonesia's education system reflects the country's diverse religious heritage, its struggle for a national identity, and the challenges of resource allocation (Frederick & Worden, 2011). The problem of providing access to basic schooling across this huge archipelago remains the main challenge (Heyward & Sopantini, 2013). Students in rural areas experience a less supportive learning environment (Wahyudi & Treagust, 2006), including inadequate teacher quality (Sari, 2019) and teacher mismatches, especially in mathematics and science (Hendayana et al., 2011). To ensure equality and equity for all Indonesians, the education system in Indonesia is the responsibility of the national government. This centralized system has received some criticism, especially regarding the balance of authority between central and local authorities (Haridza & Irving, 2017). The slogan for education in 2020 was *Merdeka Belajar* or The Freedom of Learning. Despite the challenges presented by the archipelagic structure of Indonesia and its large population, schools are expected to be the acculturation place for young Indonesians and to nurture the spirit of innovation and problem-solving. This slogan strongly empowers the education system to be more humanist and to appreciate the uniqueness of children. There is massive open recruitment of schools, teachers, lecturers, and researchers who are willing to be part of this policy. In 2021, there are 2500 schools from 34 provinces participating in this programme. The main goal is to enhance student literacy (reading, numeric, and science) to improve teachers' and schools' quality.

Schooling in Indonesia is administered through two parallel systems: The Ministry of Education and Culture administers non-religion-based schools, and the Ministry for Religious Affairs administers religion-based-schools. All Indonesians have to complete 9 years of compulsory education, including primary school and junior high school. They can then continue to senior high school (3 years) and an undergraduate degree (4 or 5 years). The schools are also divided into two main categories: public and private schools. Some of the private schools are based on a specific religion, such as Islamic or Christian schools or specific approach such as Montessori school or international school. All schools follow the national curriculum standards, and all schools provide equivalent certifications that are valid to be used as an application to a higher level of education.

Goals of K-12 Science Education in Indonesia

As Indonesia is the fourth most populous country in the world, the educational movement towards STEM education is both crucial and challenging. Based on PISA results, Indonesian students show unsatisfactory reading, science, and mathematics skills compared to other countries (Organisation for Economic Co-operation and Development [OECD], 2019). The Indonesian government is tasked with effectively educating diverse student groups to meet the country's changing needs, internally and internationally. Science education is expected to play an essential role in preparing students to be leaders in STEM fields and in improving citizens' scientific literacy (Faisal & Martin, 2019).

Science is a compulsory subject for students from grades 4 to 12. Students in grades 1 to 3 study science as a thematic topic in their reading, writing, and mathematics subjects. Primary school science is taught by class teachers who graduated from primary teacher preparation programmes. Science in junior high school is taught as integrated science, while in senior high school, it is taught as single specific subjects (physics, chemistry, and biology). Earth science is treated as part of Physics. Science teacher preparation programmes are aimed at those who wish to teach science at the high school level. The goals of science education have been translated into English and are presented in Table 13.1.

These goals are achieved through the topics covered in science learning. The topics are scientific activities and safety, living things, energy, matters and its transformation, earth and space, science-environment-technology-society. Each topic clarifies the minimum understanding and skills that students should achieve after the science learning. The goals show an awareness of balance in science learning between content knowledge, procedural knowledge, and epistemic knowledge, as in the PISA framework (OECD, 2019). Moreover, the last goal emphasizes the importance of integrating science concepts with technology, environment, and society. As a subject at the K-12 level, science is a tool with which students can develop their problem-solving skills based on scientific considerations.

Table 13.1 Goals of science education in Indonesia

No	Specific goals
1	Students live life with a positive attitude, critical thinking, creativity, innovation, and honest collaboration based on the scientific process and product
2	Students understand the natural phenomena around them based on the result of science learning through specific science subjects such as physics, chemistry, and biology
3	Students distinguish science and technology products through scientific thinking
4	Students make choices based on scientific considerations
5	Students solve real-life problems based on scientific considerations
6	Students recognize the roles of science in solving the general problems of humanity, such as food supply, health, and environmental issues
7	Students understand that the development of science leads to technology in the past and future of society and the environment

(Indonesia Ministry of Education and Culture, 2020)

The Science Teacher Preparation Programmes in Indonesia

In the following, we describe the pathways for the science teacher preparation programmes and discuss the challenges of preparing science teachers to teach effectively. The science teacher preparation programmes prepare pre-service teachers for teaching at the high school level. Science teachers are prepared to teach in junior high school (grades 7 to 9) or senior high school (grades 10 to 12). Before the senior high school level, all elementary and junior high school students learn basic science as an integrated and thematic subject with no distinct separation between physics, chemistry, and biology content. At the senior high school level, each science subject is taught separately as compulsory courses. In most schools, engineering and technology education are additional subjects for students. Engineering and technology subjects are generally taught as elective classes that students choose.

Structuring a science teacher education programme requires consideration of what teachers need to know to promote the goals of education (Olson, 2017). In Indonesia, pre-service science teachers are prepared to achieve the goals of K-12 science education (Table 13.1). National and private universities provide teacher education programmes. In 2019, 184 programmes (mathematics, science, physics, biology, and chemistry education) were available. On average, these programmes accept around 20–60 students each year (Indonesia Ministry of Education & Culture, 2021). The demography of universities that conduct these programmes on major Indonesian islands is shown in Fig. 13.1. The majority of the science teacher preparation programmes are conducted in universities on Jawa island. This island is also the most populated island in Indonesia and is where the capital city, Jakarta, is located. Around 5,000 graduates per year are shaped through these programmes, highlighting the urgency of maintaining and enhancing the programmes' quality and ensuring that they are aligned with global challenges.

There are two pathways to becoming a science teacher in Indonesia (see Fig. 13.2). The most common way is to join a science teacher preparation programme. Students

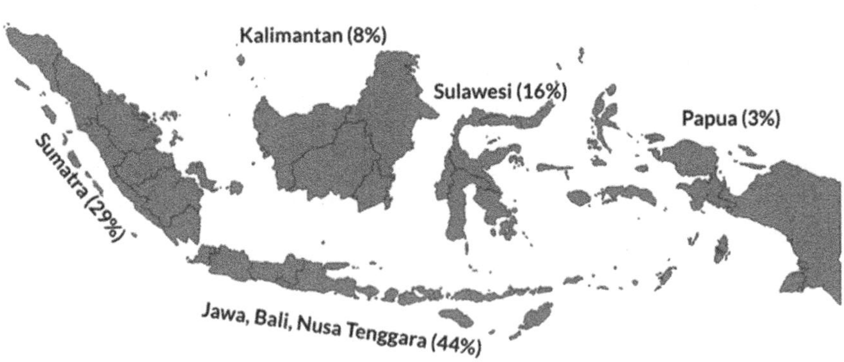

Fig. 13.1 Distribution of pre-service science teacher programmes in Indonesian universities

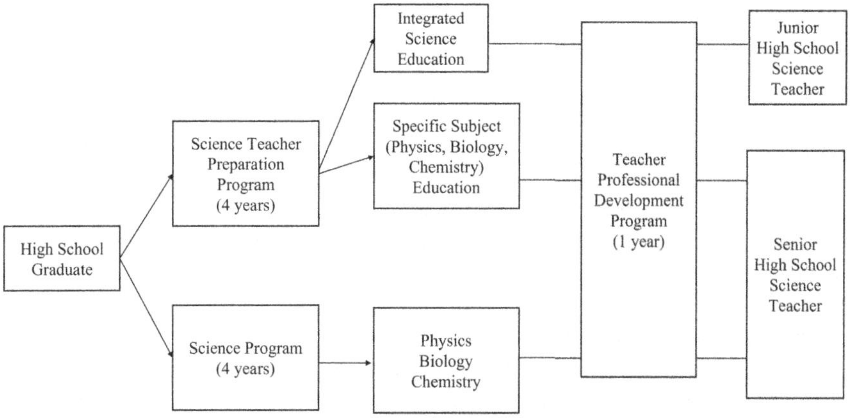

Fig. 13.2 Pathways to becoming a science teacher in Indonesia

are asked to choose one specific subject at the beginning of this programme: integrated science education, physics education, biology education, or chemistry education. After completing the programme, students are awarded a Bachelor's degree in science education. Graduates from this programme can teach at the high school level. However, they do not have a professional teaching certificate and do not qualify for certification remuneration from the Indonesian government. After finishing this programme, they can continue with a 1-year teacher professional development programme and become licensed science teachers. The second pathway to becoming a science teacher in Indonesia is completing a 4-year programme with a science major (such as physics, biology, or chemistry). After finishing this programme, students are awarded a Bachelor's degree in science. Graduates from a science major need to take the 1-year teacher professional development programme before becoming science teachers.

The Curriculum of the Science Teacher Preparation Programmes

Following national regulation (Directorate General of Higher Education, 2012), the higher education programmes (including TEIs) use the model of Outcome-Based Education (OBE). The science education programmes for teaching in Indonesia have a similar set of outcomes overall, although the details may vary across universities. The outcomes are called Programme Learning Outcomes (PLO). The outcomes consist largely of a knowledge domain and skill domain. To decide the PLOs, national regulation was discussed through the association for similar programmes. Several associations are active in Indonesia for science education, such as the Indonesian Science Teacher Association or the Physical Society of Indonesia. Moreover, insights from the international associations or accreditation boards also influence the development of PLOs. The PLOs determine the detailed outcomes for each course.

The teacher preparation programmes are divided into 4 years of coursework and 1 year of teacher professional development. The distribution of the credits across the 4 years of the science education programmes is shown in Table 13.2. Courses are divided into five main categories, with 55% related to content knowledge and 45% related to pedagogy and general courses. For the physics, biology, and chemistry education programmes, the content knowledge relates mostly to the subject that has been chosen. For an integrated science education programme, the content is integrated from each of the science domains of biology, chemistry, and physics. It is also common in Indonesia to have several elective courses that pre-service teachers choose based on their interests. Generally, the curriculum in each similar programme includes identical courses, and the Indonesian government has opened an exchange programme among these teacher preparation programmes through several programmes such as *Kampus Merdeka* or Campus Independency. These programmes allow pre-service teachers to take courses in different universities. For example, pre-service biology teachers from University A on Papua Island have opportunities to take microbiology courses at University B on Jawa island. The universities on Jawa island tend to have better facilities and quality of lecturers than others. Pre-service teachers also have opportunities to observe the uniqueness of Indonesian culture on a different island. The goal of this policy is to reduce the variation among universities in Indonesia and to nurture a sense of unity in the diversity of the country.

The content knowledge is divided into two types:

(1) Single-subject content knowledge (39 credits); for example, mathematics, fundamental physics, general biology, mechanics, astronomy
(2) Integrated subject content knowledge (42 credits); for example, biochemistry, chemical physics, basics of science, applied science, human biology.

Although the information in Table 13.2 is generally found in science education programmes, some variation can be seen in the name of courses and the number of credits awarded for each course. Moreover, each university also builds its curriculum

Table 13.2 The typical distribution of the credits in the department of science education

Group of courses	Credits	Percentage (%)
General courses (1) Religion and Indonesian nationality	6	4
(2) Language	4	3
Pedagogical knowledge	19	13
Content knowledge	81	55
Subject-specific pedagogy	26	18
Elective courses	10	7
Total	146	100

based on specific knowledge related to the university's specific knowledge expertise. Each university in Indonesia is encouraged to set up a centre of excellence based on its strengths and vision. We observed and analysed several programmes in science education at five different universities to investigate their provision of science or STEM teacher education. These programmes were chosen by considering the representation of national and private universities and the programmes in integrated science and specific science subjects.

Example A: Science Education Programme, University of Jember, National University on Jawa Island, Indonesia.

The University of Jember is located in East Jawa, the most populous island in Indonesia. It has become a centre for agriculture and medicine studies. Besides the main goals of developing science teachers' PCK, the teacher training programme also supports teachers with knowledge of agricultural issues. The curriculum of the science education programme prepares students to become science teachers in junior high school. Students must attain 144 credits (216 ECTS) through general courses, pedagogical knowledge, integrated science content knowledge, and elective courses. Additional courses that support the university's research centre are in biotechnology and agroindustry.

Example B: Physics Education Programme, Mulawarman University, National University on Kalimantan Island, Indonesia.

As the university is situated in a tropical rainforest area, this university declares itself the centre of excellence of tropical studies. A 4-year programme with a major in physics education includes 149 credits (equal to around 223 ECTS) across general courses, pedagogical knowledge, content knowledge, subject-specific pedagogy, and elective courses. After completing the minimum number of courses, students can teach physics at the high school level. The courses that support the university's centre of excellence are additional content knowledge courses in environmental physics and tropical forest environmental sciences.

Example C: Biology Education Programme, Muhamadiyah University, Private University on Jawa Island, Indonesia.

This programme is offered by a private university that is based on Islamic values. A 4-year programme with a major in biology education includes 161 credits (equal to 241 ECTS) achieved through Islamic courses, general courses, pedagogical knowledge, content knowledge, subject-specific pedagogy, and elective courses. After completing all of the required courses, students can teach biology at the high school level. The courses that support the university's Islamic values include biology and Islam, and Islam and the development of knowledge.

Teacher Professional Development Programmes

After students graduate from a 4-year science teacher preparation programme, they take the 1-year programme that prepares them to be teaching professionals. Generally, students who have graduated from teacher training faculties and those who

Table 13.3 The typical distribution of credits in the curriculum of the 1-year professional development programme

Group of courses	Credits	Percentage (%)
Content knowledge	5	42
Development of instructional design	3	25
Teacher training programme	4	33
Total	12	100

have graduated from engineering or natural science departments are able to join this programme. Students must complete two tests during the one-year programme: a comprehensive test and a performance test. The comprehensive test examines the student's mastery of content knowledge and pedagogical knowledge. The performance test examines the student's ability to teach science to an actual class of students. The typical distribution of credits in the curriculum of the 1-year teacher professional development programme is shown in Table 13.3.

Opportunities for Integration

In the earlier part of this chapter, we discussed the goals of science education at the school level (Table 13.1), the topics for science at the school level, and the curriculum for science teacher preparation programmes. While integration of science with problem-solving skills and other related subjects is emphasized at the school science level, these issues are not directly addressed in the science teacher preparation programme provision. This mismatch deserves more attention from universities. While there is a need to integrate STEM in teacher education to foster essential skills such as problem solving (Miller & Krajcik, 2019; Priemer et al., 2020), implementation in the curriculum remains a challenge. The challenge to integrating STEM education into a single subject context is a significant challenge globally (National Academy of Engineering & National Research Council, 2014), especially in the Asian context (Lee et al., 2019). It is commonly found in Asian countries, the curriculum for school level is fixed. Therefore, infusing new subject is challenging.

Previous research in the United States explained that perspectives on integrating STEM are varied. The perspectives are: (1) STEM is equal to science; (2) STEM means both science and mathematics; (3) STEM means science, but incorporates technology, engineering, and mathematics; (4) STEM equals a quartet of separate disciplines, (5) STEM means that science and mathematics are connected by either technology or engineering; (6) STEM means coordination across disciplines; (7) STEM means combining two or three disciplines; (8) STEM means complementary overlapping across disciplines; (9) STEM means transdisciplinary courses (Bybee, 2013). Of the nine possible perspectives on STEM education, integrating STEM with science and incorporating technology, engineering, and mathematics seems the

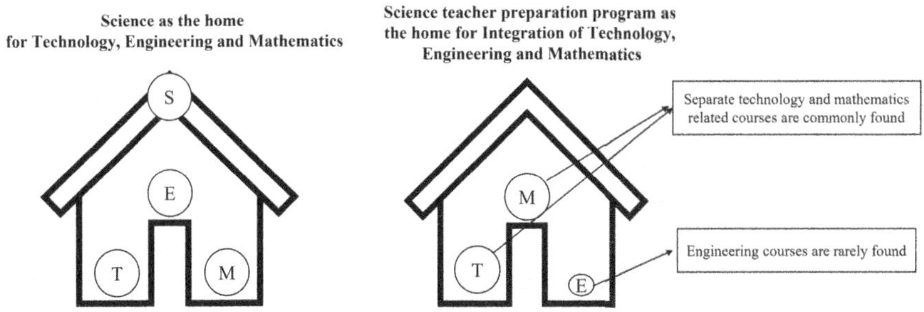

Fig. 13.3 Integration of STEM in science teacher preparation programmes in Indonesia

most feasible in the case of science teacher preparation programmes in Indonesia (Fig. 13.3). The science concepts become a 'home' that can provide a basis for the integration of technology, engineering, and mathematics elements. In the specific programmes (such as physics education, biology education, and chemistry education), the 'home' is the specific subject. The engineering component seems to be the most neglected element. This situation is also found in countries such as the United States (National Academy of Engineering & National Research Council, 2009) and Turkey (Asiroglu & Akran, 2018). The engineering element is not explicit in either the K-12 curriculum or science teacher preparation programmes in Indonesia.

Integrating STEM Education into the Science Teacher Preparation Programmes

Out of the minimum of 144 credits in the science teacher preparation programmes, mostly the integration of STEM education is considered as an additional course or additional topic in some pedagogical courses. These elements of STEM education in the current programmes should be appreciated as an initial step. However, the number of credits related to STEM education is far from sufficient. The methods of integrating STEM education vary across different universities. From our observation in five programme curricula in different universities, four approaches are currently being implemented: integrating STEM as compulsory courses, integrating STEM as part of compulsory courses, integrating STEM in elective courses, and extra-curricular activities related to STEM education.

Compulsory Courses

Some science teacher preparation programmes had STEM education courses as one of the compulsory courses. For example, at the Indonesia University of Education, STEM is a compulsory course (3 credits) that focuses on building the STEM literacy of pre-service science teachers through an integrative approach and collaborative activities. The main topics in this course are energy and new materials. This course uses inquiry- and problem-based learning in the learning process.

Part of Compulsory Courses

Although STEM education as a specific course is not part of most science teacher preparation programme curricula, our observation of several programmes revealed that STEM education topics are part of their science-specific courses. For example, at the University of Jember, there is a course called Innovative Learning Models. In this course, STEM education is discussed as an approach to learning. Another example is Mulawarman University, where STEM education is discussed in the Physics Teaching and Learning course. This course mainly examines the approaches, models, and strategies for teaching physics.

Elective Courses

As an initial process of integrating STEM education, elective courses about STEM education are offered by several universities in Indonesia. For example, in the Department of Science Education at the University of Jember, a STEM course worth two credits covers the history of the development of STEM, understanding STEM terms, perceptions of STEM learning, implementation of STEM learning, and pedagogy in STEM education. As part of this course, students develop a STEM instructional design suitable for the science curriculum in Indonesia.

STEM Education in Extra-Curricular Activities

Another approach initiated by several universities (such as Syah Kuala University) is building a research centre in STEM education. Through this centre, various activities – such as STEM camps, workshops, and seminars – help pre-service and in-service science teachers and students learn about STEM. This centre also collaborates with international non-profit organizations. Another example is the National Education Museum at the Indonesia University of Education. In this museum, there is a centre called the 4D Frame STEM Education Centre. In 2020, this centre hosted an activity called the International Students' STEM Camp for Pre-Service Teachers.

STEM Education as a Research Topic

During the 4-year programme, science teacher candidates need to conduct a research project and write up a thesis. STEM education has become a popular research topic among science teacher candidates. For example, research has been done on developing STEM learning material (Gustiani et al., 2017; Hartini et al., 2020), perceptions of STEM (Widayanti et al., 2019), and students' achievement towards twenty-first century skills in STEM education (Mutakinati et al., 2018; Putra et al., 2021). Such research shows the significant interest of science teacher candidates in exploring STEM education. After being exposed to integrated STEM during their courses, they are eager to extend their interest in STEM education to their theses.

Challenges of Integration

With the growing awareness of the complexity of real-life problems, it has become clear that a problem in physics is not solely a physics problem. Real-life problems often require the integration of science, technology, mathematics, engineering, and other related subjects. In Indonesia's school system, three possibilities for implementing STEM integration have been identified: STEM in the K-12 school system, building a STEM subject, and integrating STEM into compulsory subject (Arlinwibowo et al., 2020). The possibility to infuse STEM in school is similar to developing a magnet school that focuses on STEM or on developing STEM schools. In the history of the Indonesian school system, there is a possibility to develop a specific kind of school such as an environmental school called *Sekolah Adiwiyata* (Nomura, 2009; Parker, 2018). The school could be at the primary or secondary level and infuses environmental issues into the curriculum. The second possibility is to develop a new subject at school called STEM. The new subject would be separated from science or mathematics and would address the integration concept through problem-solving activities. The third possibility that is observable through the national curriculum is integrating STEM into the science subject curriculum.

In the science teacher preparation programmes in Indonesia, there are clear boundaries among each science subject. Therefore, out of the nine possible perspectives on STEM education (Bybee, 2013), integrating STEM with science and incorporating technology, engineering, and mathematics seems to us to be the most feasible. Here, science is the 'home' into which the other components are integrated. There is currently no possibility of developing a new programme, such as a STEM teacher programme, within the school and university system regulation.

Even though the requirement to design integrated STEM is found in the school science education curriculum, different science teacher preparation programmes are adopting different approaches to address this reform. Among these programmes, the integration of STEM is often an additional or optional part of the science teacher preparation programme. The different approaches in each programme need to be addressed, and a more organized reform towards STEM education is needed. Due to the different approaches adopted by the science teacher preparation programmes, the impacts of these changes are difficult to measure. We argue for a more comprehensive reform towards integrated STEM, setting common goals for the integration, and including integrated STEM as part of the Programme Learning Outcomes for science teacher preparation programmes.

Conclusion

Indonesia's science teacher education follows a centralized curriculum with the requirement of teaching licenses. An analysis of the curriculum shows that the programme prepares teachers in terms of pedagogical content knowledge in science. Various additional courses are offered according to local needs or specific targets or strengths of the universities. The programmes feature inspiring examples of good practice in integrating STEM education, such as efforts to initiate STEM centres, infuse STEM education into some courses (such as a micro-teaching course), and offer extra-curricular activities, such as an orientation programme. However, lasting systemic solutions remain elusive. Additionally, there is still much to be done to develop STEM Pedagogical Content Knowledge in science teacher preparation programmes in Indonesia. It is important that integrated STEM education is included in pre-service teacher programmes if these programmes are to align with the national goals for school science as presented in the school science curriculum.

References

Arlinwibowo, J., Retnawati, H., & Badrun, K. (2020). The types of STEM Education Implementation in Indonesia. *Journal of Xi'an University of Architecture & Technology, XII*(VIII), 606–613.

Asiroglu, S., & Akran, S. K. (2018). The readiness level of teachers in science, technology, engineering and mathematics education. *Universal Journal of Educational Research, 6*(11), 2461–2470. https://doi.org/10.13189/ujer.2018.061109.

Bybee, R. W. (2013). *The case for STEM education: Challenges and opportunities*. National Science Teaching Association (NSTA).

Directorate General of Higher Education. (2012). *Indonesian qualifications framework (Kerangka Kualifikasi Nasional Indonesia). Presidential decree no. 8/2012: Implication and implementation strategies*. Indonesia Ministry of Education and Culture. http://kkni.kemdikbud.go.id/asset/pdf/iqf_booklet_(english).pdf.

Faisal, & Martin, S. N. (2019). Science education in Indonesia: Past, present, and future. *Asia-Pacific Science Education, 5*(1). https://doi.org/10.1186/s41029-019-0032-0.

Frederick, W. H., & Worden, R. L. (Eds.). (2011). *Indonesia: A country study* (6th ed.). Federal Research Division, Library of Congress. https://www.loc.gov/item/2011038834/.

Gustiani, I., Widodo, A., & Suwarma, I. R. (2017). Development and validation of science, technology, engineering and mathematics (STEM) based instructional material. *AIP Conference Proceedings, 1848*. https://doi.org/10.1063/1.4983969.

Haridza, R., & Irving, K. E. (2017). The evolution of Indonesian and American Science education curriculum: A comparison study. *International Journal for Educational Studies, 9*(2), 95–110. https://journals.mindamas.com/index.php/educare/article/view/802/765.

Hartini, S., Mariani, I., Misbah, & Sulaeman, N. F. (2020). Developing of students worksheets through STEM approach to train critical thinking skills. *Journal of Physics: Conference Series, 1567*(4). https://doi.org/10.1088/1742-6596/1567/4/042029.

Hendayana, S., Supriatna, A., & Imansyah, H. (2011). Indonesia's issues and challenges on quality improvement of mathematics and science education. *Center for the Study of International Cooperation in Education (CICE), Series 4*(1), 41–51. https://ir.lib.hiroshima-u.ac.jp/ja/00034457.

Heyward, M., & Sopantini. (2013). Indonesia: The challenges of quality and equity in education. In L. P. Symaco (Ed.), *Education in South-East Asia* (pp. 71–94). Bloomsbury Academic. https://doi.org/10.5040/9781472544469.ch-004.

Indonesia Ministry of Education and Culture. (2020). *Fokus Pembelajaran SD/MI, SMP/MTs, SMA/MA.* [Learning focus for elementary school and middle school]. Badan Standar Nasional Pendidikan. https://bsnp-indonesia.org/fokus-pembelajaran/.

Indonesia Ministry of Education and Culture. (2021). *Pangkalan Data Pendidikan Tinggi* [Indonesia Higher Education Data Base]. https://pddikti.kemdikbud.go.id/.

Lee, M. H., Chai, C. S., & Hong, H. Y. (2019). STEM education in Asia Pacific: Challenges and development. *Asia-Pacific Education Researcher, 28*(1), 1–4. https://doi.org/10.1007/s40299-018-0424-z.

Miller, E. C., & Krajcik, J. S. (2019). Promoting deep learning through project-based learning: A design problem. *Disciplinary and Interdisciplinary Science Education Research, 1*(1), 1–10. https://doi.org/10.1186/s43031-019-0009-6.

Mutakinati, L., Anwari, I., & Kumano, Y. (2018). Analysis of students' critical thinking skill of middle school through STEM education project-based learning. *Jurnal Pendidikan IPA Indonesia, 7*(1), 54–65. https://doi.org/10.15294/jpii.v7i1.10495.

Nambiar, D., Karki, S., Rahardiani, D., Putri, M., & Singh, K. (2019, July). *Study on skills for the future in Indonesia.* Oxford Policy Management. https://www.unicef.org/indonesia/media/6221/file/Study%20on%20skills%20for%20the%20future%20in%20Indonesia.pdf.

National Academy of Engineering and National Research Council. (2009). *Engineering in K-12 education: Understanding the status and improving the prospects.* The National Academies Press. https://doi.org/10.17226/12635.

National Academy of Engineering and National Research Council. (2014). *STEM Integration in K-12 Education: Status, Prospects, and an Agenda for Research.* The National Academies Press. https://doi.org/10.17226/18612.

Nomura, K. (2009). A perspective on education for sustainable development: Historical development of environmental education in Indonesia. *International Journal of Educational Development, 29*(6), 621–627. https://doi.org/10.1016/j.ijedudev.2008.12.002.

Nugroho, O. F., Permanasari, A., & Firman, H. (2019). The movement of STEM education in Indonesia: Science teachers' perspectives. *Jurnal Pendidikan IPA Indonesia, 8*(3), 417–425.

Olson, J. K. (2017). Teacher preparation for science education. In K. S. Taber & B. Akpan (Eds.), *Science education, New directions in mathematics and science education* (pp. 523–537). Sense Publishers. https://doi.org/10.1007/978-94-6300-749-8_38.

Organisation for Economic Co-operation and Development. (2019). *PISA 2018 Assessment and Analytical Framework. OEDC Publishing.* https://doi.org/10.1787/b25efab8-en.

Parker, L. (2018). Environmentalism and education for sustainability in Indonesia. *Indonesia and the Malay World, 46*(136), 235–240. https://doi.org/10.1080/13639811.2018.1519994.

Permanasari, A., Rubini, B., & Nugroho, O. F. (2021). STEM education in Indonesia: Science teachers' and students' perspectives. *Journal of Innovation in Educational and Cultural Research, 2*(1), 7–16. https://doi.org/10.46843/jiecr.v2i1.24.

Priemer, B., Eilerts, K., Filler, A., Pinkwart, N., Rösken-Winter, B., Tiemann, R., & Zu Belzen, A. U. (2020). A framework to foster problem-solving in STEM and computing education. *Research in Science & Technological Education, 38*(1), 105–130. https://doi.org/10.1080/02635143.2019.1600490.

Putra, P. D. A., & Kumano, Y. (2018). Energy learning progression and STEM conceptualization among pre-service science teachers in Japan and Indonesia. *New Educational Review, 53*(3), 153–162. https://doi.org/10.15804/tner.2018.53.3.13.

Putra, P. D. A., Sulaeman, N. F., Supeno, & Wahyuni, S. (2021). Exploring students' critical thinking skills using the engineering design process in a physics classroom. *Asia-Pacific Education Researcher* (Published online: 30 November 2021). https://doi.org/10.1007/s40299-021-00640-3.

Sari, V. A. (2019). Educational assistance and education quality in Indonesia: The role of decentralization. *Population and Development Review, 45*(S1), 123–154. https://doi.org/10.1111/padr.12272.

Statistics Indonesia. (2021). *Statistical Yearbook of Indonesia 2021*. BPS-Statistics Indonesia. https://www.bps.go.id/publication/2021/02/26/938316574c78772f27e9b477/statistik-indonesia-2021.html (in Indonesian).

Steinhauer, H. (1994). The Indonesian language situation and linguistics; Prospects and possibilities. *Journal of the Humanities and Social Sciences of Southeast Asia, 150*(4), 755–784. https://doi.org/10.1163/22134379-90003070.

Suwarma, I. R., & Kumano, Y. (2019). Implementation of STEM education in Indonesia: Teachers' perception of STEM integration into curriculum. *Journal of Physics: Conference Series, 1280*, 52052. https://doi.org/10.1088/1742-6596/1280/5/052052.

Wahyudi, Y. S., & Treagust, D. F. (2006). Science education in Indonesia: A classroom learning environment perspective. In D. Fisher & M. S. Khine (Eds.), *Contemporary approaches to research on learning environments* (pp. 221–246). Worldviews. https://doi.org/10.1142/9789812774651_0009.

Widayanti, W., Abdurrahman, A., & Suyatna, A. (2019). Future physics learning materials based on STEM education: Analysis of teachers and students perceptions. *Journal of Physics: Conference Series, 1155*(1). https://doi.org/10.1088/1742-6596/1155/1/012021.

Nurul F. Sulaeman is currently an Assistant Professor in Physics Education, Faculty of Teacher Training and Education, Mulawarman University. She received the Bachelor's degree in physics education from Sebelas Maret University, Indonesia and Master's degree in Educational Research and Evaluation from Yogyakarta State University, Indonesia. She obtained the Ph.D. degree from Graduate School of Science Informatics and Technology, Shizuoka University Japan. Her research focuses on physics education, STEM education for SDGs, and STEM education in pre-service science teacher's program. She can be contacted at Email: nurul.fitriyah@fkip.unmul.ac.id.

Pramudya D. A. Putra is currently an Assistant Professor in Science Education at the University of Jember. He graduated from Surabaya State University, Indonesia with a Bachelor's degree in physics education, and Master's degree in Science Education from Sebelas Maret University Indonesia. He obtained his Ph.D. at Shizuoka University, Japan in STEM education. He is now a Leader in the Research Group for Physics Science Education at the University of Jember.

Yoshisuke Kumano is Professor Emeritus, Appointed Professor, Vice Director of STEAM Education Institute, Shizuoka University. He is also currently the Director of NPO Shizuoka STEAM Education Development Center and the Director of Shizuoka STEM Academy supported by Japan Science and Technology Agency. He is a past president of East Asian Association for Science Education (EASE). He has been Professor of Science Education, Shizuoka University for about 28 years after getting Ph.D. in Science Education at the University of Iowa through the Fulbright Program in 1993. His focus area of study is currently STEAM education.

Chapter 14
Teacher Professional Development and Education for STEM Teaching in Thailand: Challenges and Recommendations

Witat Fakcharoenphol, Chanyah Dahsah, and Tussatrin Wannagatesiri

Abstract STEM education is incorporated as part of the national policies in many countries due to the belief that it can produce the next generation of citizens with the necessary skills to raise the country's economic competitiveness. Thailand implemented a 5-year STEM Master Plan from 2015 to 2019 to increase trained personnel to serve its STEM education policies. Currently, in the Thailand Basic Core Curriculum B.E. 2551 (MOE, 2008) and revised B.E. 2560 (MOE, 2017a, 2017b), STEM Education emphasizes two subjects, science and mathematics. However, the implementation of STEM education in Thailand is encouraged to integrate other subjects to promote competencies in the Core Curriculum, for example, communication, higher-order thinking, problem-solving, and life skills. Since teachers play a pivotal role in students' learning, teachers must be able to integrate STEM into their teaching. However, integrating STEM into the classroom is a challenge in Thailand. Many studies have explored teachers' perceptions and attitudes toward STEM education. In addition, many research studies and professional development projects launched by the government and private sector have emphasized promoting teachers' competencies in STEM teaching. The teacher preparation programs also provide STEM courses and activities to encourage pre-service teachers with STEM teaching competencies. This chapter will focus on the meta-analysis of research on Thai teachers' perceptions and practices in STEM teaching and on the current trend of research and professional development projects on STEM education in Thailand. This will provide recommendations for teacher education and needed support for integrating STEM for in-service and pre-service teachers.

W. Fakcharoenphol · T. Wannagatesiri (✉)
Department of Teacher Education, Faculty of Education and Development Sciences, Kasetsart University, Nakhon Pathom 73140, Thailand
e-mail: tussatrin.k@ku.th

C. Dahsah
Science Education Center, Faculty of Science, Srinakharinwirot University, 174 Sukhuvit Soi 23 Khlong Toei, Nuea Vadhana, Bangkok, Thailand

M. M. H. Cheng et al. (eds.), *Concepts and Practices of STEM Education in Asia*,
https://doi.org/10.1007/978-981-19-2596-2_14

Keywords STEM teacher education · Classroom practices · Teacher efficacy · Teacher belief

Introduction

STEM education is being incorporated into national education policies in many countries and educational jurisdictions due to the belief that it is needed to produce the next generation of citizens with the necessary skills to raise the country's economic competitiveness. Since teachers play a pivotal role in students' learning, the effectiveness of STEM education relies in part on teachers' understanding, perceptions, and implementation of STEM. Thailand implemented a 5-year STEM Master Plan from 2015 to 2019 to increase the number of trained personnel to implement its STEM education policies. This included many professional development projects launched by the government and private sector to promote teachers' competencies in STEM teaching, mostly for in-service teachers. The teacher education programmes also provide STEM courses and activities to support pre-service teachers to develop their STEM teaching competencies. However, the integration of STEM into classrooms is a great challenge in Thailand. This chapter will focus on how recent research and professional development projects support STEM teachers in Thailand, in addition to research about teachers' perceptions of and practices in STEM teaching and learning. This chapter will also provide recommendations on how to enhance Thai teacher practices for effectively implementing STEM teaching.

Background

The Thai government has promoted the Thailand 4.0 agenda with an economic model based on creativity, innovation, new technology, and high-quality services to raise the quality of life. According to the belief that STEM education can advance the next generation of citizens with innovative thinking and other necessary work skills to raise the country's economic competitiveness, STEM education became a key educational approach in the revolution of educational policies and educational research from 2015 to 2019 (Baxter, 2017). The intention is for school students to experience STEM education which integrates the four disciplines (science, technology, engineering, and mathematics) and to develop twenty-first century learning skills, including problem solving, creativity, collaboration, communication, and critical thinking.

The Thailand Office of the Basic Education Commission, Ministry of Education (OBEC, MOE) encouraged the implementation of STEM education via the Institute for the Promotion of Teaching Science and Technology (IPST), National Center for STEM Education. 'What is STEM education?' in Thailand refers to the STEM definition conveyed by the IPST that 'Sa tĕm ṣ̄ụkṣ̄'ā khụ̄x næwthāng kār cạdkār ṣ̄ụkṣ̄'ā thī̀ būrṇ ā kār khwām rū̂ nı 4 s̄h̄ withyākār dị̂kæ̀ withyāṣ̄āṣ̄tr̒ wiṣ̄wkrrm thekhnoloyī læa khṇitṣ̄āṣ̄tr̒ doy nên kārnả khwām rū̂ pị chı̂ kæ̂ pạỵh̄ā nı chīwit cring rwm thậng kār phạt'hnā krabwnkār h̄rụ̄x p̄hlp̄hlit h̄ım̀ thī̀ pĕn prayochn̒ tx kār dảnein chīwit læa kār thảngān' [STEM education is an approach that integrates knowledge in 4 disciplines: science, engineering, technology, and mathematics, with a focus on solving real-life problems, including the development of new processes or products that benefit human life and work] (IPST, 2014). According to this definition, STEM education infuses the engineering design process into the existing science, mathematics, and technology curricula to enable the utilization of knowledge to solve actual problems and support students' future occupational undertakings (www.stemedthailand.org).

STEM education in Thailand has been strongly driven by many organizations, such as OBEC, IPST, schools and universities, and large corporations including Chevron Thailand. These organizations collectively aim to prepare science, mathematics, and technology teachers to become STEM teachers across the nation. Stages of STEM teacher professional development have evolved over the years to provide teacher with higher STEM expertise and STEM experience.

STEM Education in the Thai Basic Curriculum

Integrated STEM education brings together four disciplines: Science, Technology, Engineering, and Mathematics. However, when considering the Basic Education Core Curriculum B.E. 2551 (A.D. 2008) covering primary and secondary levels from Grade 1 to Grade 12 (7–18 years old), STEM education is directly related to three learning areas, namely, science, mathematics, and occupation-and-technology (MOE, 2008). In 2017, the curriculum was revised in mathematics, science, and geography, and the 'Design and Technology' and 'Information and Communication Technology' strands were moved from the occupation-and-technology to the science learning area. Thus, in the Basic Education Core Curriculum which is in B.E. 2551 (MOE, 2008) and revised B.E. 2560 (MOE, 2017a, 2017b), STEM education is only related to two curriculum learning areas: science and mathematics.

In the science learning area, there are four strands: life science, physical science, earth and space science, and technology. The technology strand includes design-and-technology, and computational thinking. STEM education is expected to be included in all strands; however, it is not explicit in most of the learning indicators. A few indicators clearly state expectations about STEM education, for example:

- Use separation methods for solving everyday life problems by integrating science, mathematics, technology, and engineering (Strand 2: Physical Science, Strand Sc. 2.1 Grade 8)

- Design methods to solve problems in daily life using knowledge of chemical reactions by integrating science, mathematics, technology, and engineering (Strand 2: Physical Science, Strand Sc. 2.1 Grade 9)

The engineering design process is explicit in the technology strand, since the indicators in this strand follow the steps of the engineering design process from Grade 7. For example, the indicators in Strand 4: Technology, Strand Sc. 4.1 for Grade 8 are:

- Identify problems or needs of the community or locals, summarize the scope of the problem, collect and analyse data and concepts related to the problem.
- Design problem-solving methods by analysing, comparing, and making decisions about necessary information under specific conditions and available resources, present problem-solving processes, plan working steps, and solve problems step by step.
- Test, evaluate, and explain problems or limitations under specific conditions, find out the ways for improvement, and present solutions for solving the problem.

In the mathematics learning area, there are three strands: number and algebra, measurement and geometry, and statistics and probability. Similar to science, STEM education elements are not explicit in the indicators of all the strands in mathematics. However, there are several indicators requiring the use of knowledge in mathematics to solve mathematics and real-life problems, for example:

- Understand real numbers and their relationships and use properties of real numbers to solve mathematics and real-life problems (Strand 1: Number and Algebra, Strand M. 1.1 Grade 9)
- Understand and use knowledge of proportion and trigonometry to solve mathematics and real-life problems (Strand 2: Measurement and Geometry, Strand M. 2.2 Grade 9)

For learning resources, the IPST developed several STEM lesson plans and activities for teachers of all grade levels (SciMath, 2016). These aim to support teachers to implement STEM education in their classrooms, with the intention that teachers will also use them as guidelines to create additional STEM activities to suit local contexts and careers. Additionally, many book publishers (e.g. MAC Education, SE-ED) have developed STEM activities and manuals to support students' STEM learning.

As suggested by STEM education research studies (Chamrat, 2016; Kaewklom et al., 2018; Ngaewkoodrua & Yuenyong, 2018; Pimthong & Williams, 2020), STEM education involves the promotion of generic competencies that are included in the Basic Education Core Curriculum, which emphasizes communication, thinking, problem-solving, and applying life skills and technology. STEM education is therefore clearly on the national education agenda. Thus, the education system in Thailand requires STEM education to be promoted among Thai students (National Legislative Assembly, 2014). However, the implementation of STEM education is not limited to science and mathematics but can be integrated with other subjects, such as arts, Thai, English, social sciences, geography, and health and physical education (STEM Education Thailand, 2014).

Teachers' Perceptions and Practices of STEM Teaching and Learning

'Sa tĕm ş̄ụks̄'ā khụ̄x næwthāng kār cạdkār ş̄ụks̄'ā thī̀ būrṇ ā kār khwām rū̂ nı 4 s̄h̄ withyākār dị̂kæ̀ withyāş̄āstr̒ wiş̄wkrrm thekhnoloyī læa khṇitş̄āstr̒ doy nên kārnả khwām rū̂ pị chı̂ kæ̂ pạyh̄ā nı chīwit cring rwm thậng kār phạt'hnā krabwnkār h̄rụ̄x p̄hlp̄hlit h̄ım̀ thī̀ pĕn prayochn̒ tx kār dảnein chīwit læa kār thảngān' [STEM education is an approach that integrates knowledge in 4 disciplines: science, engineering, technology, and mathematics, with a focus on solving real-life problems, including the development of new processes or products that benefit human life and work] (IPST, 2014).

The definition of STEM education by the IPST, shown above, is used throughout Thailand, but its interpretation seems to be restricted and limited in Thai practice. One interpretation is that STEM needs to integrate all four disciplines in every lesson and activity, but no one knows the minimum amount of each discipline that should be integrated. Furthermore, the examples of STEM lessons and activities usually focus on designing products to solve problems or challenges; STEM lessons without designing products might not be considered as STEM lessons. With these untold expectations, it is difficult for teachers to appreciate STEM education as an approach and then confidently and creatively design their own STEM education differently to benefit their students' learning. Initially, STEM education in Thailand has been implemented with guidance and support provided from different sources. For example, the science/mathematics learning activities through the engineering design process introduced by the IPST (Autid, 2017) uses inquiry-based learning by designing models that include encouraging students to learn relevant concepts through a hands-on inquiry-based method before drawing the design and then constructing a product with engineering design challenges. Examples include Chevron's Enjoy Science Project (Changtong et al., 2020), and STEM problem-based learning/project-based learning, which focuses on solving authentic STEM problems (Roikrong & Bongkotphet, 2019).

Ring et al. (2017) reported that teachers in the United States hold a variety of conceptions of integrated STEM, and that teachers' conceptions evolve and can be influenced by professional development experiences. Further, Holmlund et al. (2018), again in the United States, showed that although there are commonalities in understandings and views of STEM concepts among STEM personnel—such as the importance of interdisciplinary connections, the need for new, ambitious instructional practices in enacting a STEM approach, and the engagement of students in real-world problem solving—there are a variety of teaching practices and school contexts within which STEM education is enacted. Similar to the United States, STEM in Ireland seems to lack a unified definition and there are varied implementations of STEM integration (Hourigan et al., 2021). These findings align with our concern that STEM education in Thailand seems to be an 'ideal' approach to practice, and that there is a lack of unified practical goals of STEM education. Without agreement on grounded STEM concepts in practice, it is easy for teachers to find themselves lost when implementing STEM education in their classrooms.

To support our argument, we provide below a meta-analysis of the research conducted in Thailand related to in-service and pre-service teachers' perceptions of and practices in STEM teaching and learning.

According to the practices in STEM teaching and learning in Thailand, Nadelson et al. (2013) and Reeve (2015) revealed that many teachers have difficulties making connections between concepts from different STEM disciplines. They do not know how to help students see the links between inventions/products and each of the STEM disciplines, and thus cannot encourage students to learn STEM through those inventions/products. Kruatong et al. (2018) investigated Thai science student teachers' understanding of STEM-related content and found that they had difficulty identifying not only the technology, mathematics, or other content related to the given everyday inventions, but also the science content. This study demonstrated that a gap between science content knowledge and its applications may exist and could be a significant obstacle for future teaching of STEM by science student teachers.

STEM teachers first need a deep understanding of the subjects they teach before exploring the mechanisms for integration across STEM disciplines as suggested by the following studies. In a study with pre-service science teachers, Vichaidit and Faikhamta (2017) reported that participating pre-service teachers had partial understanding of STEM and characteristics of each discipline in STEM education. In another study by Pimthong and Williams (2020), they found that pre-service teachers of several majors did not have a deep understanding of STEM, such as the integration of the disciplinary concepts and skills, and their STEM focus varied based on their discipline major. In another study focused on the nature of STEM, participating science teachers and pre-service science teachers had inadequate understanding of a STEM definition, epistemological perspectives, and the impact of culture and society on STEM-related disciplines (Faikhamta, 2020; Vichaidit & Faikhamta, 2021).

Teachers seem to lack the understanding of the engineering design process and engineering practice. To investigate how teachers understand and view STEM education, Ladachart et al. (2019) asked 22 teachers who registered for a STEM education workshop and found that their understanding of STEM education emphasizing the engineering design process was limited, and that the engineering design process in some teachers' view was a fixed and rigid process with one correct solution. Another study on pre-service science teachers by Faikhamta (2020) found that their intuitive view of the engineering design process was only about making artefacts and designing things; it was not thought of as the way engineers think about and approach solving problems.

A small-scale case study with six Thai in-service teachers by Srikoom et al. (2018) investigated salient features of STEM instruction and the variations in Thai classrooms. Five categories of STEM approaches were used by the participants in the 28 STEM lessons:

- problem/project-based learning lessons were incorporated through a realistic context without any connection to other STEM content,
- the lesson addressed science and mathematics content incorporated with the engineering design process,
- the lesson followed an engineering design process but did not address appropriate content,
- the lesson did not follow an engineering design process, but included only building and testing without having the need to apply content information, for example building an air rocket without connecting it to the concept of air pressure differences,
- the lesson is a science activity but misses connections to other disciplines.

Overall, the STEM lessons designed by the participating teachers did not meet the key features of STEM educational activities.

Supporting teachers to develop a deeper understanding of the definition and key features of STEM education is needed. There are too many important features that teachers need to keep in mind as they are designing STEM lessons. From these findings, it appears that much work is still needed to improve pre-service and in-service teachers' capability in STEM education in Thailand.

Teachers' Attitudes Towards STEM Teaching and Learning

Positive attitudes can lead to positive behaviours, such as being more motivated and confident in the implementation of STEM education. Teachers' mindsets have to be transformed in order to align with the expected learning outcomes for STEM education. Many benefits and expected goals from STEM education are repeatedly mentioned by experts during workshops and by policy makers in the media. In addition, several studies have revealed that most Thai in-service teachers find the STEM teaching approach to be very interesting after professional development (Autid, 2017; Chamrat, 2016; Kaewklom et al., 2018; Nuangchalerm et al., 2020; Srikoom et al., 2017).

It seems that while teachers have positive attitudes towards STEM education, they may have a limited understanding of STEM teaching. This gap tends to appear when teachers start to put STEM education ideas into practice, as reflected by researchers in Thailand (Chamrat, 2016; Ladachart et al., 2019; Srikoom et al., 2017). This might be due to the fact that the STEM learning process is a new innovation for teachers and it comprises various complicated steps which need to be followed (Srikoom et al., 2018). Integration between disciplines is difficult, especially if teachers are still required to cover all the disciplinary content (Nuangchalerm et al., 2020). In a study by Ladachart et al. (2019), 22 primary and middle school teachers who enrolled in a professional development workshop identified that they were unclear on how to create and use STEM activities. To address these concerns, teachers might need different kinds of support or workshops.

STEM Education and Professional Development in Thailand

To effectively implement STEM education, in-service teachers and pre-service teachers need to be engaged in relevant and effective professional development (PD) or teacher education programmes. STEM PD plays an important role in preparing STEM teachers to implement STEM education. As teachers in Thailand come to know more about STEM education, the format of the PD for in-service teachers will change, from STEM introduction workshops, to coaching and mentoring workshops and developing STEM professional learning communities (PLC), lesson study, and action research, as described below.

STEM Education Workshops

Over time, STEM education workshops have been developed and differentiated to focus on teachers' different needs and levels of STEM expertise. Thus, a variety of STEM workshops have been conducted to encourage teachers' understanding and implementation of STEM activities in their classrooms. Examples include STEM challenge and activity examples, STEM learning approach and process, technology-supported STEM learning, and designing STEM lessons. These are described in more detail below, based on information that has been recorded and reported, or were part of research that was published. The details of many other PD programmes have not been published.

Examples of STEM Challenge and Activity

A three-day STEM education workshop was provided by the upper northern STEM centre with support from the IPST in 2016. The first session formed an introduction to STEM education and its relevance for developing students' twenty-first century skills via a variety of real-world and mathematical problems. The second session focused on some demonstrations of STEM classroom activities, for example, designing a bamboo rice container, tower model, and mini-helicopter; performing hands-on STEM learning activities in groups; and presenting and discussing group projects. A study by Autid (2017) investigated participants' classroom practice after the workshop and found that there were variations in STEM implementation due to teacher beliefs and interpretations of STEM teaching, teaching experiences, student context, and time available. The study also revealed that the STEM activity provided by the trainers supported the teachers to directly implement in their classrooms the examples that they had experienced in the workshop.

STEM Learning Approach and Process

Through the Coupon for Teacher Development Project supported by OBEC, a STEM workshop was run that emphasized learning by design in STEM education, where students learn science through the engineering design processes (Ladachart et al., 2019). In this workshop, participants designed several products to solve a challenge using science and mathematics. Through the workshop, the study revealed that participants lacked understanding of STEM education, specifically in relation to its definition and goals, and participating teachers' STEM lessons did not have a focus on the engineering design process as had been advocated in the workshop. The STEM smart trainer team (Yuenyong, 2019), organized by the Thailand Office of Basic Education Commission (OBEC), organized 15 workshops (for approximately 2,250 teachers) across the nation to enhance teachers' understanding and implementation of STEM education in their classrooms through communities of practice. These workshops emphasized an engineering design process comprising nine steps as follows:

- identify the problem
- analyse five capitals (physical, financial, social, human, and natural capitals)
- explore information
- share knowledge
- model a solution
- plan and develop the solution
- test and evaluate the solution
- present the solution
- reflect and revise

The participants generated ideas for developing STEM activities involving inventing and innovating for solutions related to local problems or issues, such as animal traps, machines for selecting fruit, recyclable products, watering systems for farms, and so on. Yuenyong (2019) shared his perspective as a trainer that the participants perceived STEM education as a single teaching strategy with certain steps of teaching that should be included explicitly in the national curriculum standards, and that the PD workshop needed to conceptualize assessment for STEM education.

In 2015, inquiry-based learning workshops from the Thailand STEM Education Project of Chevron, Enjoy Science, emphasized how to design and choose appropriate methods in order to enhance inquiry processes, such as think-pair-share, gallery walks, and formative assessment. Research by Ngaewkoodrua and Yuenyong (2018) investigated science teachers' existing ideas of ways of enhancing students' inventive thinking skills during the workshop. Their findings revealed that most of the participating teachers held some understanding of inquiry-based learning, which could be used as a base for developing ideas for enhancing students' inventive thinking skills. They could also design and choose the appropriate methods to be adopted in the classroom in order to enhance students' inventive thinking skills.

Technology-Supported STEM Learning

Other STEM workshops have emphasized the use of digital technology in classrooms to better support STEM instruction, for example, an adaptive personalized mobile application was used to promote TPACK for 78 in-service teachers (Kajonmanee et al., 2020). In a tablet-based STEM workshop, the programme consisted of introducing STEM education (a one-day online meeting), demonstrating STEM teaching by using iPad applications – a 'tall tower challenge' with the application *'Book creator'*; designing an edible car with *'Keynote';* and editing the video when building the instrument to collect *'Oil spill in the Ocean'* with *'iMovie'* – and developing STEM-based lessons by integrating the iPad as a learning tool. After the workshop, 150 of 240 participants (63%) identified that they had greater confidence in implementing STEM pedagogical approaches (Pitiporntapin et al., 2018). However, the limitations of time and available tablets were identified by the participants as issues that would continue to restrict their STEM instructional practices.

Designing STEM Lesson Plans

A STEM workshop designed based on Bandura's (1997) self-efficacy theory started by raising teachers' awareness of the importance, goals, and challenges of STEM education, including the meaning and nature of STEM. Then, STEM activities were demonstrated. Interdisciplinary approaches, engineering education, and learning assessment were emphasized. Lastly, participants practiced designing STEM lessons. A study of the programme (Khumwong et al., 2017) highlighted that the design of a waste water treatment system in the community, which is a problem situation connected with a real life context, helped the participating teachers clearly understand the engineering design process. The teachers' ability to make connections between STEM education and real-life contexts could be enhanced by using

STEM thinking (Reeve, 2015), a way to analyse science, mathematics, technology, and engineering concepts from the things or problems or situations in daily life.

In 2016, a 3-day face-to-face workshop followed by monthly meetings during the academic year was organized by the IPST. On the first day, the participants learned about STEM education, the nature of science, the engineering design process, integrating scientific inquiry and engineering practices, and applying mathematics and technology with the STEM approach. The second day focused on micro-teaching of STEM activities, analysing lessons, and sharing guidelines for designing STEM lesson plans. On the last day, the participants shared their own STEM lessons, received feedback, and improved the lessons. Throughout the academic year, teachers met on a monthly basis to share and reflect on their implementation and outcomes of their action research. Srikoom et al. (2018) found that the participants who attempted to teach STEM lessons needed more support related to asking essential questions, linking STEM activities to careers or life in the real world, understanding of the engineering design process, and integrating engineering into the lessons.

Coaching, Mentoring and Developing Professional Learning Communities (PLCs)

In addition to STEM workshops, teachers need different kinds of support that can guide them in their own specific contexts. Coaching and mentoring can be helpful to teachers. Coaches and mentors can give advice on lesson plans and feedback from class observations. Specific guidance for each classroom may help individual teachers to improve their STEM lessons. Coaching and mentoring can also improve the management of STEM lessons (Boonsong et al., 2017).

Another kind of support comes from peer teachers, that is, professional learning communities (PLCs) or, interchangeably, communities of practice (CoPs). The STEM smart trainer project has developed a community of practice for STEM in Thailand, resulting in school partnerships and cooperation among teachers (Yuenyong, 2019). PLCs can empower teachers to change their understanding of and beliefs about STEM teaching (Vasinayanuwatana et al., 2021). Intalapaporn (2019) presented a PLC process that started with participating teachers sharing problems during the implementation of STEM, brainstorming for developing STEM classroom activities, trying out the ideas, and providing feedback. Through the PLC activities, primary teachers' abilities to design learning activities for STEM education were enhanced. The PLC activities also increased teacher's self-efficacy in teaching science, personal efficacy, and outcome expectancy efficacy (Krainara & Chatmaneerungcharoen, 2019). Through PLCs, teachers can share their vision for STEM education, STEM activities, and best practice (Thana et al., 2018). PLCs can create a friendly and supportive atmosphere for teachers as well as support friendly learning atmospheres for students (Wetwiriyasakun et al., 2021).

Lesson Study and Classroom Action Research

To effectively and appropriately implement STEM education in the classroom, teachers need to understand the reasons behind the implementation of STEM activities. A lesson study on high school physics teachers was used to design, implement, critique, and develop good STEM learning activities (Teevasutornsakul et al., 2015). Teachers can improve their lessons based on reflections of peer teachers and students' feedback.

Classroom action research, with a cycle of planning, teaching, and guided reflection, have also been used to develop STEM lessons: teachers compare and contrast research articles, design a STEM lesson and receive feedback, and, lastly, implement the STEM lesson and analyse their students' learning data (Faikhamta et al., 2020). Lastly, metacognitive reflections have been used to develop and shape teachers' pedagogical reasoning for STEM lessons (Park & Prommas, 2017). Through reflective writing, teachers can reflect on their teaching practices and pedagogical reasoning.

Initial Teacher Education

For pre-service teachers, STEM education workshops for both extracurricular activities and classroom lessons have been included in the Initial Teacher Education programme. In a study reported by Kruatong et al. (2017), pre-service teachers in a science method course were provided with 18 hours of learning activities (3 hours per week) about a 5E inquiry-based STEM education programme. Analysis of the pre-service teachers' lesson plans revealed that they were able to create their own STEM activities (based on their own experiences with STEM activities), for example, designing and building an earthworm condo and electromagnet door, designing inventions using local materials, and solving everyday problems.

Saratapan et al. (2019) studied ways to enhance pre-service teachers' integration of STEM education into home economics lessons. The programme included an introduction to STEM, analysing examples of STEM education in classrooms, and developing lesson plans in STEM education. The lesson can incorporate real-world experiences and problems, immerse students in hands-on inquiry and open-ended exploration, and involve students in productive teamwork, which are guided by the engineering process, integrate content from both mathematics and science courses, and allow for failure.

A STEM programme for science pre-service teachers based on pedagogical content knowledge (PCK) was introduced by Srisawasdi (2012). This programme sought to change the perceptions of the participants regarding STEM education and the implementation of STEM education. The programme adopted a case-based learning approach with a series of workshops integrating digital technologies into teaching in science and mathematics. Findings collected after the programme suggested that the pre-service teachers showed positive gains in TPACK competency.

From the above review of STEM education PD, we found that each PD workshop has its own objectives and limitations. Improvements in science teachers' knowledge and attitudes towards STEM were observed, and teachers were generally motivated and became more confident in the implementation of STEM education after the PD workshops. However, a combination of coaching and mentoring, lesson study, promoting reflection after implementing STEM lessons, and PLCs are necessary to sustain changes in instructional practices. These initiatives helped teachers integrate science with other subjects, design learning objectives related to twenty-first century skills, and encourage students to apply knowledge to their daily life (Chatmaneerungcharoen, 2019). Through coaching and peer support, teachers and teacher communities can develop their own STEM lessons with more confidence. To deeply understand and be able to design effective STEM lessons, teachers need to reflect on their teaching. This can be done through lesson study and classroom action research.

Discussion and Recommendations

With the push for STEM education at the national level in Thailand, a large budget and workforce has been allocated to preparing in-service teachers through STEM professional development by the IPST, OBEC, universities, and many other organizations. With this huge effort, STEM education PD should have a positive effect on teachers' performance and on students' learning. However, to implement STEM education in a real classroom in Thailand is challenging, and STEM teachers share many concerns. For example, some research studies have reported that teachers in STEM workshops cannot design STEM lessons to suit the subject, context, and students' potential (Pantu et al., 2019) and that teachers who attended STEM workshops were not proficient in all of the essential elements of STEM teaching (Lomarak et al., 2019). Moreover, teachers' understanding of STEM is tied to the definitions of the four STEM disciplines. This makes teachers focus too much on identifying science, technology, engineering, and mathematics in their activities, or on trying to make clear connections between disciplines. These concerns weigh heavily on teachers and result in them losing focus on the real emphasis of their STEM activities, that is, to capture students' interest, and to evaluate the workability or the design of the invention (examples can be found in Autid, 2017 and in Pimthong & Williams, 2020).

Similar to other countries, challenges and obstacles of implementing integrated STEM education include lack of knowledge and lack of time for collaborative preparation and instruction (Shernoff et al., 2017). For example, it has been shown that teachers in Hong Kong need articulated professional development, pedagogic support, and curricular resources for empowering them to implement STEM education in practice (Geng et al., 2019). Even within classrooms where integrated

STEM education is being implemented, it is difficult to sustain the integration of STEM concepts, and some disciplines seem to disappear during different phases of learning (Estapa & Tank, 2017). Teachers need support from peers, experts, and administrators to manage their workloads as well as to reflect on their STEM education practice.

STEM education is not only concerned with the learning of STEM content knowledge, but also the development of twenty-first century skills. However, another concern of teachers is the national standardized examinations, which prioritize subject matter and knowhow over assessing the twenty-first century skills. Stronger emphasis on authentic assessment of students' learning outcomes in STEM education must be included in PD workshops (Chamrat, 2016; Kaewklom et al., 2018; Nuangchalerm, 2020). Moreover, to further promote the quality of STEM education in Thailand, more research studies are needed to identify changes in teacher efficacy, outcome expectancies, and awareness of STEM careers. For example, research on changes in teacher practices after the adaptation of STEM education would provide grounded evidence for Thai STEM education reform, as has been done in Australia by Anderson and Tully (2020) whose study showed a significant increase in the use of small group problem-solving through inquiry, engagement, and opportunities for student reasoning.

To support STEM teachers in implementing STEM education in their classrooms, STEM materials such as STEM lessons and activities, STEM challenges, and STEM authentic assessment would be helpful. These materials can be provided by experts or by peer STEM teachers. However, due to the variety of learning objectives and different school contexts, there is no STEM material that can fit all classrooms. After extensive effort and resources have been poured into STEM education workshops, everyone expects STEM teachers to be able to immediately and automatically design their own STEM material. However, that is not the case. STEM teachers require after-workshop support to experiment and to gain experience before becoming great STEM teachers. PLCs offer scope for providing ongoing support in terms of developing educative materials and moral support. As has been demonstrated elsewhere, communities of practice have been shown to support STEM teachers by learning within the community, engaging in STEM practices, and using STEM knowledge to solve real-world problems (Kelley et al., 2020; Weinberg et al., 2021).

References

Anderson, J., & Tully, D. (2020). Designing and evaluating an integrated STEM professional development program for secondary and primary school teachers in Australia. In J. Anderson & Y. Li (Eds.), *Integrated approaches to STEM education. Advances in STEM education* (pp. 403–425). Cham: Springer.

Autid, W. (2017). *Mathematics teachers' classroom instruction after participating in a STEM education workshop in Thailand* [Unpublished doctoral dissertation]. Pennsylvania State University. https://etda.libraries.psu.edu/files/final_submissions/15672.

Bandura, A. (1997). *Self-efficacy: The exercise of control.* W. H. Freeman and Company.
Baxter, W. (2017). *Thailand 4.0 and the future of work in the Kingdom.* International Labour Organization. http://tinyurl.com/mkhf5r5.
Boonsong, K., Petsom, N., & Ngerntong, C. (2017). Teacher development by coaching and mentoring process and manual for development of learning management by STEM education to develop the thinking process of learners. *Academic Journal of Phetchaburi Rajabhat University, 7*(2), 134–141 (in Thai).
Chamrat, S. (2016). STEM education on the road of socially-engaged scholarship: Game changer for future learning. *Kasetsart Educational Review, 31*(3), 34–47 (in Thai).
Changtong, N., Maneejak, N., & Yasri, P. (2020). Approaches for implementing STEM (Science, Technology, Engineering & Mathematics) activities among middle school students in Thailand. *International Journal of Educational Methodology, 6*(1), 185–198. https://doi.org/10.12973/ijem.6.1.185.
Chatmaneerungcharoen, S. (2019). Improving Thai science teachers' TPACK through an innovative continuing professional development program. *Journal of Physics: Conference Series, 1340*(1). https://doi.org/10.1088/1742-6596/1340/1/012017.
Estapa, A. T., & Tank, K. M. (2017). Supporting integrated STEM in the elementary classroom: A professional development approach centered in an engineering design challenge. *International Journal of STEM Education, 4*(1). https://doi.org/10.1186/s40594-017-0058-3.
Faikhamta, C. (2020). Pre-service science teachers' views of the nature of STEM. *Science Education International, 31*(4), 356–366. https://doi.org/10.33828/sei.v31.i4.4.
Faikhamta, C., Lertdechapat, K., & Prasoblarb, T. (2020). The impact of a PCK-based professional development program on science teachers' ability to teaching STEM. *Journal of Science and Mathematics Education in Southeast Asia, 43*(December), 1–22.
Geng, J., Jong, M. S. Y., & Chai, C. S. (2019). Hong Kong teachers' self-efficacy and concerns about STEM education. *Asia-Pacific Education Researcher, 28*(1), 35–45. https://doi.org/10.1007/s40299-018-0414-1.
Holmlund, T. D., Lesseig, K., & Slavit, D. (2018). Making sense of "STEM education" in K-12 contexts. *International Journal of STEM Education, 5*, Article 32. https://doi.org/10.1186/s40594-018-0127-2.
Hourigan, M., O'Dwyer, A., Leavy, A. M., & Corry, E. (2021). Integrated STEM—A step too far in primary education contexts? *Irish Educational Studies*. https://doi.org/10.1080/03323315.2021.1899027.
Institute for the Promotion of Teaching Science and Technology. (2014). *Rūcạk s̄a tĕm* [know stem]. STEM Education Thailand. Retrieved April 22, 2021, from http://www.stemedthailand.org/?page_id=23.
Intalapaporn, C. (2019). Teacher development model for professional learning community to promote STEM education learning activity design ability for primary teachers: Case study of Don Yai Hom School, Muang District, Nakhon Pathom Province. *Journal of Research and Curriculum Development, 9*(2), 46–67 (in Thai).
Kaewklom, W., Khumwong, P., & Dahsah, C. (2018). Current situation, problem and needs for STEM learning of primary science teacher. *Veridian E-Journal, Silpakorn University, 11*(3), 2092–2112 (in Thai).
Kajonmanee, T., Chaipidech, P., Srisawasdi, N., & Chaipah, K. (2020). A personalized mobile learning system for promoting STEM discipline teachers' TPACK development. *International Journal of Mobile Learning and Organization, 14*(2), 215–235. https://doi.org/10.1504/IJMLO.2020.106186.
Kelley, T. R., Knowles, J. G., Holland, J. D., & Han, J. (2020). Increasing high school teachers' self-Efficacy for integrated STEM instruction through a collaborative community of practice. *International Journal of STEM Education, 7*, Article 14. https://doi.org/10.1186/s40594-020-00211-w.
Khumwong, P., Pruekpramool, C., & Phonphok, N. (2017). The impact of STEM education professional development workshop on secondary teachers' STEM teaching efficacy. *Journal*

of Education, Mahasarakham University, 11(3), 108–121. http://www.edu.msu.ac.th/journal/home/journal_file/365.pdf (in Thai).

Krainara, S., & Chatmaneerungcharoen, S. (2019). Building a professional learning community with team endeavors while creating elementary-focused STEM-integrated lesson plans. *Journal of Physics: Conference Series, 1340*, 012015. https://doi.org/10.1088/1742-6596/1340/1/012015.

Kruatong, S., Kruea-In, N., Nugultham, K., & Wannagatesiri, T. (2017). Inquiry-based STEM activities created by science student teachers in a teaching method course, Nov 13–17. *Proceeding of the 7th International Conference on Science & Mathematics Education*. Seameo Recsam, Penang, Malaysia. November 13 – 17.

Kruatong, S., Kruea-In, N., Nugultham, K., & Wannagatesiri, T. (2018). Science student teachers' confidence teaching and understanding STEM-related content at the middle school level. *The International Journal of Science, Mathematics and Technology Learning, 24*(3), 19–32. https://doi.org/10.18848/2327-7971/CGP/v24i03/19-32.

Ladachart, L., Phothong, W., Rittikoop, W., & Ladachart, L. (2019). Teachers' understandings and views about STEM education and engineering design. *Silpakorn University Journal, 39*(3), 133–149. https://so05.tci-thaijo.org/index.php/sujthai/article/view/122924 (in Thai).

Lomarak, T., Nuansai, B., Promden, W., & Sangsila, A. (2019). Assessment of in-service science teachers' teaching competencies in integrated STEM program. *Journal of Physics: Conference Series, 1340*, 012009. https://doi.org/10.1088/1742-6596/1340/1/012009.

Ministry of Education. (2017a). *Tạw chī̂ wạd læa s̄āra kār reīyn rū̂ kæn klāng klùm s̄āra kār reīyn rū̂ khṇitṣ̄āstr̒ (c̄hbạb prạbprung ph.Ṣ. 2560) Tām h̄lạks̄ūtr kæn klāng kār ṣ̄ụks̄'ā k̄hận phụ̄nṭhān phuthṭhṣ̄ạkrāch 2551* [Indicators and core learning areas mathematics (Revised B.E. 2560) based on the basic education core curriculum B.E. 2551]. Office of the Basic Education Commission, MOE. https://drive.google.com/file/d/1F4_wAe-ZF13-WhvnEAupXNiWchvpcQKW/view.

Ministry of Education. (2017b). *Tạw chī̂ wạd læa s̄āra kār reīyn rū̂ kæn klāng klùm s̄āra kār reīyn rū̂ withyāṣ̄āstr̒ (c̄hbạb prạbprung ph.Ṣ. 2560) Tām h̄lạks̄ūtr kæn klāng kār ṣ̄ụks̄'ā k̄hận phụ̄nṭhān phuthṭhṣ̄ạkrāch 2551* [Indicators and core learning areas science (Revised B.E. 2560) based on the basic education core curriculum B.E. 2551]. Office of the Basic Education Commission, MOE. https://drive.google.com/file/d/1_ALwE9xuCL3Fjet3XI4gYjBj8p_1zLaA/view.

Ministry of Education. (2008). *The basic education core curriculum B.E. 2551 (A.D. 2008)*. http://academic.obec.go.th/images/document/1525235513_d_1.pdf.

Nadelson, L. S., Callahan, J., Pyke, P., Hay, A., Dance, M., & Pfiester, J. (2013). Teacher STEM perception and preparation: Inquiry-based STEM professional development for elementary teachers. *The Journal of Educational Research, 106*(2), 157–168. https://doi.org/10.1080/00220671.2012.667014.

Ngaewkoodrua, N., & Yuenyong, C. (2018). The Teachers' existing ideas of enhancing students' inventive thinking skills. *The Turkish Online Journal of Educational Technology, 17*(2), 169–175. http://www.tojet.net/articles/v17i2/17216.pdf.

Nuangchalerm, P., Prachagool, V., Prommaboon, T., Juhji, J., Imroatun, I., & Khaeroni, K. (2020). Views of primary Thai teachers toward STREAM education. *International Journal of Evaluation and Research in Education, 9*(4), 987–992. https://doi.org/10.11591/ijere.v9i4.20595.

Pantu, R., Kemwimoottiwong, C., Ratchawet, A., & Chaiworn, P. (2019). The model of STEM education leader teacher development for the school under Chiang Mai primary educational service areas office 2. *Journal of Humanities and Social Science Thonburi University, 13*(3), 133–144. https://so03.tci-thaijo.org/index.php/trujournal/article/view/214073/148942 (in Thai).

Park, D.-Y., & Prommas, C. (2017). Development of Thai teachers' pedagogical reasoning by utilizing metacognitive reflections in STEM professional development. *Journal of Education and Human Development, 6*(3), 137–152. https://doi.org/10.15640/jehd.v6n3a15.

Pimthong, P., & Williams, J. (2020). Preservice teachers' understanding of STEM education. *Kasetsart Journal of Social Sciences, 41*, 289–295. Retrieved from https://kasetsartjournal.ku.ac.th/downloadFile.aspx?param=YXJ0aWNsZUlEPTY0MTl8bWVkaWFJRD02Njk0.

Pitiporntapin, S., Chantara, P., Srikoom, W., Nuangchalerm, P., & Hines, L. M. (2018). Enhancing Thai in-service teachers' perceptions of STEM education with tablet-based professional development. *Asian Social Science, 14*(10), 13–20. https://doi.org/10.5539/ass.v14n10p13.

Reeve, E. M. (2015). STEM Thinking! *Technology and Engineering Teacher, 74*(4), 8–16.

Ring, E. A., Dare, E. A., Crotty, E. A., & Roehrig, G. H. (2017). The evolution of teacher conceptions of STEM education throughout an intensive professional development experience. *Journal of Science Teacher Education, 28*(5), 444–467. https://doi.org/10.1080/1046560X.2017.1356671.

Roikrong, W., & Bongkotphet, T. (2019). STEM problem-based learning on exploration and production of petroleum topic to enhance of scientific literacy of the 9th grade students. *Social Sciences Research and Academic Journal, 14*(3), 135–148 (in Thai).

Saratapan, N., Pitiporntapin, S., & Hines, L. M. (2019). Enhancing pre-service teachers' integration of STEM education into home economics lessons through a professional development program. *International Education Studies, 12*(8), 11–20. https://doi.org/10.5539/ies.v12n8p11.

SciMath. (2016). *Xekšār kickrrm ša tĕm lœa khūmụ̄x kickrrm ša tĕm chwng chận thī 1–4* [STEM education document: STEM activities and manuals for level 1 – 4]. Institute for the Promotion of Teaching Science and Technology, MOE. https://www.scimath.org/stem.

Shernoff, D. J., Sinha, S., Bressler, D. M., & Ginsburg, L. (2017). Assessing teacher education and professional development needs for the implementation of integrated approaches to STEM education. *International Journal of STEM Education, 4*, Article 13. https://doi.org/10.1186/s40594-017-0068-1.

Srikoom, W., Faikhamta, C., & Hanuscin, D. (2018). Dimensions of Effective STEM Integrated Teaching Practice. *K-12 STEM Education, 4*(2), 313–330.

Srikoom, W., Hanuscin, D., & Faikhamta, C. (2017). Perceptions of in-service teachers toward teaching STEM in Thailand. *Asia-Pacific Forum on Science Learning and Teaching, 18*(2), 1–15.

Srisawasdi, N. (2012). Fostering pre-service STEM teachers' technological pedagogical content knowledge: A lesson learned from case-based learning approach. *Journal of the Korean Association for Science Education, 32*(8), 1356–1366. https://doi.org/10.14697/jkase.2012.32.8.1356.

STEM Education Thailand. (2014). *Sa tĕm ṣ̄ụkš'ā khụ̄x xarị?* [What is STEM education?] http://www.stemedthailand.org/?post_type=faq.

Teevasutornsakul, C., Aunsaard, W., & Taowannee, C. (2015). *The model for development the teacher of physics in higher school to designs the learning activities of STEM education with lesson study*. Burapha University. http://dspace.lib.buu.ac.th/bitstream/1234567890/3571/1/2563_021.pdf (in Thai).

Thana, A., Siripun, K., & Yuenyong, C. (2018). Building up STEM education professional learning community in school setting: Case of Khon Kaen Wittayayon School. *AIP Conference Proceedings, 1923*, 030067. https://doi.org/10.1063/1.5019558.

The National Legislative Assembly. (2014). *Rāyngān k̄hx šenx cheing nyobāy ša tĕm ṣ̄ụkš'ā (STEM Education) nyobāy cheing ruk pheụ̄x phạt'hnā yeāwchn lœa kảlạng khn d̄ān withyāṣ̄āštr' thekhnoloyī wiṣ̄wkrrmṣ̄āštr' lœa khṇitṣ̄āštr'* [STEM education policy proposal review report, proactive policy to develop youth and manpower in science technology engineering and mathematics]. http://61.19.241.96/w3c/senate/comm.php?url=view&comm_id=1547&content_id=5162&size=12.

Vasinayanuwatana, T., Teo, T. W., & Ketsing, J. (2021). Shura-infused STEM professional learning community in an Islamic School in Thailand. *Cultural Studies of Science Education, 16*, 109–139. https://doi.org/10.1007/s11422-020-09990-8.

Vichaidit, C., & Faikhamta, C. (2017). Exploring orientations toward STEM education of pre-service science teachers. *Rajabhat MahaSarakham University Journal, 11*(3), 165–174. https://so05.tci-thaijo.org/index.php/rmuj/article/view/144783/107071 (in Thai).

Vichaidit, C., & Faikhamta, C. (2021). Science teachers' perceptions on pedagogical content knowledge for STEM. *Journal of Education Naresuan University, 23*(2), 152–168. https://so06.tci-thaijo.org/index.php/edujournal_nu/article/view/160452/168007 (in Thai).

Weinberg, A. E., Balgopal, M. M., & Sample McMeeking, L. B. (2021). Professional growth and identity development of STEM teacher educators in a community of practice. *International Journal of Science and Mathematics Education, 19*, 99–120. https://doi.org/10.1007/s10763-020-10148-9.

Wetwiriyasakun, P., Kaeomani, C., & Suwanmanee, S. (2021). PLC model for science teacher development at the basic educational level. *Journal of Physics: Conference Series, 1835*, 012064. https://doi.org/10.1088/1742-6596/1835/1/012064.

Yuenyong, C. (2019). Lesson learned of building up community of practice for STEM education in Thailand. *AIP Conference Proceedings, 2081*(1), 020002–1–6. https://doi.org/10.1063/1.5093997.

Witat Fakcharoenphol is an Assistant Professor in Science Education at Kasetsart University, Kamphaeng Saen Campus, Thailand. He teaches in the Ph.D. of Science Education, the Master of Education and the Bachelor of Education programmes. His research focuses on educational technology in science education such as educational games, instructional media, and online learning systems. He received his B.A. in Physics from Cornell University and Ph.D. in Physics (Physics Education Research) from University of Illinois at Urbana-Champaign.

Chanyah Dahsah is an Assistant Professor of Science Education and director of Science Education Center at the Faculty of Science, Srinakharinwirot University, Thailand. She received a Ph.D. in Science Education from Kasetsart University, Thailand in collaboration with the University of Waikato in 2007. In 2013–2015, she was a researcher scholar at CREATE for STEM Institute, Michigan State University. Her research interests include K-12 STEM learning, assessment, and professional development. Her current projects involve the Little Scientists' House Project, STEM-Entrepreneur, and STEM2 curriculum development and implementation.

Tussatrin Wannagatesiri is an Associate Professor in Education and head of the Teacher Education Department at Kasetsart University, Kamphaeng Saen Campus, Thailand. She teaches in the Ph.D. of Science Education, the Master of Education and the Bachelor of Education programmes. She received a Ph.D. in Science Education from Kasetsart University, Thailand in collaboration with the University of Waikato in 2007. Her research interests include K-12 learning, assessment, and professional development.

Chapter 15
STEM Teacher Professional Development for Primary School Teachers in Hong Kong

May May Hung Cheng and Fang-Yin Yeh

Abstract This study draws on the findings from a STEM education project to examine Hong Kong in-service primary school teachers' perceived challenges in implementing STEM education, the support they received in STEM teacher professional development (STEM TPD), and their needs for future STEM TPD. The study engaged teacher professional development through a school-university partnership and adopted a practitioner research approach that aimed at enhancing primary school teachers' professional capacity of designing STEM activities relevant to the Hong Kong curriculum, with an emphasis on the learning of crosscutting concepts and inquiry-based teaching. STEM TPD is aimed at collaborative curriculum development as an opportunity to foster active learning through co-creating curriculum materials among teachers and university facilitators. Twelve primary school teachers from various subject teaching backgrounds were interviewed. Semi-structured interviews were carried out to collect the teachers' experiences with the STEM TPD and their views on the integration of cross-cutting concepts in designing STEM lessons. Findings of the study revealed teachers' perceived challenges during the preparation and teaching phases related to STEM instruction and lesson planning, limited resources, and other concerns embedded in broader contextual situations. While teachers reported to have received different types of support from the STEM TPD relating to the pedagogical knowledge of STEM, future directions for STEM TPD were expressed in relation to content components of STEM TPD, opportunities for authentic learning and first-hand experiences, and coherence. Finally, this chapter discusses areas that need to be resolved before a further enhancement in terms of quality and quantity of STEM lessons could be expected.

Keywords STEM teacher professional development · Primary teacher · Hong Kong · Practitioner research · Collaborative curriculum development

---

M. M. H. Cheng (✉) · F.-Y. Yeh
Department of Curriculum and Instruction, The Education University of Hong Kong, 10 Lo Ping Road, Hong Kong SAR, China
e-mail: maycheng@eduhk.hk

F.-Y. Yeh
e-mail: fyyeh@eduhk.hk

M. M. H. Cheng et al. (eds.), *Concepts and Practices of STEM Education in Asia*,
https://doi.org/10.1007/978-981-19-2596-2_15

Introduction

In Hong Kong, STEM education has become a major policy priority in 2015 after the Curriculum Development Council (CDC) published the document titled 'Promotion of STEM Education—Unleashing Potential in Innovation'. In the Hong Kong curricular context, STEM education is promoted through the science, technology, and mathematics education key learning areas (KLAs) in secondary schools, and through mathematics and general studies (GS, which incorporates science, technology and personal, social, and humanities education) in primary schools. Thus, GS teachers with non-STEM related backgrounds are sometimes required to teach STEM in primary schools. Renewing the curricula of the KLAs and GS has been proposed as one of the main strategies for promoting STEM education, in addition to organizing STEM co-curricular activities (i.e. education fairs, outside classroom learning, competitions), enhancing learning and teaching resources, and targeting the professional development of KLAs and GS curriculum leaders (CDC, 2017). Based on the directional measures of the CDC (2017), curriculum content for KLAs and the GS curriculum were updated to foster a shift towards inquiry-based learning, cross-disciplinary integration, application of knowledge and skills, and hands-on and minds-on activities.

In 2020, the Task Force on Review of School Curriculum (Task Force School Curriculum), set up by the Hong Kong Education Bureau (EDB) to holistically review the primary and secondary curricula in Hong Kong, further suggested a developmental priority to define STEM education and develop a handbook for school-based STEM education to clarify expectations across the primary and secondary school levels. STEM curricular attention shifted from co-curricular activities to the formal curriculum and co-coordination among related KLAs and GS curricula. Additionally, initiatives for STEM professional development programmes further emphasized strengthening the professional capacities of frontline teachers, specifically on teaching strategies and technological and pedagogical content knowledge of STEM topics (Task Force School Curriculum, 2020).

Continuous teacher professional development is emphasized and supported by the Hong Kong government, especially under the ongoing renewal of the curriculum. The Task Force on Professional Development of Teachers (Task Force Professional Development) was set up by the EDB in 2017 to further study the development, promotion, and implementation of teachers' professional growth. As part of teachers' continuous professional development (CPD) stipulated by the Hong Kong government, there is a basic requirement for every teacher to complete no less than 150 hours of professional development activities (e.g. overseas visits, exchanges, learning circles, seminars, workshops, etc.) in a three-year cycle. In addition, to be eligible for promotion, relevant training courses (e.g. refresher courses and management training courses) are also required for teacher CPD, guided by professional development guidelines, such as the 'T-standard^{+}' launched by the Committee on Professional Development of Teachers and Principals (COTAP) (Task Force Professional Development, 2019). STEM education has been stipulated as a major initiative for school-based development (Task Force Professional Development, 2019), and has also been a primary focus of the professional development programmes offered by the Curriculum Development Institute of the EDB for primary school teachers (especially for GS teachers), covering topics relating to STEM knowledge, the application of science process skills in STEM education, STEM curriculum planning and implementation, STEM teaching and assessment, and coding and computational thinking (Curriculum Development Institute, 2021).

Research, especially Asian-based STEM education research, is still in an emerging state (Lee et al., 2019), and there has been less research on Asian teachers' perceived readiness to implement STEM (see Margot & Kettler, 2019) and STEM TPD (see Chai, 2019; Geng et al., 2019). In Hong Kong, only a few and very recent studies on STEM TPD have examined STEM pedagogy (e.g. project-based learning, robot-based pedagogy, design-based pedagogy) (Chiu et al., 2021; Kong et al., 2020; Szeto et al., 2021) and teachers' attitudes towards designing STEM curriculum activities (Lin et al., 2021). In their quantitative study of Hong Kong teachers' responses to STEM education, Geng et al. (2019) found that only around 6% of school teachers perceived themselves as being 'well prepared' for STEM education, and suggested further professional development, pedagogical support, and curricular resources relating to 'information' (e.g. how to perform STEM instruction, share information through learning circles), 'management' (e.g. receiving supports in time-related concerns, such as administrative tasks, class hours), and 'consequences' (e.g. hoping to optimize STEM pedagogical approaches, access to promoted STEM resources to further empower STEM teaching).

Against the research background and the nascent STEM education directional measures in Hong Kong for the STEM curriculum, instruction and teacher professional development, a STEM TPD project for Hong Kong primary teachers was conducted.

Engaging Teachers in Practitioner Research as a Form of Teacher Professional Development

The nature of the recent development of teacher professional development (TPD) has seen a move away from workshops and courses to workplace and professional learning communities (Campbell et al., 2004), with school-university partnerships as one prevalent approach to facilitating teachers' professional learning (Cheng & So, 2012; Day & Smethen, 2010). One form of the school-university partnership focuses on the generation of educational knowledge through practitioner research (PR). PR is a paradigm of educational research that involves teachers' dual roles as practitioner and researcher and is considered as a strategy to foster TPD by enhancing teachers' research skills and teaching effectiveness in their work contexts (Burns, 2009; Cain & Milovic, 2010; Cheng & Li, 2020).

In the context of new curriculum directives and reforms, studies have shown that collaborative curriculum development — as one form of professional development in which teachers work together with university facilitators to create curriculum materials — fosters teachers' active professional learning and provides opportunities for in-service teachers to examine their beliefs and classroom teaching practices and to increase their subject matter and content knowledge (Drits-Esser & Stark, 2015; Handelzalts, 2019).

Literature relating to teacher professional development has recognized the impact of constructivist learning theory on the conceptions, organization, and structure of professional development (Cheng et al., 2009; Harfitt & Chan, 2017; Keiny, 1994; Sparks, 2002), and an array of studies have claimed successful implementation of curriculum changes involving TPD programmes underpinned by the constructivist framework (e.g. Fung, 2000; Howe & Stubbs, 1997; Zehetmeier et al., 2015). The key feature of these programmes is that they helped teachers to construct their own learning experiences, reflect, and take more responsibility for and control over their own learning. PR engages practitioners in a 'systematic and intentional inquiry' (Cochran-Smith & Lytle, 2009, p. 142) of one's own professional practices. The stance of PR, situating teachers as learners, connects PR to constructivist and inquiry-based approaches, and serves as a way to promote different aspects of teachers' professional learning through reflective teaching practices and teacher collaboration (Atay, 2007; Stavroula et al., 2015).

Challenges and the Professional Development Support Needed by School Teachers to Implement STEM Instruction

International studies on school teachers' perceived challenges in implementing STEM education and the professional development needs for STEM education are limited, especially studies involving primary school teachers (Chai, 2019). However, the studies have shown that school teachers perceive different challenges in STEM teaching. Relating to STEM pedagogical content knowledge, the concerns include insufficient understanding of subject content and integrated STEM instruction (Shernoff et al., 2017), concerns on more efficient and optimized STEM pedagogical approaches (Geng et al., 2019), the need for professional learning in teaching inquiry-based STEM (Nadelson, et al., 2012, 2013) and knowledge on implementing engineering-designed based STEM curriculum units in mathematics or science lessons (Guzey et al., 2016; Guzey et al., 2014).

Other concerns on teaching resources, including the lack of appropriate instructional materials and technological resources (Shernoff et al., 2017), were also voiced by teachers when implementing STEM instruction. Additionally, organizational and contextual-related concerns were also expressed, including the lack of opportunities for teacher collaboration and development on STEM practices, concerns about additional workload and time spent on collaboration and administrative issues, and the availability of class hours to perform STEM activities (Geng et al., 2019). Students' lack of understanding or lack of motivation to learn in different ways was also one of the many classroom challenges encountered by school teachers when implementing STEM instruction (Shernoff et al., 2017).

Other literature on school teachers' perceptions of the effective STEM TPD components highlighted teachers' needs in terms of content, process, and contexts of TPD in STEM education. For content, school teachers stressed the practicality and relevance of the TPD content to directly connect to student STEM learning. STEM TPD contents would include STEM content knowledge, STEM activities that meet the curriculum outcomes, pedagogy (e.g. instructions for teaching diverse learners) and assessment techniques (Goodnough et al., 2014). In terms of the process of STEM TPD, school teachers emphasized planned opportunities for collaboration and active learning (Goodnough et al., 2014; Shernoff et al., 2017), opportunities to access good examples and models (e.g. review video recordings of experienced teachers), first-hand experiences of structured lesson plans and materials for key pedagogical approaches (such as project-based learning) (Goodnough et al., 2014; Shernoff et al., 2017), and space for individualization and autonomy in the application (Goodnough et al., 2014). For contexts of STEM TPD, concerns about organizational supports (e.g. release days to collaboration) or programme supports (e.g. specialist supports, supports within school, technological supports) (Goodnough et al., 2014) were also expressed.

Broader literature on teacher professional development concluded different core features that have a powerful effect on learning and changes in classroom practice (Campbell et al., 2004; Garet et al., 2001; Llinares & Krainer, 2006). Whilst espousing managerial supports, the studies highlighted TPD core features on subject-matter knowledge (content), creating opportunities for active learning among teachers (process), and fostering coherence in relation to individuals' previous experiences and alignment with wider curricular framework and assessment (context) to have significantly affect teacher learning. The few studies reviewed above on school teachers' perceptions of the ideal STEM TPD components (e.g. Goodnough et al, 2014; Shernoff et al., 2017) echoed these findings.

Considering the current need for further understanding of Hong Kong primary school teachers' perceived challenges, supports, and future directions for STEM TPD, the following research questions were investigated in this STEM TPD study in the Hong Kong context:

RQ 1: What are the challenges, as perceived by the primary school teachers, when implementing STEM education?
RQ 2: What are the supports received in the STEM TPD through the school-university partnership, as perceived by the primary school teachers?
RQ 3: What are the future directions for STEM TPD, in terms of content and process of professional development, as perceived by the primary school teachers?

The Design of the Study

The project engaged teacher professional development through a school-university partnership that aimed at enhancing primary school teachers' professional capacity of designing STEM activities relevant to the Hong Kong curriculum, with an emphasis on the learning of crosscutting concepts. The study makes reference to the Next Generation Science Standards (National Research Council [NRC], 2013), a multi-dimensional standard for STEM learning currently implemented in the United States for K-12 that emphasizes combining core ideas, practices, and cross-cutting concepts, and STEM research conducted in Asia suggesting the lack of attention given to the integration or evaluation of cross-cutting concepts in STEM programmes (see Cheng & Yeh, Chap. 2). As stated by NRC (2013), cross-cutting concepts are to 'bridge disciplinary core boundaries' for explaining the core disciplinary knowledge. The cross-cutting concepts are observed patterns, cause and effect, the structure of phenomena, system models, limitations of the system, function (e.g. interaction of humans and nature), and change (e.g. growth, changes in states of matter, energy) (NRC, 2013).

Two key cross-cutting concepts of 'change' and 'human and nature' are the focus of the construction of the STEM learning framework for STEM TPD for this study. The two cross-cutting concepts were chosen to reflect the emphasis of the primary school curriculum in Hong Kong. In particular, 'human and nature', a cross-cutting concept developed by the project team, echoes one of the key strands, 'People and Environment', in the General Studies curriculum for primary schools (CDC, 2017). Apart from cross-cutting concepts, the 6E teaching cycle was also introduced to the teachers such that they may consider emphasizing the 'engineering' component in designing STEM lessons. The 6E instructional cycle for STEM (Burke, 2014) is a modification of the 5E instructional model for science instruction (Bybee et al., 2006) which comprised the inquiry cycle of engage, explore, explain, expand, and evaluate. The 6E instructional cycle added a practical element of 'engineering' to emphasize meaning making and the construction of knowledge from hands-on experiences and learning by doing. The 6E teaching cycle for STEM comprises the main steps: engage, explore, explain, engineer, enrich, and evaluate (Burke, 2014).

The research adopted a PR approach to working with primary school teachers. The two primary investigators of the project are researchers with backgrounds in science and teacher education from a university in Hong Kong that specializes in teacher education. A total of 36 primary school teachers from six Hong Kong schools that follow the local curriculum framework were recruited. Each participating school involved four to eight primary school teachers in the school-based STEM TPD (Teachers were identified in the research through an alphabet and number system).

Research members and the participating teachers co-constructed and designed school-based lessons that demonstrated the integration of cross-cutting concepts, skills, and attitudes. Additionally, active professional learning was further encouraged through purposefully designed opportunities for teachers to provide insights to improve instructional materials, peer observations, and expert feedback on STEM teaching trials during the STEM TPD. Teaching trials were implemented in each school by one to three teachers, with teaching trial lessons ranging from three to six class periods depending on the school-based, co-constructed lesson designs. Altogether 12 teachers were involved in conducting the teaching trials. This chapter reports findings from investigating the STEM TPD experiences of these 12 teachers from six different local primary schools.

The 12 primary school teachers comprised six male and six female teachers. Five teachers have bachelor's degrees in STEM-related majors (computer science, mathematics, biology), two reported having a background in GS, four reported neither a STEM nor a GS background, and one did not specify their academic background. Participants' teaching experiences in core STEM-related subjects, that is, in GS and in mathematics, differed. The majority of the 12 teachers had 3–10 years of teaching experience in GS, while most participants had either no or 3 to 5 years of experience of mathematics teaching (see Table 15.1). The more experienced teachers (6 years experiences and above) that participated in the teaching trials were usually senior teachers or subject coordinators of the schools. The school-based STEM TPD was implemented for a duration of one school term (about 4 months).

Semi-structured interviews (Creswell, 2002) with participating teachers were carried out to examine the effectiveness of STEM TPD on the pedagogical skills and awareness of the development of cross-cutting concepts, and to collect their observations on teachers' professional development and views on the integration of cross-cutting concepts in designing STEM lessons. Interviews were carried out in participants' schools before, during, and after the teaching intervention, which lasted around 30 min each. The study examined the data set from the 12 teachers' post teaching intervention interviews.

The audio-recorded interviews were transcribed. When analysing the interview data, teachers' perceived challenges were categorized with reference to three aspects of challenges, namely pedagogical content knowledge, resources, and contextual/organizational level concerns. Furthermore, three core features of effective professional development, that is, content, active professional learning, and coherence, served as a typology to categorize the qualitative interview data related to teachers' perceived received support, and future directions for STEM TPD were examined.

The research team obtained ethical approval from the Human Research Ethics Committee of the university before the research project began. Informed consent

Table 15.1 Participants' teaching experiences in general studies and mathematics ($N = 12$)

	No experience	1–2 years	3–5 years	6–10 years	11–15 years	16–20 years	n/a
General studies	1	1	5	2	1	1	1
Mathematics	4	0	5	1	0	1	1

was obtained from the teacher and students in the teaching trials. Designated labels were used for the teachers in reporting the interview findings.

Findings and Discussion

Teachers' Perceived Challenges

Participating in the project has proved to be a challenging experience for the teachers. Three main types of challenges were reported, including those during the preparation and teaching phases, relating to students' needs, and collaborating with other teachers. Challenges during the preparation and teaching phases were reported by 10 teachers, including the lack of time as STEM lessons require extra teaching time and careful preparation, addressing technical issues, conducting trials, teaching outside textbooks, finding relevant resources and teaching materials, considering safety precautions and possible safety issues, ensuring a smooth flow of the lesson, and handling unexpected results during the teaching.

These challenges during the preparation and teaching phases are related to STEM instruction and lesson planning, limited resources, and other concerns embedded in broader contextual situations. For example, some teachers talked about the lack of access to structured lesson plans and instructional materials, and the extra time needed for lesson preparation.

> 'I need to find more extra time for preparation, from curriculum design, to figure out the relevant scientific principle for teaching. [I] will need to spend extra time ... textbooks provided sufficient information to teach the topic in the past... and with this [STEM] the publisher [textbooks] may not be providing the information, teachers have to prepare it...' (T14)

> 'The teacher has to put in more effort in lesson preparation. We used to rely on textbooks and it was a lot easier. I now feel pressure as I have to produce a booklet; being not very familiar with the topic, I have to find a lot of information.' (T46)

One teacher reflected on the challenges and opportunities of lesson preparation within the broader concept of integrated STEM lesson planning (i.e. better linkage between topics and subjects) and collaborative teaching.

> 'In this project, I am teaching with Mr Chan (pseudonym) but our teaching topics are different.... Each of us has to consider our topic, we know the general direction, but I have to figure out the details and try... It is better to involve all the teachers at the same level, and each class can try and improve after the previous class finishes the teaching... The topics can actually be linked, for example, we can link the discussion of hydrogen vehicles to energy problems and the environment, etc.' (T54)

Other teachers expressed concerns about lack of access to adequate space and environment for experiments (laboratory)-based STEM integration.

> 'There were some safety concerns in conducting experiments. [In the classroom,] it was not very convenient, for example, having water or hot water, students may hurt themselves.' (T14)

> 'The problem is to find the location. We need to test the parachutes, there are safety concerns. We have spent a lot of time considering the right timing and location. When will the sports field not be occupied by PE lessons?' (T21)

Time-scheduling and completing the curriculum were also challenges perceived by the teachers, especially with the government guidance on COVID-19 preventions and social distancing measures for schools.

> 'Colleagues who figure out the class timetable arrangement have a hard time. The number of lessons has been reduced, and we have less time to complete the curriculum. There is additional challenge due to the half-day school arrangement during COVID-19.' (T46)

Other than challenges related to resources and contextual limitations, managing students' behaviours, learner differences, and engaging students in group learning were also challenging aspects during the STEM lesson implementation. Two of the teachers reflected on how to handle students' behaviour during STEM lessons, such as 'managing the difference in responses from the boys and girls' (T45), 'young children having difficulties using scissors and they cannot control the precision, [and] how to stimulate students to discuss among themselves, work out solutions instead of waiting for teachers to provide them with answers' (T23). One of the teachers found it challenging to engage students to achieve the learning objectives (in prediction and explanations) during STEM activities,

> 'There are differences among students. Some would want to "play and try" as soon as possible, and were impatient as the teacher said, "you need to predict", they may think "why do we need to predict? Can we just try now!" …. We need to strike a balance, this is what we put importance on and we need to explain; however, too much talking will make them feel bored.' (T54)

Teachers' Perceptions of the Support Received and Their Own Learning

Teachers participating in the project reflected on having received four main types of support from the STEM TPD relating to the pedagogical knowledge of STEM teaching, namely having a clearer idea of planning and implementing STEM lessons; realizing how cross-cutting concepts may work as a framework; and realizing how alternative content and ideas could be included in STEM lessons.

Three of the teachers, T23, T21, and T51, found themselves having a clearer idea of how to plan and implement STEM lessons having received the examples and reading materials from the project team. They reported 'knowing what to do', although the project team provided flexibility for the teachers to plan their lessons. Pedagogically, the teachers reported gains as they learned to 'teach outside the classroom' (T46), experienced 'more systematic lesson preparation, for example, using 2E in Primary 3 and 6E at the upper primary level' (T51), 'know more about 6E' (T63), and realizing 'the importance of pedagogical theories underpinning the design of the activities, such that teachers are not "implementing the activities for the sake of doing them' (T67). The implementation of 6E has benefited student learning, as suggested by one of the teachers as follows:

> 'We used to adopt POE (predict, observation, explain) when we conducted experiments in the past. This time we are using 6E and I think we learnt more. … 6E has facilitated students' development of higher order thinking.' (T32)

Three of the teachers realized 'the possibility of using cross-cutting concepts to frame STEM lessons' (T51), and 'with this framework, we know how to add new activities' (T63). The introduction of cross-cutting concepts in the STEM TPD like 'nature and people' has provided teachers with a framework for lesson planning that is coherent with their previous teaching.

> 'When we planned STEM teaching or the curriculum in our school, key words like 'nature and people' never appeared. However, nature is related to other aspects of knowledge, and is related to the activities covered in our previous STEM curriculum. Realizing this has helped us to prepare STEM lessons and make the purpose of our activities more explicit.' (T51)

> 'The most obvious difference is the consideration of cross cutting concepts; this impacts on our preparation, to consider the key concepts and directions, how to guide students to think. This is very different.' (T21)

Three teachers, T46, T54, and T14, appreciated the provision of resources and alternative content and ideas. One of the teachers (T54) felt that the most significant benefit of participation was having received 'new ideas and directions', and 'an alternative way of teaching'. Furthermore, two of the teachers realized a different role for teachers as they reflected,

> 'I realized that STEM is not about assembling a simple machine but to adjust [the activity] to their level, related to daily life experience, let them consider the problem and suggest solutions. Teachers are there to support the process.' (T16)

> 'This is inquiry teaching. I used to tell them answers, they copy, remember and go to exams. This time, I explain the objectives, tell them what they have to do, they find out their answers and there are no standard answers; it is up to them to find out and learn.' (T45)

With the content support from the STEM TPD programme, teachers also reported changes in their attitudes towards teaching STEM. The change in teachers' attitudes towards STEM were all positive including having 'a more open attitude towards teaching STEM' (T23), agreeing that 'teaching STEM is a good direction to take' (T46), and both T54 and T14 found themselves more receptive of or open to new ideas about STEM teaching.

Directions for Professional Development

Teachers participating in the project provided six directions as their future professional development needs. Among these six directions, three are content-focused, including lesson planning (e.g. template or framework for planning STEM learning), input related to pedagogical considerations (e.g. relating to cross-cutting concepts, diverse student needs), and enriching their content knowledge, which in turn supports their teaching. The fourth direction is related to active professional learning in STEM TPD, such as engaging teachers in the STEM learning experience. Finally, the fifth and sixth directions are to further address the coherence of STEM TPD relevant to teachers' background (e.g. changing teachers' conceptions or mindset) and the STEM curricular context in Hong Kong.

Content Components of STEM TPD: Lesson Planning, Pedagogical Skills, Content Knowledge

Three teachers mentioned their need for further professional development on lesson planning. They would like to find ways to organize a STEM lesson, how to handle previous suggestions on lesson or curriculum planning when they are including new ideas about STEM and need help to identify focus in lesson planning. For example,

> 'How to plan a lesson or the consideration of the organization of a lesson?' (T14)

> 'I would like to have some expert views on how to focus and arrange given there are so many skills, six thinking hats, etc. How do we change our practice? Does it mean giving up old practices? Would there be more suggestions on curriculum planning and can they let us know how many to focus on? There are many ways related to creative teaching. Professional Development could provide teachers with a clear direction and how to focus.' (T54)

The teachers mentioned professional development (PD) needs on specific pedagogical skills, including ways to teach the cross-cutting concepts, for example, 'change', guiding students through a design cycle, ways to distinguish a 'good' STEM lesson, catering for diverse student needs and assessing students' learning outcomes/performance. For example,

> 'It is particularly useful to find out new ways to teach, especially related to "change" and "humans and nature".' (T14)

> 'Teachers need to know how to guide students to go through a design cycle, fair tests, etc., not just completing the experiments as instructed.' (T23)

> 'To provide teachers with more lesson examples, how to conduct STEM lessons, and what is a good STEM lesson.' (T67)

> 'We will need some training related to science concepts, how to assess if students have these concepts. There is a worry about whether the content is too difficult or too easy, do we need to teach them? Some content is not covered in the textbooks, some classes have higher ability, and do we need to teach them certain concepts?' (T32)

There is some confusion as to whether some STEM-related concepts which are not covered in the textbooks have to be taught, and it is deemed a challenge for teachers to pitch the content to match students' ability levels.

Active Professional Learning in STEM TPD: Opportunities for First-Hand Experiences

Planned opportunities for active learning, specifically opportunities to experience STEM learning in the STEM TPD was also mentioned. Teachers expressed the importance of having first-hand STEM learning experience themselves before they plan lessons for their students. One of the teachers concurred that 'it will be useful to let teachers experience STEM learning themselves. We then know how to organize the lesson' (T46).

Coherence: Addressing the Challenges Relevant to STEM Curricular, Instructional Context, and Teachers' Backgrounds

Teachers expressed their weaknesses in relation to the wider curriculum focus on coding, experiments, and connecting daily life experiences to the formal curriculum in the Hong Kong STEM contexts. More relevant knowledge and teachers' own experience in learning more STEM-related knowledge, technology, coding and updated knowledge related to daily life experiences will then be translated or incorporated into their STEM lessons, and the teachers will find themselves more confident in designing activities for STEM teaching. For example,

> 'Teachers are weaker in technology and coding, and do not pay as much attention to data analysis in the M(mathematics) part.' (T32)

> 'This is new to primary teachers; we didn't have "Coding" in our curriculum before. There is a lot of pressure on primary school teachers.' (T51)

> 'To know more about STEM knowledge, which is the experiments that may be suitable and the new ideas for integrating into the lesson.' (T63)

> 'It is important for teachers to update their STEM knowledge, consider how to introduce it to the students, relate to daily life experiences and design a good lesson.' (T16)

In a similar vein, one of the teachers realized a shift in pedagogical practices, for example, from 5 to 6E: 'How to lead students to complete STEM activities, linking STEM lessons. There are a lot of variations for STEM teaching, I would find how to write lesson plans useful. It used to be 5E and POE when I was a university student and it is now 6E' (T45). The need for further input on how to develop linkage between STEM lessons is reflected.

The coherence of the STEM TPD for primary school teachers was also expressed in relation to the broader curriculum connection between school levels. Two of the teachers found it important to consider the linkage between STEM lessons and even the linkage or development of a framework describing STEM learning between primary and secondary levels. The teacher suggested, 'The linkage and framework showing the relationship between primary and secondary school. For example, students may need to learn micro:bit, mBot and Scratch at the primary school level, and learn Python and coding at the secondary level.' (T16)

Within the Hong Kong context, teachers with non-STEM related backgrounds are sometimes required to teach STEM. This means that STEM teaching is very different from other primary school subjects like Chinese or mathematics, and teaching STEM means leaving their comfort zone. One direction for coherence in STEM TPD is to address the needs of teachers with different subject backgrounds. A change of mindset on the importance of the process as compared with covering the content as planned was well reflected by one of the teachers as follows:

> 'Teachers have different needs, I have a General Studies background, and my preparation is fine. The situation is different for Math and Chinese teachers.... STEM is different from traditional subjects, there are no standard answers and many colleagues are not accustomed to this. Many colleagues hope to see some answers, share with their students and they can follow the next time. This is a change in mindset and needs a long way to go. This is different from teaching Chinese and Math. The process is more important for STEM.' (T21)

Conclusion and Implications

Many efforts were made in the last few years, since the STEM guideline was published, to offer TPD providing input for teachers to implement STEM lessons in schools. For example, the Task Force on Review of School Curriculum (2020) set directions to 'further enhance STEM-related professional development programmes and equip teachers with necessary knowledge and skills to further promote STEM education in schools' (p. v), and numerous opportunities were provided including dissemination of good practices. STEM is no longer a new endeavour for primary school teachers in Hong Kong.

Findings suggest that researchers may need to move away from assuming that teachers need professional development for STEM because they have little idea of how to conduct STEM lessons. With the support and guidance provided by the education community including the government, universities, and other organizations, primary teachers in Hong Kong have been planning and implementing STEM lessons. With these experiences, there are some issues to be resolved before a further enhancement in terms of quality and quantity of STEM lessons could be realized. The teachers have identified some challenges that need to be addressed including access to structured lesson plans and instructional materials, extra time involved in preparing and implementing STEM instruction and teacher collaboration, access to adequate space and environments for experiments (laboratory)-based STEM integration, and timetabling issues in schools.

For future STEM TPD, a few directions are pertinent, including addressing the needs of the teachers to provide them with authentic STEM learning experiences. Taking a constructivist point of view for TPD (Zehetmeier et al., 2015), teachers themselves need to construct their own successful learning experience, and this experience will form a basis upon which they can design STEM learning for their students.

As reflected by one of the teachers in the study, teachers need to adopt a 'more open attitude' towards STEM teaching. This may be the most fundamental basis as teacher educators consider STEM TPD opportunities. Teachers need to reconstruct their conceptions of teaching, that is, having to provide standard answers, not allowing failures, covering and following textbooks, and guiding students to a perfect solution. STEM teaching may be seen as challenging their beliefs and practices which they have adopted for years in other subjects such as mathematics and languages. STEM TPD opportunities will have to address these changes in conceptions and work with the teachers to change their fundamental beliefs.

Teachers participating in this study were keen to improve the quality of their STEM teaching. There were questions like 'What does good STEM teaching look like?', 'How can we implement 6E?', and 'How can we develop a linkage between primary and secondary STEM teaching?'. Similar to STEM learning, there are no fixed or standard answers to these questions. Quality STEM teaching will need to be adapted to the needs of the students.

In the design of TPD, Llinares and Krainer (2006) underlined 'reflection', which is the attitude towards, and competence in (self-)criticism of one's own action, and 'networking', which is the attitude towards, and competence in communicative and cooperative work (p. 12) as key interventions in TPD programmes. The stance of PR, in the form of school-university partnership, engages teachers in authentic STEM professional learning through systematic and intentional reflection of their own professional practices and teacher cooperation and communication. PR which situates teachers as practitioners, researchers, and learners serves as a way to foster STEM professional learning. It is, therefore, proposed that future STEM TPD adopt a practitioner research model and engage teachers in researching into their own teaching. Teachers will then be able to collect evidence of student learning, analyse the impact of their teaching, and adjust and re-adjust their teaching strategies to enhance the quality of their teaching. Promoting active professional learning is crucial if the STEM education community is keen to ensure the quality of STEM lessons in classrooms.

Acknowledgements The project reported in this chapter was funded by the Central Reserve Allocation Committee at the Education University of Hong Kong (EdUHK). The authors would also like to acknowledge the support and contribution of Dr. Li Wai Chin at the EdUHK to the project.

References

Atay, D. (2007). Teacher research for professional development. *English Language Teaching (ELT) Journal, 62*(2), 139–147. https://doi.org/10.1093/elt/ccl053.

Burke, B. N. (2014). The ITEEA 6E learning by design model: Maximizing informed design and inquiry in the integrative STEM classroom. *Technology and Engineering Teacher, 73*(6), 14–19.

Burns, A. (2009). Action research in second language teacher education. In A. Burns & J. C. Richards (Eds.), *The Cambridge guide to second language teacher education* (pp. 289–297). Cambridge University Press.

Bybee, R. W., Taylor, J. A., Gardner, A., Van Scotter, P., Powell, J. C., Westbrook, A., & Landes, N. (2006). *The BSCS 5E instructional model: Origins and effectiveness*. Biological Sciences Curriculum Studies (BSCS). https://media.bscs.org/bscsmw/5es/bscs_5e_full_report.pdf.

Cain, T., & Milovic, S. (2010). Action research as a tool of professional development of advisers and teachers in Croatia. *European Journal of Teacher Education, 33*(1), 19–30.

Campbell, A., McNamara, O., & Gilroy, P. (2004). *Practitioner research and professional development in education*. Sage Publications Inc. https://doi.org/10.4135/9780857024510.

Chai, C. S. (2019). Teacher professional development for science, technology, engineering and mathematics (STEM) education: A review from the perspectives of technological pedagogical content (TPACK). *Asia-Pacific Education Researcher, 28*, 5–13. https://doi.org/10.1007/s40299-018-0400-7.

Cheng, M. M. H., Chan, K.-W., Tang, S. Y. F., & Cheng, A. Y. N. (2009). Pre-service teacher education students' epistemological beliefs and their conceptions of teaching. *Teaching and Teacher Education, 25*(2), 319–327. https://doi.org/10.1016/j.tate.2008.09.018.

Cheng, M. M. H., & Li, D. D. Y. (2020). Implementing practitioner research as a teacher professional development strategy in an Asia-Pacific context. *Journal of Education for Teaching, 46*(1), 55–70. https://doi.org/10.1080/02607476.2019.1708627.

Cheng, M. M. H., & So, W. W. M. (2012). Analysing teacher professional development through professional dialogue: An investigation into a university-school partnership project on enquiry learning. *Journal of Education for Teaching, 38*(3), 323–341. https://doi.org/10.1080/02607476.2012.668331.

Chiu, T. K. F., Chai, C. S., Williams, P. J., & Lin, T.-J. (2021). Teacher professional development on self-determination theory–based design thinking in STEM education. *Educational Technology & Society, 24*(4), 153–165.

Cochran-Smith, M., & Lytle, S. L. (2009). *Inquiry as stance: Practitioner research for the next generation*. Teachers College Press.

Creswell, J. W. (2002). *Educational research: Planning, conducting, and evaluating quantitative*. Merrill Prentice Hall.

Curriculum Development Council. (2017). *General studies curriculum guides for primary schools (primary 1 – primary 6)*. Hong Kong SAR: Education Bureau. https://www.edb.gov.hk/attachment/en/curriculum-development/cross-kla-studies/gs-primary/GSCG_2017_Eng.pdf.

Curriculum Development Institute. (2021). *Professional development programmes for heads and teachers (web version) 2021/22: Primary*. Hong Kong SAR: Education Bureau. https://www.edb.gov.hk/attachment/en/curriculum-development/resources-support/booklet_on_professional_development_programmes/pdp_for_heads_and_teachers_web_2022/PDP_Booklet_2021-22_Pri_Eng_Final.pdf.

Day, C., & Smethem, L. (2010). Partnerships between schools and higher education. In P. L. Peterson, E. Baker, & B. McGaw (Eds.), *International encyclopaedia of education* (pp. 757–763). Elsevier Ltd.

Drits-Esser, D., & Stark, L. A. (2015). The impact of collaborative curriculum design on teacher professional learning. *The Electronic Journal of Science Education, 19*(8), 15129.

Fung, Y. (2000). A constructivist strategy for developing teachers for change: A Hong Kong experience. *Journal of in-Service Education, 26*(1), 153–167. https://doi.org/10.1080/13674580000200108.

Garet, M. S., Porter, A. C., Desimone, L., Birman, B. F., & Yoon, K. S. (2001). What makes professional development effective? Results from a national sample of teachers. *American Educational Research Journal, 38*(4), 915–945. https://journals.sagepub.com/doi/pdf/10.3102/00028312038004915.

Geng, J., Jong, M. S. Y., & Chai, C. S. (2019). Hong Kong teachers' self-efficacy and concerns about STEM education. *The Asia-Pacific Education Researcher, 28*(1), 35–45. https://doi.org/10.1007/s40299-018-0414-1.

Goodnough, K., Pelech, S., & Stordy, M. (2014). Effective professional development in STEM education: The perceptions of primary/elementary teachers. *Teacher Education and Practice, 27*(2/3), 402–423.

Guzey, S. S., Moore, T. J., & Harwell, M. (2016). Building up STEM: An analysis of teacher-developed engineering design-based STEM integration curricular materials. *Journal of Pre-College Engineering Education Research (J-PEER), 6*(1), Article 2. https://doi.org/10.7771/2157-9288.1129.

Guzey, S. S., Tank, K., Wang, H.-H., Roehrig, G., & Moore, T. (2014). A high-quality professional development for teachers of grades 3–6 for implementing engineering into classrooms. *School Science and Mathematics, 114*(3), 139–149. https://doi.org/10.1111/ssm.12061.

Handelzalts, A. (2019). Collaborative curriculum development in teacher design teams. In J. Pieters, J. Voogt & N. P. Roblin (Eds.), *Collaborative curriculum design for sustainable innovation and teacher learning* (pp. 159–173). Springer, Cham. https://doi.org/10.1007/978-3-030-20062-6_9.

Harfitt, G., & Chan, C. (2017). Constructivist learning theories in teacher education programmes: A pedagogical perspective. In D. J. Clandinin & J. Husu (Eds.), *The SAGE handbook of research on teacher education* (pp. 545–560). SAGE Publications Ltd.

Howe, A. C., & Stubbs, H. S. (1997). Empowering science teachers: A model for professional development. *Journal of Science Teacher Education, 8*(3), 167–182. https://doi.org/10.1023/A:1009487417488.

Keiny, S. (1994). Constructivism and teachers' professional development. *Teaching and Teacher Education, 10*(2), 157–167. https://doi.org/10.1016/0742-051X(94)90010-8.

Kong, S. C., Lai, M., & Sun, D. (2020). Teacher development in computational thinking: Design and learning outcomes of programming concepts, practices and pedagogy. *Computers & Education, 151*, 103872. https://doi.org/10.1016/j.compedu.2020.103872.

Lee, M.-H., Chai, C. S., & Hong, H.-Y. (2019). STEM education in Asia Pacific: Challenges and development. *Asia-Pacific Education Researcher, 28*(1), 1–4.

Lin, P.-Y., Chai, C. S., & Jong, M. S.-Y. (2021). A study of disposition, engagement, efficacy, and vitality of teachers in designing science, technology, engineering, and mathematics education. *Frontiers in Psychology, 12*. https://doi.org/10.3389/fpsyg.2021.661631.

Llinares, S., & Krainer, K. (2006). Mathematics (student) teachers and teacher educators as learners. In A. Gutiérrez & P. Boero (Eds.), *Handbook of research on the psychology of mathematics education: Past, present and future* (pp. 429– 459). Sense Publishers. https://doi.org/10.1163/9789087901127_016.

Margot, K. C., & Kettler, T. (2019). Teachers' perception of STEM integration and education: A systematic literature review. *International Journal of STEM Education, 6*, Article 2. https://doi.org/10.1186/s40594-018-0151-2.

Nadelson, L. S., Callahan, J., Pyke, P., Hay, A., Dance, M., & Pfiester, J. (2013). Teacher STEM perception and preparation: Inquiry-based STEM professional development for elementary teachers. *The Journal of Educational Research, 106*(2), 157–168. https://doi.org/10.1080/00220671.2012.667014.

Nadelson, L. S., Seifert, A., Moll, A. J., & Coats, B. (2012). I-STEM summer institute: An integrated approach to teacher professional development in STEM. *Journal of STEM Education: Innovation and Outreach, 13*(2), 69–83.

National Research Council. (2013). *Next generation science standards: For States, by States*. Washington, DC: The National Academies Press. https://doi.org/10.17226/18290.

Shernoff, D. J., Sinha, S., Bressler, D. M., & Ginsburg, L. (2017). Assessing teacher education and professional development needs for the implementation of integrated approaches to STEM education. *International Journal of STEM Education, 4*, Article 13. https://doi.org/10.1186/s40594-017-0068-1.

Sparks, D. (2002). *Designing powerful professional development for teachers and principals*. National Staff Development Council.
Stavroula, P., Chrystalla, P.-K., & Loucas, L. (2015). 'The exchange of ideas was mutual, I have to say': Negotiating researcher and teacher 'roles' in an early years educators' professional development programme on inquiry-based mathematics and science learning. *Professional Development in Education, 41*(2), 382–400. https://doi.org/10.1080/19415257.2014.999381.
Szeto, E., Sin, K., & Leung, G. (2021). A cross-school PLC: How could teacher professional development of robot-based pedagogies for all students build a social-justice school? *Professional Development in Education, 47*(1), 141–155. https://doi.org/10.1080/19415257.2020.1787201.
Task Force on Professional Development of Teachers. (2019). *Task force on professional development of teachers—Report*. Hong Kong SAR: Education Bureau. https://www.edb.gov.hk/attachment/en/teacher/report-tftpd/TF%20Final%20Report_en_final.pdf.
Task Force on Review of School Curriculum. (2020). *Optimise the curriculum for the future: Foster whole-person development and diverse talent*. Hong Kong SAR: Education Bureau https://www.edb.gov.hk/attachment/en/curriculum-development/renewal/taskforce_cur/TF_CurriculumReview_FinalReport_e.pdf.
Zehetmeier, S., Andreitz, I., Erlacher, W., & Rauch, F. (2015). Researching the impact of teacher professional development programmes based on action research, constructivism, and systems theory. *Educational Action Research, 23*(2), 162–177. https://doi.org/10.1080/09650792.2014.997261.

May May Hung Cheng obtained her Bachelor's and Master's degrees from the University of Hong Kong and her Ph.D. from the University of Waikato, New Zealand. She is currently Chair Professor of Teacher Education in the Department of Curriculum and Instruction at the Education University of Hong Kong. She was previously a Reader in Professional Education at the Department of Education, University of Oxford. Her main areas of research are teacher education and science education with a focus on teacher learning and teacher professional development. She was the president of the East Asian Association for Science Education (EASE) between 2016 to 2019.

Fang-Yin Yeh obtained her Bachelor's degree from the National Taiwan University of Science and Technology, Master's degree from the University of Kansas, USA, and Ph.D. from the University of Hong Kong. She is currently a Post-Doctoral Fellow in the Department of Curriculum and Instruction at the Education University of Hong Kong. Her research focuses on school curriculum in East Asia. She is also interested in teacher professional development.

The manufacturer's authorised representative in the EU is Springer Nature Customer Service Centre GmbH, Europaplatz 3, 69115 Heidelberg, Germany. If you have any concerns regarding our products, please contact ProductSafety@springernature.com

Printed and bound by CPI Group (UK) Ltd, Croydon, CR0 4YY
30/09/2026
02225320-0010